Biology and ecology of mangroves

Tasks for vegetation science 8

Series Editor

HELMUT LIETH

University of Osnabrück, F.R.G.

OTHER BOOKS IN THE TASKS FOR VEGETATION SCIENCE SERIES

1 Box, E.O. *Macroclimate and Plant Forms.* An introduction to predictive modelling in phytogeography. ISBN 90 6193 941 0
2 Navin Sen, D. & Singh Rajpurohit, K. *Contributions to the Ecology of Halophytes.* ISBN 90 6193 942 9
3 Ross, J. *The Radiation Regime and Architecture of Plant Stands.* ISBN 90 6193 607 1
4 Margaris, N.S. & Mooney, H.A. (eds) *Components of Productivity of Mediterranean – Climate Regions.* ISBN 90 6193 944 5
5 Müller, M.J. *Selected Climatic Data for a Global Set of Standard Stations for Vegetation Science.* ISBN 90 6193 945 3
6 Roth, I. *Stratification in Tropical Forests as seen in Leaf Structure.* ISBN 90 6193 946 1
7 Steubing, L. & Jäger, H.J. *Monitoring of Air Pollutants by Plants: Methods and Problems.* ISBN 90 6193 947 X
9 Teas, H.J. *Physiology and Management of Mangroves.* ISBN 90 6193 949 6
(For Contents of Volume 9 see page VII of this book)

1983 **SPRINGER-SCIENCE+BUSINESS MEDIA, B.V.**

Biology and ecology of mangroves

edited by

H.J. TEAS

1983 **SPRINGER-SCIENCE+BUSINESS MEDIA, B.V.**

Library of Congress Cataloging in Publication Data

Main entry under title:

Biology and ecology of mangroves.

(Tasks for vegetation science ; 8)
1. Mangrove swamp ecology. 2. Mangrove swamps.
I. Teas, Howard J. (Howard Jones), 1920-
II. Series.
QH541.5.M27B56 1983 574.5'26325 83-6186
ISBN 978-90-481-8526-9 ISBN 978-94-017-0914-9 (eBook)
DOI 10.1007/978-94-017-0914-9

Cover design: Max Velthuijs

Contents

CONTENTS OF

Physiology and management of mangroves

H.J. TEAS (ed.)

To be published as Volume 9 of the series Tasks for Vegetation Science

IN MEMORIAM

VALENTINE JACKSON CHAPMAN
1910–1980

Val Chapman received his Doctorate from Cambridge, was a Postdoctoral Fellow at Harvard, and served as Chair Professor of Botany at Auckland University from 1946 until his retirement as Emeritus Professor in 1974.

Val Chapman was known to most of us at the Mangrove Symposium for his books on mangrove vegetation, coastal vegetation, his chapters on mangroves in wet coastal ecosystems, and his papers on Jamaican mangroves, respiration and physiology of mangrove seedlings, mangrove phytosociology, mangrove biogeography, and mangroves of New Zealand. His scientific side that we knew less well extended to studies of marine and freshwater algae, seashore ecology, weeds and eutrophication of freshwater lakes, salt deserts, salt marshes, and conservation of marshes and swamps. Further afield and even less known to his scientific colleagues was Val Chapman's role in University, in civic and religious affairs. For example, he chaired an array of University of Auckland committees, and was two years Dean of Sciences; he served four years as an elected member of the Auckland City Council, was awarded the O.B.E. in 1974 for his contributions to science, civic life and conservation; and he was active in the Anglican Church and served 17 years as a member of St. John's Theological College Board of Governors.

In science or public service Val Chapman will be remembered by those who knew him for his creativity, energy and the enthusiasm he inspired in those around him.

Howard J. Teas

CHAPTER 1

Development of mangrove forests from a geological perspective

COLIN D. WOODROFFE*

Department of Geography, University of Auckland, Private Bag, Auckland, New Zealand

* Present address: The Australian National University Casuarina Northern Territory Australia

Introduction

Mangrove swamps develop only where coastal physiography and energy conditions are favourable. They are most extensive where there is a low shore gradient, and occupy a broader belt on shorelines which have a large tidal range (Walsh, 1974). Sheltered habitats are essential for mangrove development, and on coasts which are exposed, mangroves are localised in the lee of other coastal landforms (Davies, 1972).

Mangrove swamps are tropical analogues of the temperate salt marshes (Chapman, 1975) and many of the adaptations which the plants show to the intertidal environment are similar in the two plant formations. These formations are found in association at the mid-latitudinal limits to their distributions. On salt marshes plant species may show a zonation parallel to the shore and this zonation has been recognised as an expression of a temporal succession of species, whereby each zone will, in time, be replaced by a more landward plant assemblage. Within the salt marsh, accretion of sediment is promoted by the vegetation and the level of the substrate is raised, triggering the replacement of species (Chapman 1960).

Zonation of species parallel to the shore has also been widely identified on mangrove shorelines of the tropics, and Chapman (1970, 1975) has suggested that it is possible to separate the mangrove coasts of the world into nine schemae according to the species present and their distribution. The recognition of zonation need not imply succession of species but may simply reflect plant response to some factor varying normal to the coast, such as salinity or freqency of tidal inundation (Lugo and Snedaker, 1974; Zahran, 1977); however, it has been proposed by several authors (e.g. Richards, 1952; Macnae, 1968) that zonation on mangrove shorelines does reflect temporal succession of those species. It is believed that the mangrove vegetation also promotes sediment accretion and so one species 'prepares the way' for another (Watson, 1928; Rosevear, 1947; Banijbatana, 1958; Kuenzler, 1974).

Associated with the concept of species replacement through time is the belief that the seaward species has an ability to prograde into shallow water. The viviparous nature of seedlings of many species of mangroves appears to support this contention and several early workers regarded mangroves as 'makers of land' (Curtis, 1888; Hitchcock, 1891; Phillips, 1903; Vaughan, 1909; Davis, 1938; Stephens, 1962) and 'trees that walk to the sea' (Hodge, 1956). Particularly influential in this respect was the study of the geologic role of mangroves in southwestern Florida by Davis (1940). Davis attributed a substantial progradational ability to the seaward species *Rhizophora mangle*, which he considered to be progressively colonising the marine shoals of Florida Bay and Biscayne Bay.

Mangroves are often associated with fine-grained sediments, but they are able to grow on a wide range of substrate types (Salomon, 1978). Sedi-

Tasks for vegetation science, Vol. 8 ed. by H.J. Teas

 ISBN 978-90-481-8526-9

ments accumulating beneath a mangrove swamp may be of two types. (i) Allochthonous sediments are those which are derived from outside the swamp itself and are brought into the swamp either from landward or from seaward. In deltaic areas much terrigenous sediment may be imported by the river system; in estuaries both terrigenous and marine sediments may accumulate; and on the broad mud flats of shallow-water carbonate banks, such as the Bahama Banks, mangroves may grow on intertidal calcareous flats in which the sediment is almost totally marine in origin. (ii) Autochthonous sediments are those which are deposited *in situ* and include mangrove-derived peats, with a high organic content, which may underlie mangroves in areas where allochthonous sediment supply is low. In addition, mangroves can grow on sandy, reefal or rocky substrates (Chapman, 1975).

Mangrove swamps which grow in areas receiving large supplies of allochthonous sediments may show rapid coastal progradation. In Sumatra seaward migration of mangrove margins has been recorded at rates of up to 125 m/yr, and in Java rates of up to 200 m/yr have been observed (Macnae, 1968). In parts of Malaysia, inland peat swamp forests may be established overlying mangrove mud deposits which have been shown by radiometric dating to be Holocene in age (Anderson, 1964; Hewitt, 1967), evidence of coastal progradation of several kilometres in 3–4000 years.

Deltaic areas are prone to rapid adjustments in the patterns of sedimentation in response to sediment inputs, and many studies on the dynamics of mangrove shorelines have been concentrated in such deltaic areas (Fosberg, 1964; Allen, 1965). Studies of mangroves in the Atrato Delta (Vann, 1959) and on the Pacific coast of Colombia (West, 1956) suggested that vegetation was not an important factor in determining sedimentation patterns, as had previously been supposed, but that the dynamics of the mangrove vegetation may in many cases be a response to, and not a cause of, coastal progradation (van Steenis, 1958; Gill, 1971).

On these geomorphologically active shorelines the patterning of mangrove species can be seen to reflect the distribution of landforms and sedimentary environments (Carter, 1959). Thom has demonstrated the control that microtopography exercises on the distribution of mangrove species in the coastal plain of Tabasco, Mexico (Thom, 1967, 1975). This is an area where the pattern of landforms is dynamic and constantly changing, and mangroves are opportunistic in their colonisation of the various landform habitats. Geomorphic change in habitat characteristics induces vegetational change (Thom, 1975).

Similar conclusions were reached from several studies of harbour and estuarine environments in Australia; vegetation on mangrove shorelines has been shown to be dynamic, but movements of the mangrove margins are found to occur subsequent to sedimentation on seaward flats (Bird, 1971, 1973; Bird and Barson, 1977; Thom *et al.*, 1975; Butler *et al.*, 1977). In these geomorphologically active areas, where allochthonous sediment inputs can be large, the general view appears to be that mangroves can modify the rate at which sedimentary processes take place but generally do not change the pattern of landform evolution (Carlton, 1974).

Mangrove swamps do not only grow in areas that are geomorphologically active, or in areas which receive large supplies of allochthonous sediment. Mangroves may often be associated with coral reefs, an association which may have been in existence since the Cretaceous (McCoy and Heck, 1976). It has been suggested by Welch (1963) that there is a succession from a coral reef to a *Thalassia*-covered plain to mangrove forest and eventually to an island habitat. Thus, although a coral reef is not an area of rapid sediment accumulation on the scale that a delta or an estuary may be, reef growth and consolidation can result in a mobile substrate, which it is implied may be colonised by mangroves.

Recently there has been an appreciation of the interdependence of reefal substrates and sea level movements. It has been illustrated, using radiometric dating, that the development of reefs has been controlled by the rate of sea level rise (Easton and Olson, 1976; Montaggioni, 1976; MacIntyre and Glynn, 1976; Focke, 1978) and a model of the nature of the response of the growth potential of reefs to sea level movements and resultant energy conditions has been proposed by Adey (1978). The work of the 1973 expedition to the northern Great

Barrier Reef illustrates well the relationship between reef morphology and sea level change; here the sea level appears to have reached a level close to its present one about 6000 years ago (Polach *et al.*, 1978; McLean *et al.*, 1978), and since that time vertical reef growth has been superseded by horizontal reef flat consolidation and the formation of microatolls (Stoddart *et al.*, 1978a).

Coral reefs grow in predominantly high energy environments, while mangrove forests are found in low energy environments. However, mangrove forests can be established in the low energy leeward areas of the reef flat, where they are dependent on the reef materials as a substrate. Steers (1937) considered that on the low wooded islands of the Great Barrier Reef mangroves could, given time, cover the entire reef flat. On the best studied of these islands, Low Isles, there has been some dispute as to what extent the mangrove has expanded (Fairbridge and Teichert, 1948; Macnae, 1966). Stoddart (1980) has shown that the area of continuous mangrove has increased at the expense of the mangrove park, a trend not reflected on adjacent islands, and has inferred that the spread of mangroves on these islands appears to be dependent upon the evolutionary stage of the reefal substrate and its suitability for colonisation by mangroves (Stoddart *et al.*, 1978b).

In the same way that dating of coral material can show that reef development has been controlled by sea level movements, accumulations of mangrove sediments which exceed in thickness the range within which these sediments can now be deposited indicate that these deposits have formed during a time of change in the relationship between land and sea (Ebanks, 1967; Bloom, 1970). This was recognised by Davis (1940) in his classic work on the mangroves of Florida, and he believed that mangrove peat had developed beneath the mangroves during a period of sea level rise, but that the mangroves had also prograded seaward over marine shoals while the area was being submerged.

The suggestion that the mangroves of Florida were active progradational geologic agents was challenged by Egler (1952), who considered that the mangroves would generally have moved landward as sea level had risen, invading the freshwater Everglades environments, and that the seaward margin would also have receded, or at best might have remained static. Subsequent stratigraphic work by Scholl (1964a, b, 1969) revealed that the sedimentary sequence is indeed transgressive, and that the basal calcareous sediments are freshwater calcitic muds and not marine deposits as identified by Davis. These basal muds are overlain by fibrous peats and record the incursion of marine waters into the freshwater environment.

More recent work in the Everglades-mangrove complex has confirmed that the area was under the influence of a transgressing sea for the last 4–5000 years (Riegel, 1965; Spackman *et al.*, 1966, 1969) but reveals that the lower unit within the peat profile is in many cases a freshwater peat, similar to those now forming in the Everglades (Cohen, 1968; Cohen and Spackman, 1977; Gleason *et al.*, 1974), and indicating that the transgression occurs in some instances within the peat profile, and not at its base (Smith, 1968).

A similar sequence of transgressive sediments, recording a rise in sea level, has been recovered in cores in Florida Bay (Turmel and Swanson, 1976; Enos and Perkins, 1979). However in this instance calcareous marine sediments overlie the mangrove peat, forming islands, and representing the final stage of the marine transgression. Locally mangroves have recolonised these islands and may have promoted lateral extension of the islands, so producing a regressive sequence of sediments overlying the transgressive sequence (Wanless, 1974; Enos and Perkins, 1979). Similar complex transgressive and regressive sedimentary sequences have been reported from the islands on the Belize shelf, central America (Ebanks, 1967, 1975).

In both geomorphologically active areas, such as deltas and estuaries, and in geomorphologically less active areas, such as sheltered reef flats, substrate characteristics and patterns of sedimentation or substrate adjustment are a major control on the morphology of mangrove swamps. The form and dynamics of the mangrove forest are also a function of the varying relationship between land and sea and suitable energy conditions. Where allochthonous sediments are accumulating, landforms evolve, and development of intertidal flats can promote the progradation of mangroves. In reefal environments, where allochthonous sediments are limited, material may accumulate *in situ* beneath the swamp; the forests are less dynamic but may respond to stages in the growth and consolidation of the reef, as a function of the level of land and sea, and wave and current energy. Elsewhere mangroves may colonise substrates which are immobile and in this case the swamp morphology will be primarily controlled by the pre-existing topography, but change can result from movements of sea level.

This paper describes the morphology and development of areas of mangrove swamp on Grand Cayman, a low-lying coralline island, which provides a good example of extensive swamplands developed in an area which is not geomorphologically active. On Grand Cayman there is no terrestrial drainage system, and marine sediments on the island are largely calcareous and are not accumulating in the swamps. The mangrove forests are underlain by autochthonous deposits of mangrove-derived peat which have evidently formed during a period of marine transgression. The morphology of the swamps has been controlled by pre-existing bedrock topography, which reflects depositional environments in the Pleistocene rocks, and by resultant energy conditions.

Area of study

The Cayman Islands form emergent peaks on the Cayman Ridge which runs from the Sierra Maestra of Cuba to the Misteriosa Banks. Grand Cayman (19° 20′N 81° 20′W) is the largest of the three islands and has the most extensive cover of mangroves. The island is composed of a core of Tertiary rocks of the Bluff Limestone Formation (Matley, 1926), around which have been deposited rocks of the Ironshore Formation of Pleistocene age (Brunt *et al.*, 1973). It is on the lower-lying rocks of the younger formation that mangrove swamps are most extensive.

The island is microtidal with a maximum tidal

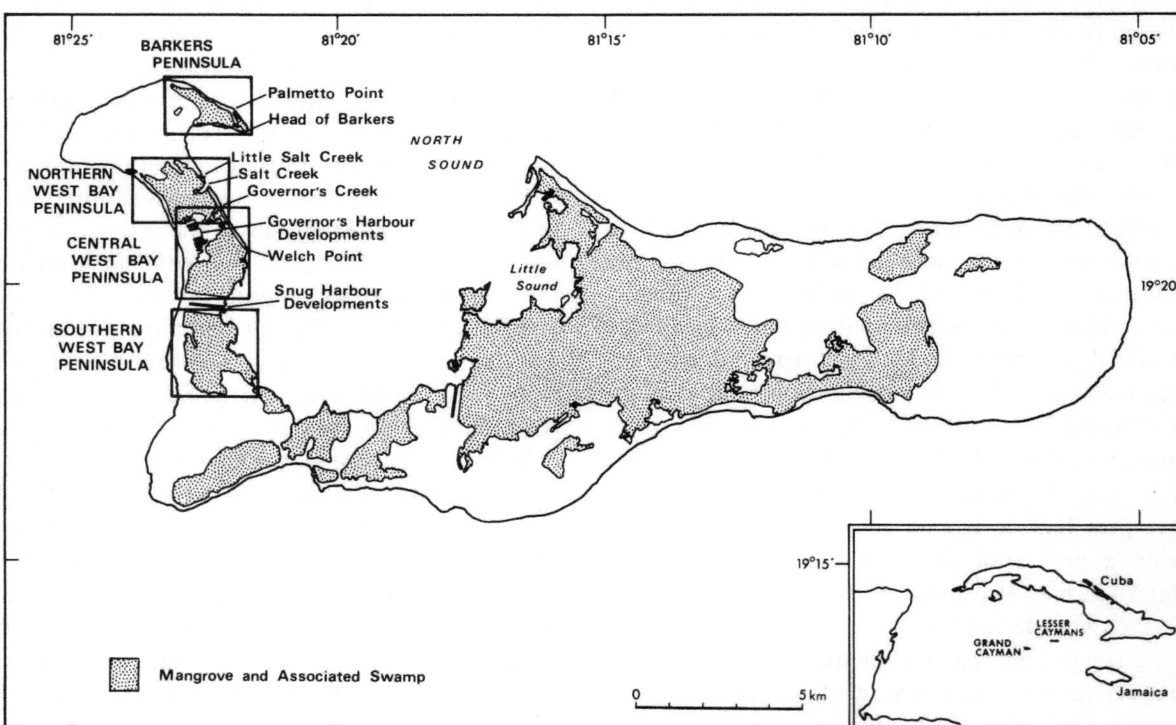

Fig. 1. Distribution of mangrove and associated swamp on Grand Cayman, and location of areas described.

range of 60 cm (Rigby and Roberts, 1976). The mangrove swamps are composed of *Rhizophora mangle* and *Avicennia germinans*; both can grow in monospecific stands, but over much of the swamplands are found together. *Laguncularia racemosa* occurs within the swamps but forms a less important element of the vegetation. *Conocarpus erectus* exists on the margin of the swamp, often in association with *Laguncularia*, and also can be found growing monospecifically in inland basins.

The West Bay peninsula, to the west of the island (Fig. 1) is covered by mangrove swamp. The morphology and development of this peninsula is discussed below, with particular reference to the northern West Bay peninsula, in the vicinity of Salt Creek.

Northern West Bay peninsula, Grand Cayman

Physiography and vegetation

The area is bounded to the west by a beach ridge of coralgal sand, reaching a height of 3 m or more, and dominated by shrubs, especially *Coccoloba uvifera* and *Chrysobalanus icaco*. To the east there is a

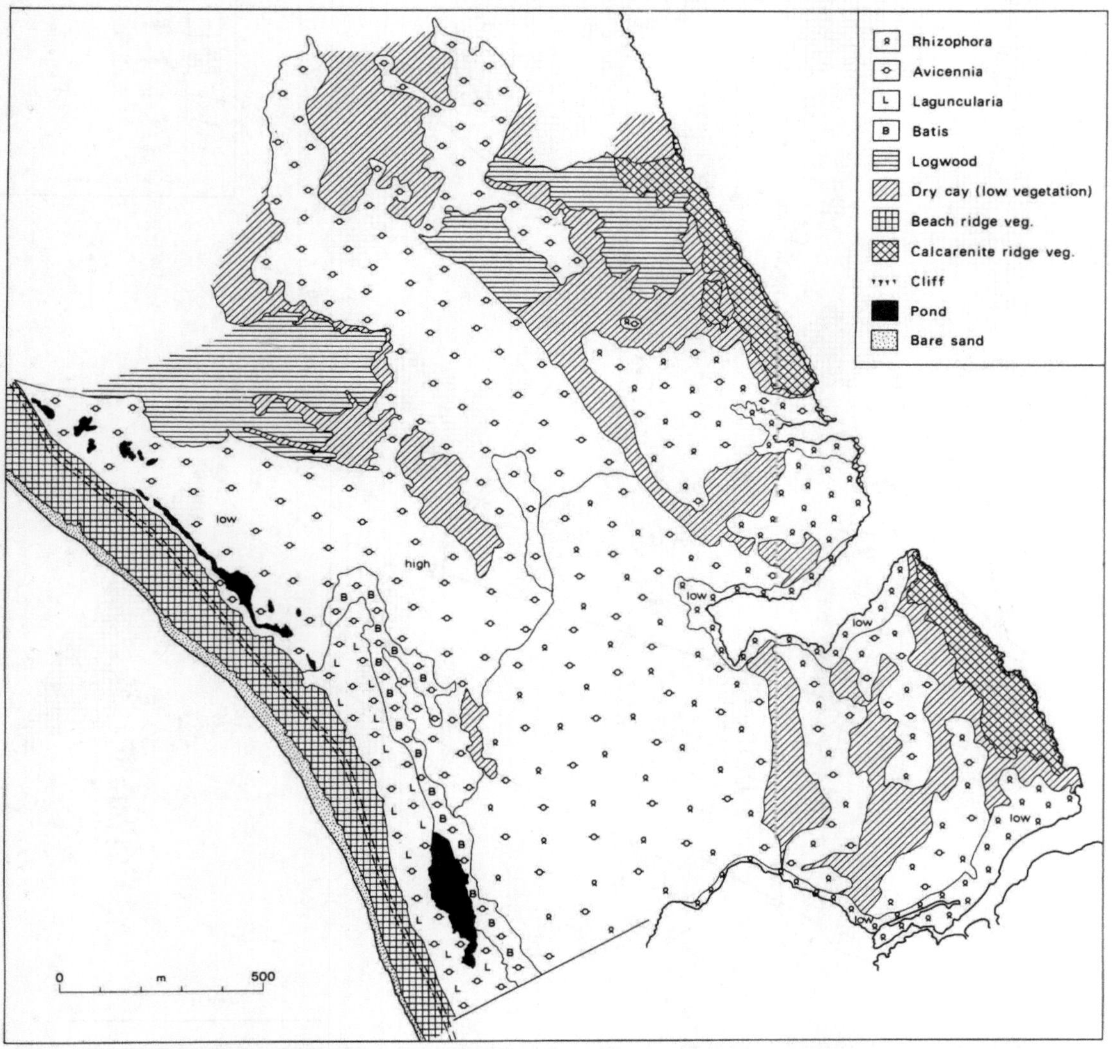

Fig. 2. Vegetation of the northern West Bay peninsula.

calcarenite ridge rising to a maximum height of 7 m, and dominated by a dense scrub of *Coccoloba uvifera*, *Borrichia arborescens* and *Mallotonia gnaphalodes* (Fig. 2).

There are low ridges of bedrock, with a northwest-southeast trend, exposed within the mangrove swamp and carrying either a low tree cover of *Haematoxylum campechianum* and *Thespesia populnea*, or low herbaceous vegetation of *Batis maritima*, *Sporobolus virginicus*, *Salicornia peruviana* and *Sesuvium portulacastrum*. In addition small bedrock protrusions, locally termed 'dry cays', occur within the swamp and these also carry a similar non-swamp vegetation (Woodroffe *et al.*, 1980).

Tidal water inundating the northern West Bay peninsula flows through breaches in the calcarenite ridge, Little Salt Creek, Salt Creek, and Governor's Creek, or from Governor's Harbour, and it is around these inlets that there is a low fringe of seaward *Rhizophora*. At the margin the *Rhizophora* can be as low as 1–2 m tall, but the trees increase in stature to 15 m or more to landward. Further from the tidal inlets, *Rhizophora* and *Avicennia* occur in

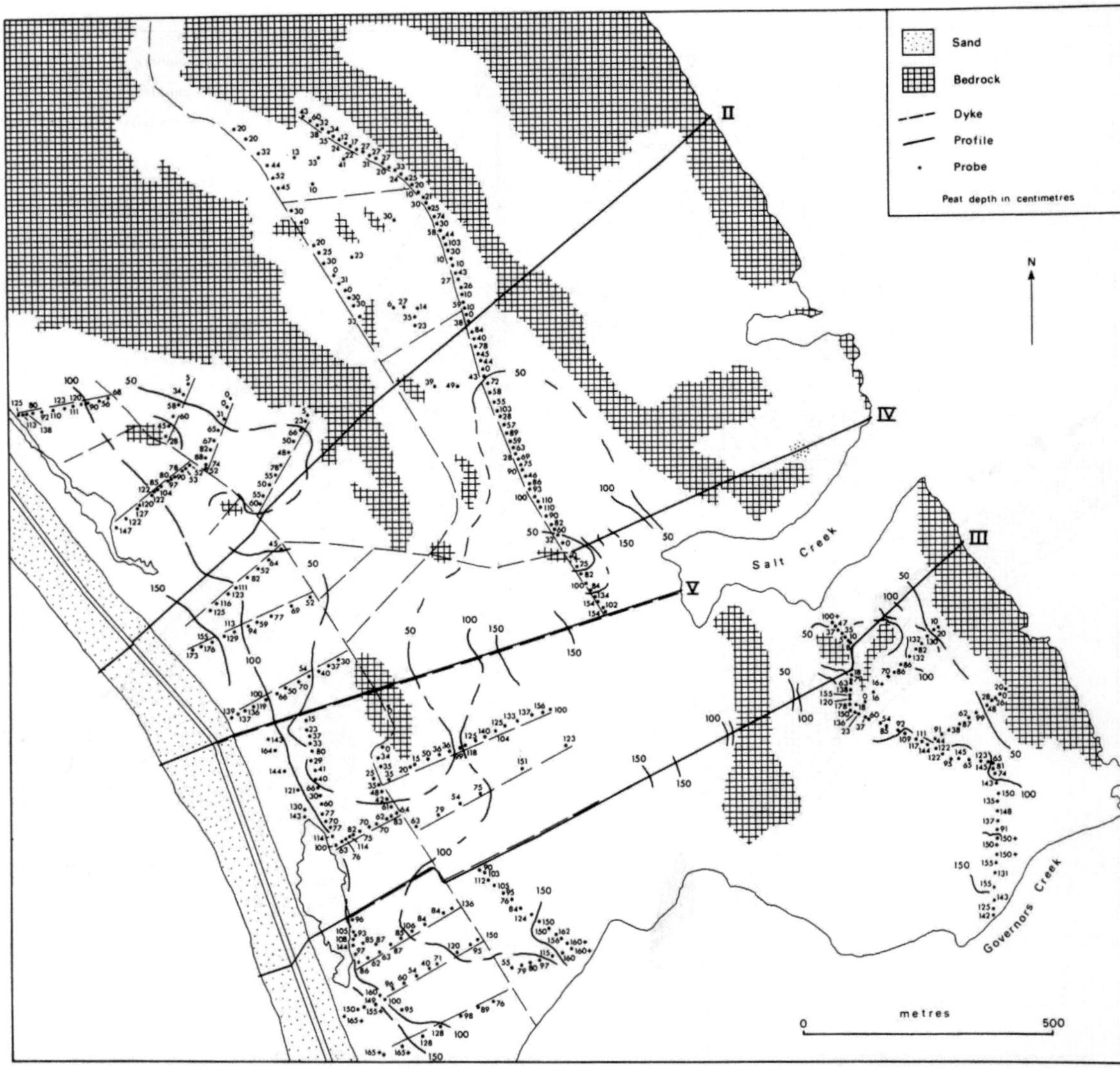

Fig. 3. Sediment depths, northern West Bay peninsula.

association, and *Rhizophora* becomes less frequent until there is a pure *Avicennia* forest. *Avicennia* grows to heights of 15 m in the centre of the peninsula, but decreases to less than 3 m to the northwest. In places *Batis maritima* may form a dense ground cover. In a sense there exists a crude zonation of species away from the source of tidal inundation, which parallels the zonation of species observed elsewhere in the West Indies (Davis, 1940; Chapman, 1944; Dansereau, 1947).

Sediment depth and characteristics

The depth of unconsolidated sediment overlying bedrock was probed throughout the northern West Bay peninsula (Fig. 3) and shows that the physiography of the underlying bedrock is complex. Much of the area to the north is covered by sediment depths of less than 50 cm, but thicknesses increase to the south. Several basins in the centre of the peninsula are deeper than 150 cm, and sediments in these were evidently deposited in isolated depressions in bedrock. The two ridges of emergent bedrock continue in form to the south, but are submerged by Holocene sediments with only occasional exposures.

The nature of the sediments was examined in a series of cores (Fig. 4). Most of the swamp is underlain by mangrove peat. The basal unit in shallow cores to the north is an orange plastic mud which resembles that now accumulating beneath a low herbaceous cover on the emergent bedrock ridges and the 'dry cays' exposed within the swamp. These areas are seasonally-flooded and the mud indicates that the northern part of the peninsula was similarly seasonally-flooded prior to the establishment of mangroves (Woodroffe *et al.*, 1980). A less distinctive green mud, also with a high degree of plasticity, was recorded at depths of more than 130 cm in the isolated basins in the centre of the peninsula, and is also considered to have been deposited in a seasonally-flooded habitat.

Development of the northern West Bay peninsula

The surface of the peat substrate upon which mangroves are presently established lies within the upper part of the tidal range. Where the surface was levelled it was found to be 15–30 cm above mean sea level. (Mean sea level is the datum for the benchmark system of the Directorate of Overseas Survey. It only approximates true mean sea level, ±10 cm, M. Giglioli, pers. comm.) The existence of peats more than 130 cm deep (Fig. 5) indicates that these peats must have been initiated by mangroves growing at a lower elevation than present, and as mangroves are intertidal this must have occurred when sea level was relatively lower. Mangroves were established in those areas which were first flooded by tidal waters, i.e. the lowest-lying parts of the peninsula, and subsequently encroached on previously terrestrial or seasonally-flooded environments.

The calcarenite ridge to the east, however, has acted as a threshold. At present the ridge flanks the east coast of the peninsula and allows tidal water access to the swamp behind only through the breaches, the three creeks, Little Salt Creek, Salt Creek and Governor's Creek. When sea level was lower these breaches were shallower and at some stage the ridge was continuous and prevented inundation of water into the central peninsula.

The morphology of each of these breaches is shown in Fig. 6, illustrating the depth of the creek below still water level and the thickness of sand on the creek floor. The eastern lip of each inlet was examined, though this need not be the shallowest point and erosion may have occurred to deepen the inlet since it became tidal. It appears that Salt Creek was breached when the sea was approximately 200 cm below its present level; but tidal waters evidently did not flood the centre of the peninsula at this stage as sediments around Salt Creek are generally shallow (Fig. 3) and bedrock must have been exposed. Sea level rose and breached Governor's Creek which before dredging was about 175 cm deep, and more recently breached Little Salt Creek which has a maximum depth of 120 cm.

Detailed lithological and palaeontological examination of the bedrock of the area reveals that the present configuration of the surface reflects depositional environments within the calcareous rocks. The Pleistocene rocks have been classified into five facies by Brunt *et al.* (1973): reef, back-reef,

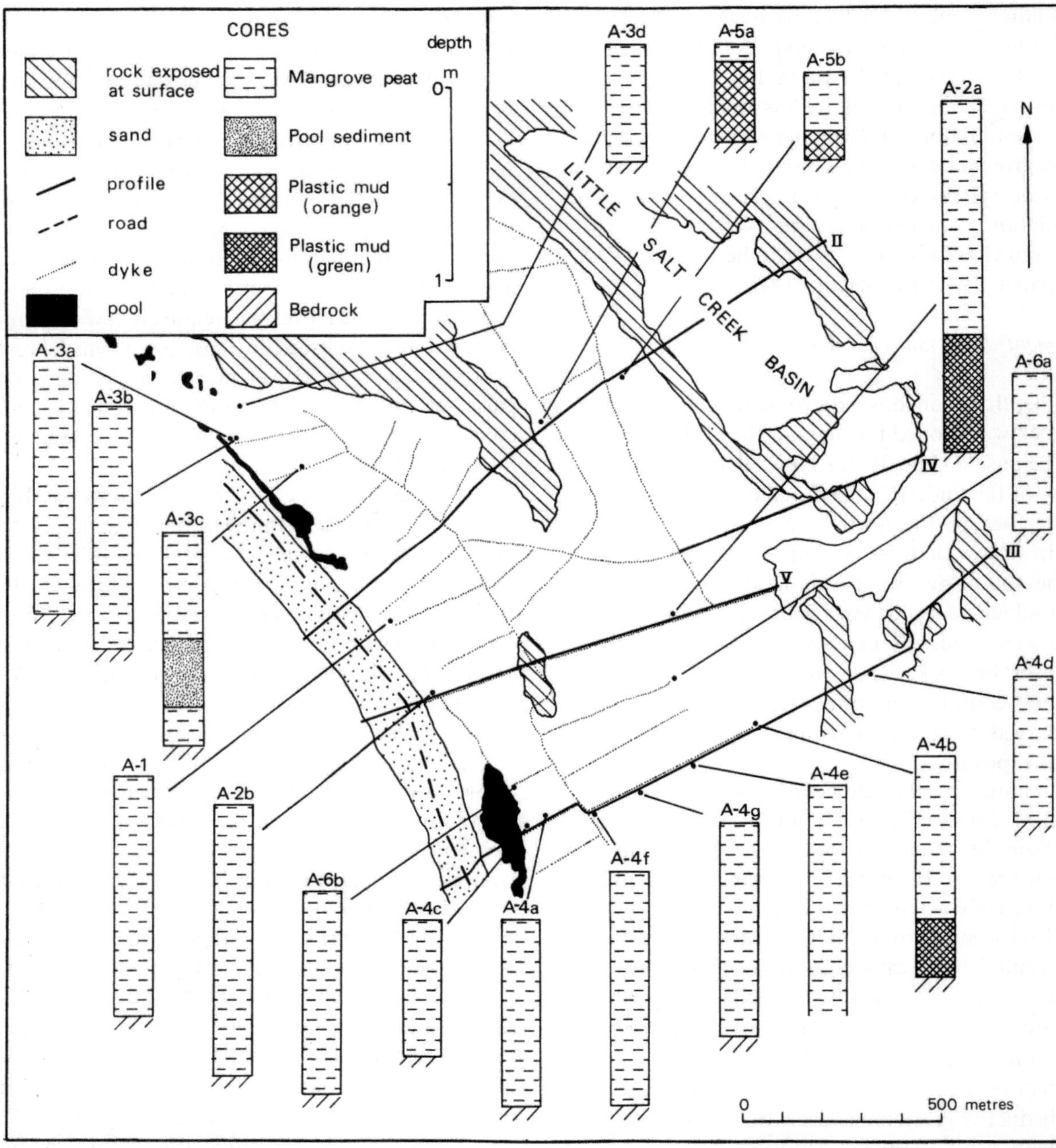

Fig. 4. Stratigraphy of cores, northern West Bay peninsula.

lagoonal, shoal and beach ridge facies. Exposures of rock and excavations for mosquito-control canals by the Mosquito Research and Control Unit, Georgetown, and the facies occurring within the northern West Bay peninsula were mapped (Fig. 7). The calcarenite ridge is part of the shoal facies of cross-bedded oolitic deposits, and this can be seen to extend over much of the east of the area. Most of the peninsula is composed of the lagoonal facies, consisting of poorly-consolidated silts and sands and containing a diverse molluscan fauna. However, the emergent peaks of bedrock, the 'dry cays',

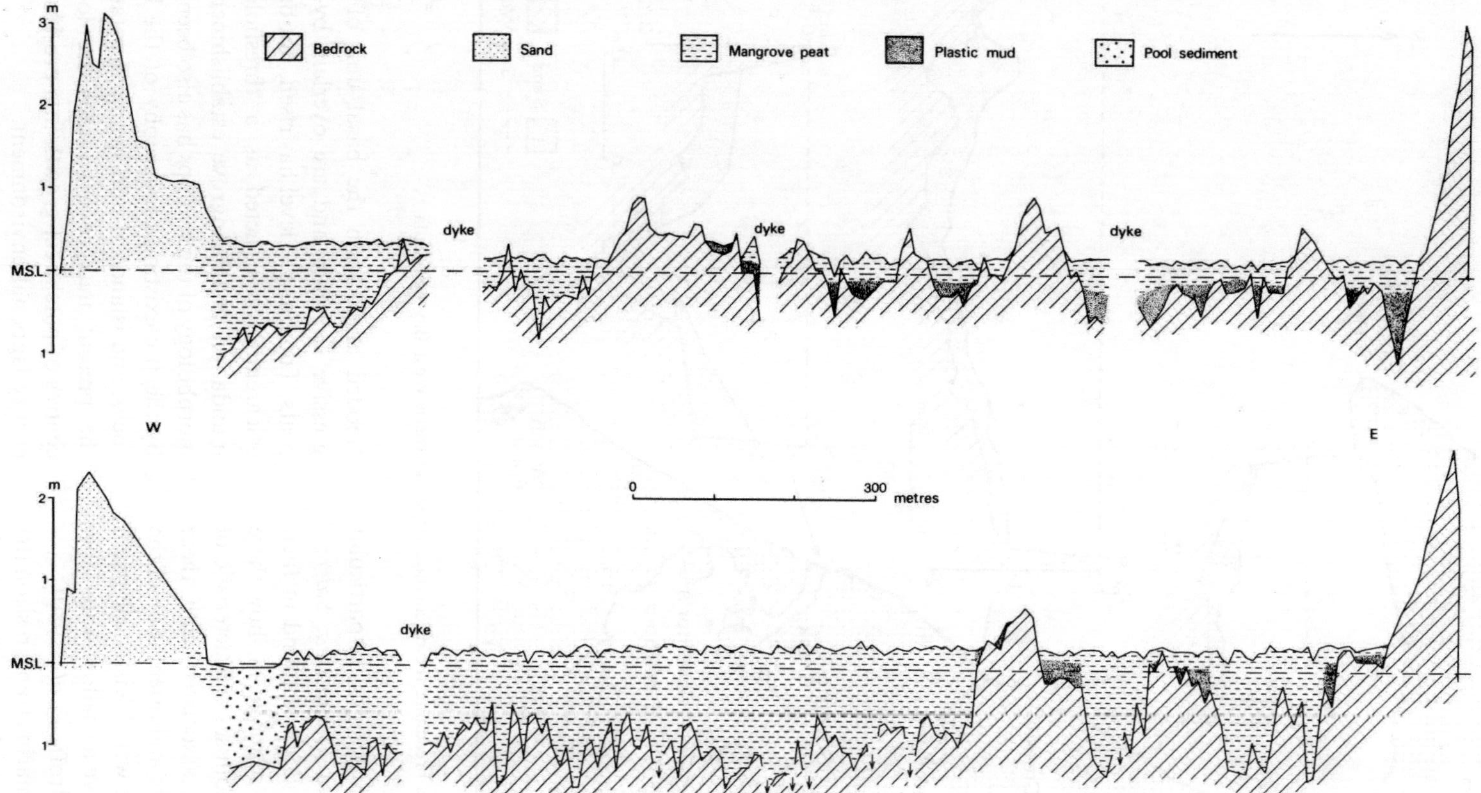

Fig. 5. West-east profiles across the northern West Bay peninsula.

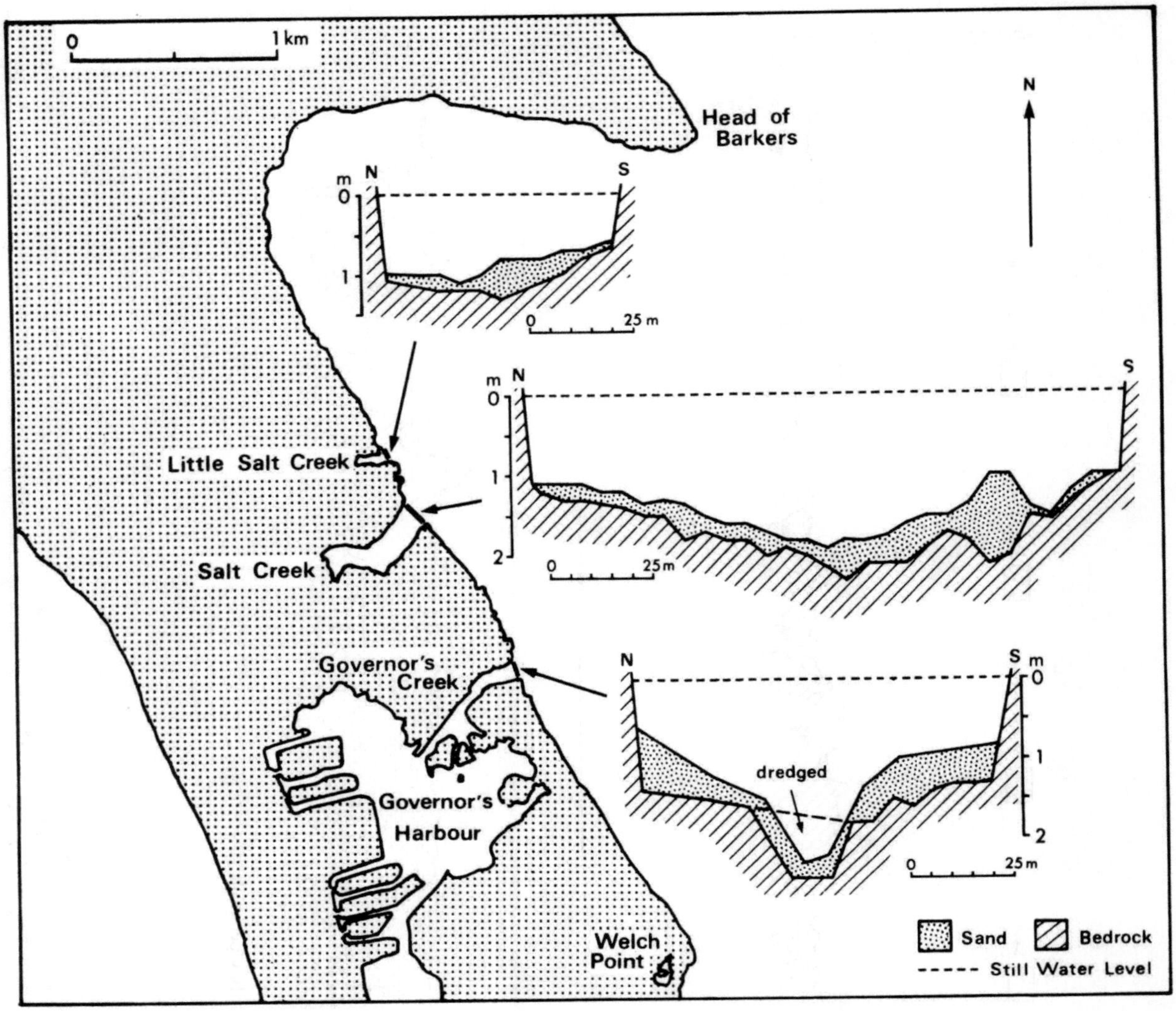

Fig. 6. Morphology of tidal inlets breaching the calcarenite ridge, northern West Bay peninsula.

are composed of coral bearing rocks. In particular massive corals of the genera *Montastrea*, *Siderastrea*, *Diploria* and *Colpophyllia* are found in their position of growth. Counts of molluscs show there to be a predominance of molluscs characteristic of rock and coral substrates associated with these enclaves of coral, and the size and orientation of the outcrops suggests that these were Pleistocene patch reefs which were raised above a shallow-water, low energy lagoon floor (Woodroffe *et al.*, 1980).

During the course of the marine transgression the swamp sediments have encroached on these fossil patch reefs; sediments characteristic of seasonally-flooded areas form the basal unit of the transgressive sequence and are overlain by mangrove peats. Thus, as sea level has risen, not only has the calcarenite ridge acted as a threshold to tidal inundation and mangrove establishment, but the morphology of the swamps has also been controlled by the pre-existing topography of the Pleistocene rocks, and islands of non-swamp vegetation within the present mangrove swamps are localised on centres of patch reef growth in a pre-Holocene low energy lagoonal environment.

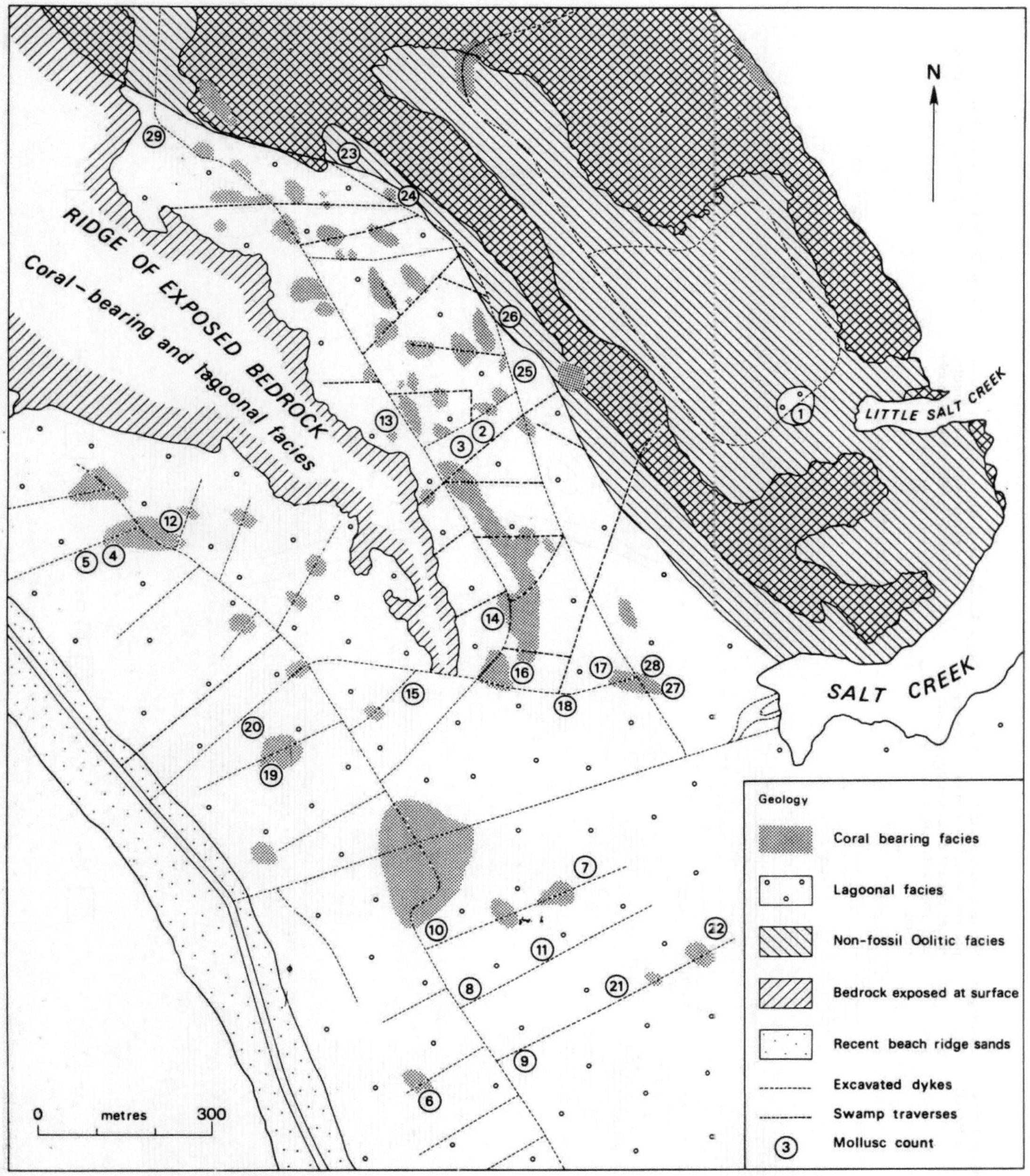

Fig. 7. Distribution of facies of Pleistocene rocks, northern West Bay peninsula.

Development of West Bay peninsula as a whole

The rest of the West Bay peninsula has developed in much the same manner as the northern West Bay peninsula. Sediment depths were probed and sediments were examined in a series of cores in the central and southern part of the peninsula and in the Little Salt Creek basin, and the course of the evolution of the region as a whole has been interpreted (Fig. 8).

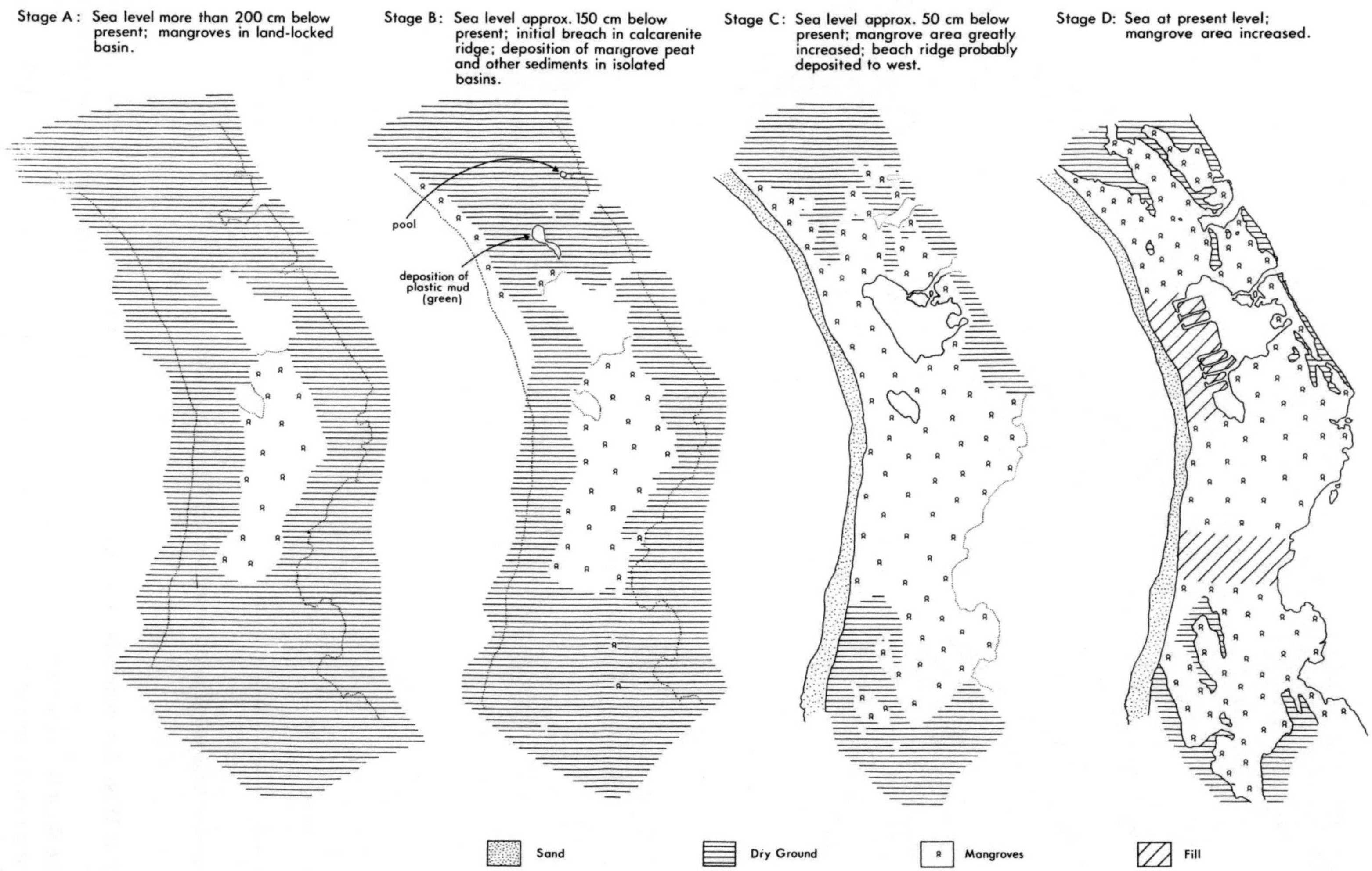

Fig. 8. Development of West Bay peninsula.

When the sea level was 200 cm below present the area of dry ground was more extensive than now. Much of the peninsula was then dry; however, there is a basin of deep peat in the centre, south of Governor's Harbour. The area has a cover of stunted *Rhizophora* scrub, like other dwarf mangrove described from Florida (Davis, 1940), Jamaica (Chapman, 1944) and Barbuda (Stoddart *et al.*, 1973). In places the mangrove peats in this basin exceed 4 m in depth and the mangrove evidently first grew here when sea level was considerably lower. It is not clear how the swamp in this inland depression was initially inundated by tidal water, as sea water presumably did not flow through Governor's Creek which was not breached until the sea reached a level about 175 cm below present (Fig. 5), and no evidence of another direct link with open water has been found.

As the sea rose to 150 cm below its present level, Governor's Creek was breached, and much more of the peninsula was flooded by tidal water with the result that the area of mangrove expanded. However some areas were still not inundated, particularly those behind the calcarenite ridge. A pool existed in the vicinity of Little Salt Creek and basins in the centre of the northern West Bay peninsula were flooded only seasonally and by freshwater, and muds were being deposited.

At a later stage in the marine transgression the mangroves encroached still further on the terrestrial vegetation, as sea level rise made more landward substrates available for colonisation.

Radiometric dating of mangrove peats

The continuous nature of the peat record in cores on Grand Cayman implies that the submergence of the area has been gradual and has not taken place as a rapid increase in the level of the sea concentrated in a short period of time. Mangroves generally grow only in the tidal range (though some mangroves were found growing outside the tidal range above the highest tide in non-marine swamps on Grand Cayman), and in this respect mangrove sediments provide more information on the rate of sea level rise than shallow water evidence, such as corals, which though limited by the level of the sea will continue to grow upwards whether submerged gradually or rapidly.

Radiocarbon dating of samples of peat, laid down at the peat/bedrock interface, is reported in Table 1. The peat contains remains of *Rhizophora* root material and dates for peat from different depths generally confirm that the swamps have developed during a period of sea level rise. Samples BA-3 and BA-9 appear anomalously young, otherwise the relationship between depth and age seems to imply that sea level was about 185 cm below present at least 2100 years ago. Unfortunately radiocarbon dates on mangrove peat are subject to a number of sources of error (Smith and Coleman, 1967), and in particular to contamination by penetration of younger root material from mangroves from above. For this reason dates on mangrove peat can only be regarded as minimum ages and it is not possible to construct a sea level curve for Grand Cayman using only these dates.

Figure 9 shows the sea level curve constructed by Scholl *et al.* (1969), based on dates on mangrove peat, calcitic mud, freshwater gastropods, marine and brackish-water shells from Florida. Also shown on Fig. 9 are dates on mangrove peat from Grand Cayman, Florida, Bahamas and Belize. Almost all the dates (except those from Belize) fall below the line of sea level change proposed by these authors as

Table 1. Radiocarbon dates on samples of mangrove peat from the peat/bedrock interface, Barkers peninsula.

Sample no.	Lab no.	Depth below surface (cm)	Depth below M.S.L. (cm)	^{14}C years B.P.
BA-1	SRR-1002	205	184	2132 ± 40
BA-2	SRR-1003	55	47	842 ± 50
BA-3	SRR-1004	237	209	928 ± 80
BA-4	SRR-1005	170	140	2160 ± 75
BA-5	SRR-1006	133	113	1404 ± 50
BA-6	SRR-1007	78	50	567 ± 55
BA-7	SRR-1008	146	116	1549 ± 70
BA-8	SRR-1009	139	114	1856 ± 60
BA-9	SRR-1010	155	130	878 ± 60

^{14}C half life 5570 ± 30 years. Samples dated at Scottish Universities Research and Reactor Centre.

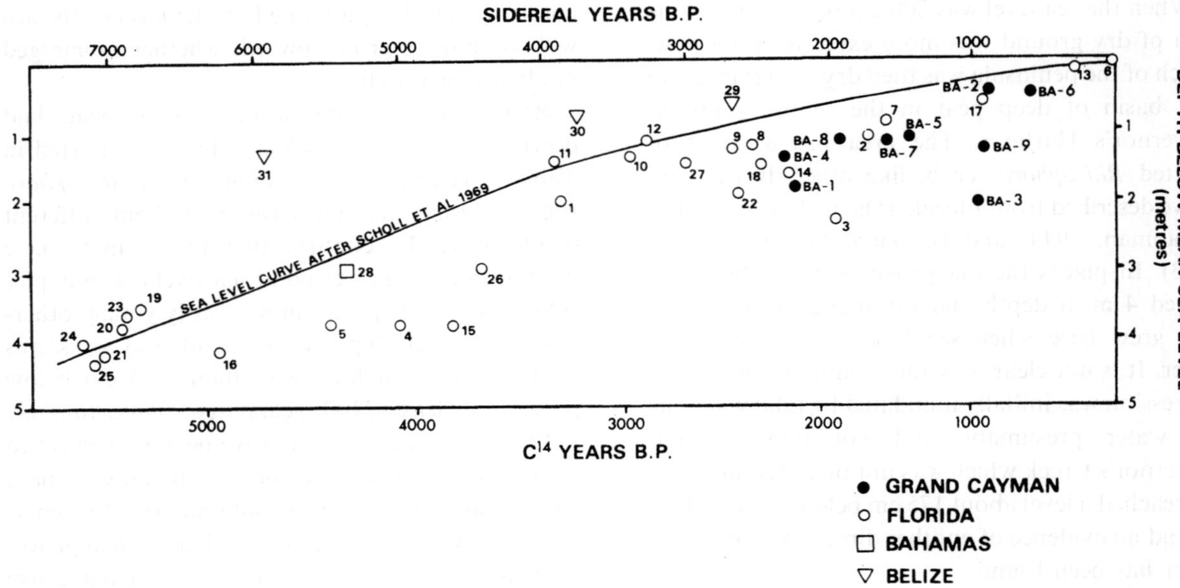

Fig. 9. Radiocarbon dates on mangrove peat from Florida, Bahamas, Belize and Grand Cayman, and their relationship to the sea level curve of Scholl et al. (1969).

would be expected for peats which had been contaminated by younger roots. The Grand Cayman peat samples for the depth from which they were recovered show very similar ages to mangrove peat samples from Florida, and are consistent with the sea level curve of Scholl *et al.* (1969). This suggests that gradual Holocene submergence of Grand Cayman may have occurred at a very similar rate to that recorded in Florida.

Conclusion

On Grand Cayman mangrove swamps receive little allochthonous sediment, but have colonised a stable substrate, the pre-Holocene bedrock surface, and have deposited *in situ* mangrove peats. The initial colonisation of the bedrock has occurred during a period of marine transgression, and the morphology of the swamp has been controlled by the topography of the substrate, which is shown to reflect Pleistocene depositional environments, and resultant energy conditions. Radiometric dating of the mangrove peats confirms that the mangrove forest has expanded, encroaching on previously terrestrial or seasonally-flooded habitats, as submergence has allowed incursion of marine waters into more landward areas.

Grand Cayman is not an area that is geomorphologically active, as are deltas and estuarine environments where landform habitats may evolve and change rapidly, but nevertheless it can be shown that it is the availability of a suitable substrate that is the major control on the establishment of mangroves, and that this can change over a longer time scale, as a function of the rate and stage of marine transgression and consequent energy conditions. In the more active sedimentary environments rates of landform evolution may be rapid enough to mask the influence that changes in sea level have on mangrove swamp development.

It is concluded that mangrove forests can develop in different geomorphological and sedimentological environments and be subject to different processes. The dynamics of mangrove shorelines is dependent upon patterns of sedimentation or substrate adjustment where allochthonous sediment inputs are high or the substrate is evolving, but in stable areas change in the morphology of mangrove swamps occurs in relation to pre-existing topography as a response to movements of sea level.

The geomorphological processes influencing a mangrove swamp in different parts of the world may therefore be different not only because sedimentary characteristics differ in relation to supply of allochthonous sediment and *in situ* production of sediment, but also because the history of sea level change may have been different, and the relationship between land and sea can vary. Study of the course of the development of a mangrove forest may then throw light on the processes presently acting within that swamp.

Acknowledgements

Research on mangrove swamps in the Cayman Islands was undertaken during the tenure of a Natural Environment Research Council Research Studentship. The work was directed by Dr. D.R. Stoddart. Fieldwork was carried out at the Mosquito Research and Control Unit, Grand Cayman. I am grateful to the Cayman Islands Government for provision of accommodation and transport facilities on Grand Cayman, and in particular to Dr. M.E.C. Giglioli for much assistance in the field.

Literature cited

Adey, W.H. 1978. Coral reef morphogenesis: a multidimensional model. Science, N.Y. 202:831–837.

Allen, J.R.L. 1965. Coastal geomorphology of Eastern Nigeria: Beach-ridge barrier islands and vegetated tidal flats. Geologie Mijnb. 44:1–21.

Anderson, J.A.R. 1964. The structure and development of the peat swamps of Sarawak and Brunei. J. Trop. Geogr. 18:7–16.

Banijbatana, D. 1958. Mangrove forest in Thailand. Proc. 9th Pacif. Sci. Congr., Bangkok. 22–34.

Bird, E.C.F. 1971. Mangroves as land-builders. Victoria Nat. 88:189–197.

Bird, E.C.F. 1973. Mangroves and coastal morphology in Cairns Bay, North Queensland. J. Trop. Geogr. 35:11–16.

Bird, E.C.F. and M.M. Barson. 1977. Measurement of physiographic changes on mangrove-fringed estuaries and coastlines. Mar. Res. Indonesia. 18:73–80.

Bloom, A.L. 1970. Paludal stratigraphy of Truk, Ponape and Kusaie, Eastern Caroline Islands. Bull. Geol. Soc. Am. 81:1895–1904.

Brunt, M.A., M.E.C. Giglioli, J.D. Mather, D.J.W. Piper and H.G. Richards. 1973. The Pleistocene rocks of the Cayman Islands. Geol. Mag. 110:209–221.

Butler, A.J., A.M. Depers, S.C. McKillup and D.P. Thomas. 1977. Distribution and sediments of mangrove forests in South Australia. Trans. Roy. Soc. S. Austr. 101:35–44.

Carlton, J.M. 1974. Land-building and stabilization by mangroves. Environ. Conserv. 1:285–294.

Carter, J. 1959. Mangrove succession and coastal change in South-West Malaya. Trans. Inst. Br. Geogr. 26:79–88.

Chapman, V.J. 1944. 1939 Cambridge University Expedition to Jamaica, I. A study of the botanical processes concerned in the development of the Jamaican shore-line. Jour. Linn. Soc. Bot. 52:407–447.

Chapman, V.J. 1960. Salt Marshes and Salt Deserts of the World. Plant Science Monographs, London, Leonard Hill, 392 pp.

Chapman, V.J. 1970. Mangrove phytosociology. Trop. Ecol. 11:1–19.

Chapman, V.J. 1975. Mangrove Vegetation. Vaduz, J. Cramer, 447 pp.

Cohen, A.D. 1968. The petrology of some peats of southern Florida (with special reference to the origin of coal). Ph.D. thesis Pennsylvania State University.

Cohen, A.D. and W. Spackman. 1977. Phytogenic organic sediments and sedimentary environments in the Everglades-Mangrove Complex. Pt II. The origin, description and classification of the peats of southern Florida. Palaeontographica B 162:71–114.

Curtis, A.H. 1888. How the mangroves form islands. Garden and Forest. 1:100.

Dansereau, P. 1947. Zonation et succesion sur la restinga de Rio de Janeiro I La holosere. Revue Can. Biol. 6:448–477.

Davies, J.L. 1972. Geographical Variation in Coastal Development. Edinburgh, Oliver and Boyd, 204 pp.

Davis, J.H. Jr. 1938. Mangroves, makers of land. Nature Mag. 31:551–553.

Davis, J.H. Jr. 1940. The ecology and geologic role of mangroves in Florida. Pap. Tortugas Lab. 32:303–412.

Easton, W.H. and E.A. Olson. 1976. Radiocarbon profile of Hanauma Reef, Oahu, Hawaii. Bull. Geol. Soc. Am. 87:711–719.

Ebanks, W.J. Jr. 1967. Recent carbonate sedimentation and diagenesis, Ambergris Cay, British Honduras. Ph.D. thesis, Rice University.

Ebanks, W.J. Jr. 1975. Holocene Carbonate Sedimentation and Diagenesis, Ambergris Cay, Belize. In: Belize (British Honduras) Shelf-Carbonate Sediments, Clastic Sediments and Ecology, 234–296.

Egler, F.E. 1952. Southeast saline Everglades vegetation, Florida, and its management. Vegetatio 3:213–265.

Enos, P. and R.D. Perkins. 1979. Evolution of Florida Bay from island stratigraphy. Bull. Geol. Soc. Am. 90:59–83.

Fairbridge, R.W. and C. Teichert. 1948. The Low Isles of the Great Barrier Reef: a new analysis. Geogr Jour. 111:67–88.

Focke, J.W. 1978. Holocene development of coral fringing reefs, Leeward off Curaçao and Bonaire (Netherlands Antilles). Marine Geology, 28:M31–M41.

Fosberg, F.R. 1964. Vegetation as a geological agent in tropical deltas. In: UNESCO, Les problems scientifiques des deltas de la zone tropicale humide et leurs implications, 227–233.

Gill, A.M. 1971. Mangroves: is the tide of opinion turning? Fairchild Trop. Gard. Bull. 26:5–9.

Gleason, P.J., A.D. Cohen, W.G. Smith, H.K. Brooks, P.A. Stone, R.L. Goodrick and W. Spackman. 1974. The environmental significance of Holocene sediments from the Everglades and saline tidal plain. In: Environments of South Florida: present and past. Miami Geol. Soc. Mem. 2:287–341.

Hewitt, B.R. 1967. The occurrence, origin and vegetation of lowland peat in Malaya. Proc. Linn. Soc. N.S.W. 92:58–66.

Hitchcock, A.S. 1891. A visit to the West Indies. Bot. Gaz. 1891:130–141.

Hodge, W.H. 1956. The trees that walk to the sea. Nature Mag. 49:456.

Kuenzler, E.J. 1974. Mangrove swamp systems. In: Coastal Ecological Systems of the United States, the Conservation Foundation, Washington D.C. 1:346–371.

Lugo, A.E. and S.C. Snedaker. 1974. The ecology of mangroves. Ann. Rev. Ecol. Syst. 5:39–64.

McCoy, E.D. and K.L. Heck Jr. 1976. Biogeography of corals, sea-grasses, and mangroves: an alternative to the centre of origin concept. Syst. Zool. 25:201–210.

MacIntyre, I.G. and P.W. Glynn. 1976. Evolution of modern Caribbean fringing reef, Galeta Point, Panama. Bull. Am. Ass. Petrol. Geol. 60:1054–1072.

McLean, R.F., D.R. Stoddart, D. Hopley and H. Polach. 1978. Sea level change in the Holocene on the northern Great Barrier Reef. Phil. Trans. Roy. Soc. A. 291:167–186.

Macnae, W. 1966. Mangroves in eastern and southern Australia. Aust. Jour Bot. 14:67–104.

Macnae, W. 1968. A general account of the fauna and flora of mangrove swamps and forests in the Indo-West-Pacific region. Advances Marine Biol. 6:73–270.

Matley, C.A. 1926. The geology of the Cayman Islands (British West Indies) and their relation to the Bartlett Trough. Q. Jour. Geol. Soc. Lond. 82:352–386.

Montaggioni, L. 1976. Holocene submergence on Réunion Island (Indian Ocean). Ann. S. Afr. Mus. 71:69–75.

Phillips, O.P. 1903. How the mangrove tree adds new land to Florida. Jour. Geogr. N.Y. 2:10–21.

Polach, H.A., R.F. McLean, J.R. Caldwell and B.G. Thom. 1978. Radiocarbon ages from the northern Great Barrier Reef. Phil. Trans. Roy. Soc. A. 291:139–158.

Richards, P.W. 1952. The Tropical Rain Forest: an Ecological Study. Cambridge Univ. Press, 450 pp.

Riegel, W. 1965. Palynology of environments of peat formation in southwestern Florida. Ph.D. thesis, Pennsylvania State University.

Rigby, J.K. and H.H. Roberts. 1976. Grand Cayman Island; Geology, Sediments and Marine Communities: Brigham Young University Geology Studies, Spec. Publ. 4:122 pp.

Rosevear, D.R. 1947. Mangrove swamps. Farm and Forest. 8:23–30.

Salomon, J.N. 1978. Contribution à l'étude écologique et géographique des mangroves. Revue Géomorph Dyn. 27:63–80.

Scholl, D.W. 1964a. Recent sedimentary record in mangrove swamps and rise in sea level over the southwestern coast of Florida. I. Marine Geology. 1:344–366.

Scholl, D.W. 1964b. Recent sedimentary record in mangrove swamps and rise in sea level over the southwestern part of Florida II. Marine Geology. 2:343–364.

Scholl, D.W. 1969. Modern coastal swamp stratigraphy and the ideal cyclothem. In: Environments of Coal Deposition, Geol. Soc. Am. Spec. Pap. 114:37–61.

Scholl, D.W., F.C. Craighead and M. Stuiver. 1969. Florida submergence curve revised: its relation to coastal sedimentation rates. Science, N.Y. 163:562–564.

Smith, W.G. 1968. Sedimentary environments and environmental change in the peat-forming area of South Florida. Ph.D. thesis: Pennsylvania State University.

Smith, W.G. and J.M. Coleman. 1967. Recent submergence of southern Florida: Discussion. Bull. Geol. Soc. Am. 78:1191–1194.

Spackman, W., C.P. Dolsen and W. Riegel. 1966. Phytogenic organic sediments and sedimentary environments in the Everglades-mangrove complex. Part 1 – Evidence of a transgressing sea and its effect on environments of the Shark River area of southwest Florida. Palaeontographica B. 117:135–152.

Spackman, W., W.L. Riegel and C.P. Dolsen. 1969. Geological and biological interactions in the swamp-marsh complex of southern Florida. In: Environments of Coal Deposition, Geol. Soc. Am. Spec. Pap. 114:1–35.

Steenis, C.G.G.J. van 1958. Tropical shore formations. Proc. Symp. Humid Trop. Vegetation (Tijiawi, Indonesia) 215–217.

Steers, J.A. 1937. The coral islands and associated features of the Great Barrier Reefs. Geogr. Jour. 89:1–28, 119–139.

Stephens, W.M. 1962. Trees that make land. Sea Front. 8:219–230.

Stoddart, D.R. 1980. Mangroves as successional stages, inner reefs of the northern Great Barrier Reef. J. Biogeogr. 7:269–284.

Stoddart, D.R., G.W. Bryan and P.E. Gibbs. 1973. Inland mangroves and water chemistry, Barbuda, West Indies. Jour. Nat. Hist. 7:33–46.

Stoddart, D.R., R.F. McLean, T.P. Scoffin, B.G. Thom and D. Hopley. 1978a. Evolution of reefs and islands, northern Great Barrier Reef: synthesis and interpretation. Phil. Trans. Roy. Soc. Lond. B. 284:149–160.

Stoddart, D.R., R.F. McLean, T.P. Scoffin and P.E. Gibbs. 1978b. Forty-five years of change on low wooded islands, Great Barrier Reef. Phil. Trans. Roy. Soc. B 284:63–80.

Thom, B.G. 1967. Mangrove ecology and deltaic geomorphology: Tabasco, Mexico. Jour. Ecol. 55:301–343.

Thom, B.G. 1975. Mangrove ecology from a geomorphic viewpoint. Proc. Int. Symp. Biology Management Mangroves 2:469–481.

Thom, B.G., L.D. Wright and J.M. Coleman. 1975. Mangrove ecology and deltaic-estuarine geomorphology: Cambridge Gulf-Ord River, Western Australia. Jour. Ecol. 63:203–232.

Turmel, R.J. and R.G. Swanson. 1976. The development of Rodriguez Bank, a Holocene mudbank in the Florida Reef Tract. J. Sedim. Petrol. 46:497–518.

Vann, J.H. 1959. Landform-vegetation relationships in the Atrato delta. Ann. Ass. Am. Geogr. 49:345–360.

Vaughan, T.W. 1909. The geologic work of mangroves in southern Florida. Smithson, Misc. Collns. 52:461–464.

Walsh, G.E. 1974. Mangroves: a review. In: Ecology of Halophytes. Academic Press New York and London, 51–174.

Wanless, H.R. 1974. Mangrove sedimentation in geologic perspective. In: Environments of South Florida: present and past. Miami Geol. Soc. Mem. 2:190–200.

Watson, J.G. 1928. Mangrove forests of the Malay Peninsula. Malay. Forest Rec. 6:1–275.

Welch, B.L. 1963. From coral reef to tropical island via *Thalassia* and mangrove. Va. Jour. Sci. 14:213–214.

West, R.C. 1956. Mangrove swamps of the Pacific coast of Colombia. Ann. Ass. Am. Geogr. 46:98–121.

Woodroffe, C.D., D.R. Stoddart and M.E.C. Giglioli. 1980. Pleistocene patch reefs and Holocene swamp morphology, Grand Cayman Island, West Indies. Jour. Biogeogr. 7:103–113.

Zahran, M.A. 1977. Africa A. Wet formations of the African Red Sea coast. In: Wet Coastal Ecosystems: Ecosystems of the World Elsevier, Amsterdam 215–231.

CHAPTER 2

Evidence for an Upper Carboniferous mangrove community

ANNE RAYMOND*† and TOM L. PHILLIPS**

* *Department of Geophysical Sciences, University of Chicago, Chicago, IL, U.S.A.*
** *Botany Department, University of Illinois, Urbana, IL, U.S.A.*

† Present address: Dept. Geology, Texas A and M University, College Station, Texas 17843

Abstract. The morphological and paleogeographic evidence suggesting that cordaitean-dominated swamps from the early Westphalian D (= 295 m.y.b.p.) of Iowa grew in salt water swamps is reviewed. Cordaitean trees, *Psaronius* tree ferns, and *Medullosa* seed ferns contributed 95% of the peat in coal-ball samples taken from three Iowa mines. Ordination analysis shows that associations dominated by *Psaronius* and *Medullosa* succeeded an association dominated by cordaitean trees at each site. Based on comparisons with Upper Carboniferous compression flora deposits, *Psaronius* and *Medullosa* seem to have grown in freshwater or only slightly brackish water habitats. The root morphology of cordaitean trees from Iowa and the paleogeographic position of these swamps suggest that cordaitean trees which produced *Nucellangium glabrum* and *Cardiocarpus spinatus* seeds grew in salt water. Results of the ordination analysis suggest that cordaitean trees which produced *Mitrospermum* seeds grew in freshwater or only slightly brackish habitats. The methods used to assess these deposits as mangrove peats can be applied to any structurally preserved peat. Most other putative mangrove communities in the fossil record have been described from compression-flora deposits. The non-morphological criteria for assessing these floras as mangrove swamp deposits are reviewed.

Introduction

Genera of extant mangrove communities can be traced back to the early Tertiary (55 m.y.b.p.) largely by means of palynology, and there is a gradual expansion in the diversity and association of mangrove genera from earliest Eocene time. Pollen from the back mangrove genera, *Nypa* and *Brownlowia*, occur together in the lower Eocene of Borneo, while mangroves with representatives adapted to more coastal environments appear in successive epochs: *Rhizophora* pollen in the early Oligocene; *Sonneratia* pollen in the early Miocene and *Avicennia* in the middle Miocene (Muller, 1964). From the lower Eocene Deccan Intertrappean Beds of India and deposits of similar age near Kutch, India, Lakhanpal (1974) has described pollen assigned to *Rhizophora*, *Sonneratia* and *Nypa*. From the middle Eocene Plantagenet Group of Australia, Churchill (1973) has described pollen of *Nypa*, *Avicennia*, the *Rhizophora* tribe and a possible *Sonneratia* type. Apart from inferences based on similarities of microfossil assemblages to those of modern mangroves where habitats are known, the task of establishing mangrove habitats earlier in the fossil record is heavily dependent on geological information. The oldest putative mangrove community of angiosperms has been described from the Late Cretaceous Dakota Sandstone of Kansas by Retallack and Dilcher (1981) who suggest that such an environment played a major role in the dispersal of early angiosperms.

Morphological adaptations do not directly separate freshwater and mangrove swamp plants

Tasks for vegetation science, Vol. 8 ed. by H.J. Teas
 ISBN 978-90-481-8526-9

because of the numerous similar adaptations (Duddington, 1974). Many modern mangroves have viviparous reproduction, whereas vivipary occurs in few freshwater swamp plants (Duddington, 1974). However in pre-angiosperm times, the reproductive adaptations of mangroves were necessarily different. We do not know if sexually reproductive lower vascular plants could reproduce in mangrove habitats because of the salinity, and it is uncertain how the gymnosperms were adapted reproductively to such habitats since none of them grow in salt water today. Nevertheless, vast coastal areas have been available in the geologic past for mangrove habitats and one of the most extensive such intervals is the Late Carboniferous or Age of Coal when much of Europe and North America was located in tropical and subtropical paleolatitudes. The cordaitean-rich peat swamps of the Western Interior Coal Region of the United States, particularly those in Iowa, have been suggested as the oldest putative mangrove swamps (Cridland, 1964; Phillips, 1979).

The objectives of this preliminary report on the fossil swamp vegetation of Iowa are to quantitatively document the composition of the cordaitean-rich peat beds and to discuss the criteria for placing some of these plants in a mangrove habitat as well as criteria for other kinds of fossil beds that may represent mangrove environments.

Most of the *in situ* and anatomically preserved fossil vascular plants of the geological past occur as partially preserved peat beds in coal-ball concretions of Upper Carboniferous coal seams in the United States and Europe. They provide the most direct evidence of swamp vegetation available from subtropical and tropical climates of the past (Phillips, 1980). These occurrences are sufficiently scattered stratigraphically and geographically so as to provide an overview of the vegetational patterns in swamps throughout the Late Carboniferous (Phillips *et al.*, 1974).

The coal fields of the midcontinent region of the United States are divided into Western and Eastern Coal Regions; the Iowa coal balls come from the Forest City Basin of the Western Region. Arborescent lycopods generally dominate the coal-ball floras of the Eastern Coal Region and in the western European coal fields during the Early and Middle Pennsylvanian (Phillips, 1980). The composition of Iowa coal balls differs markedly from the quantitative composition of these lycopod-dominated coal-ball floras, both in the abundance and diversity of cordaitean trees.

The concept that some gymnospermous cordaitean trees had a mangrove growth habit goes back to the study of Cridland (1964) in which he suggested that the root systems known as *Amyelon iowense*, which belonged to some cordaitean trees, possessed many anatomical traits also found in modern mangroves. Cridland (1964) also noted that such an interpretation indicated the presence of a marine influence within the Western Coal Region. our evidence lends support to such a mangrove interpretation for some of the cordaitean trees. It seems likely that some of the Iowa coal swamps represent an extreme environment, even for the Western Coal Region.

American coal balls were first recognized as such from Iowa by Stopes and Watson (1908) and some of the earliest morphological studies of the anatomically preserved plants were begun by Darrah (1939) and by Andrews (1951) who have provided reviews. The most unusual feature of the vegetation from the classic localities of the Shuler and Urbandale Mines near Des Moines, Iowa is the abundance and diversity of cordaitean trees and the dearth of lycopods (Darrah, 1941).

Materials and methods

The coal-ball samples used in this study were collected from working coal mines between 1935 and 1945 and were donated to the paleobotanical collection of Harvard University. No data as to the relative positions of individual coal balls within the coal seam accompanied these samples. The samples come from the following mines: the Shuler Mine and the Urbandale Mine located near the city of Des Moines in Dallas Co., Iowa; and the Williamson Mine located in Lucas Co., Iowa in the south central portion of the state. Darrah (1941) gives locations for these and other Iowa coal-ball localities.

Unlike most of the coal-ball sources from other Euramerican coals, the identification of the coal or coals yielding the coal balls in Iowa has been lacking, as is a precise stratigraphic placement of these Iowa coals within the Desmoinesian Series. However, recent biostratigraphic studies in Iowa coals by Ravn (in press) and by R.A. Peppers (personal communication, 1980) indicate that the Iowa coals represented in this study are approximately equivalent to the Buffaloville Coal Bed in the Illinois Basin and early Westphalian D (= 295 m.y.b.p.) in age. Ravn (in press) treats this interval as interval 3 and the coals are rich in *Florinites* pollen from cordaitean trees.

Fifteen coal balls from each mine were examined in the study. The contents of ten of the coal balls from each mine had never been inspected prior to this study. Five coal balls from each locality were taken from the previously sectioned coal balls in the paleobotanical collections. Selected coal balls from museum collections are likely to introduce some bias into the quantitative analysis of the vegetation.

Two methods used in the study of the coal-ball vegetation of the three mines are quantitative volumetric analysis of floral constituents and ordination analysis. In order to measure the volumetric contribution of coal-ball plants to the peat, uncut coal balls were sectioned and cellulose acetate peels of the cut surfaces of all coal balls were made according to techniques given in Phillips, Avcin and Berggren (1976). An attempt was made to identify each constituent on the surface of the peels. The surface area of each constituent on the peel was measured with a 2 mm^2 grid. Approximately 85% of the plant constituents in each coal-ball peel was identified. The sum of surface areas of floral constituents belonging to each genus was tallied. This sum reflects the volumetric contribution of each plant type to the peat, and presumably the importance of that plant type in the original coal swamp. The sample sizes used for this study are considerably smaller than those recommended for quantitative volumetric analysis by Phillips *et al.* (1977). However, the method used to investigate these samples also differs from that outlined in the earlier study in that all recognizable peat constituents were measured and counted. Measuring all the constituents on a peel surface maximized the amount of information available from a small sample.

Because the relative positions of coal balls within the coal is not known for these samples, the occurrence of peat constituents within coal balls from the same level of the coal seam can not be used to determine paleoecological associations or successions of plants. Instead, a form of ordination analysis, reciprocal averaging analysis, was used to associate genera on the basis of their co-occurrence on the surfaces of the sampled coal-ball peels. Hill (1973) discusses reciprocal averaging analysis which is an iterative ordination technique similar to weighted averaging. The main advantages of reciprocal averaging analysis are mathematical simplicity and ability to deal with data containing genera with irregular, non-normal distributions (Gauch *et al.*, 1977). The results of a reciprocal averaging analysis are most straightforward when the data contain a single gradient (e.g. samples going from wet to dry soil). However, this technique can also uncover multiple gradients when they occur (Gauch *et al.*, 1977).

Results

Quantitative volumetric analysis

Results of the quantitative volumetric analyses of floral constituents in the three coal-ball samples appear in Table 1. The generic level was used in this analysis. Quantitative volumetric analyses of the three samples show that detritus derived from cordaitean trees, *Psaronius* tree ferns and *Medullosa* seed ferns compose nearly 95% of the peat in these deposits. Cordaitean trees are coniferophyte gymnosperms with large strap-like leaves; they contributed from 51 to 71% of the peat. Three genera and four species of cordaitean seeds occur in the samples. The principal components of the cordaitean peat are roots and leaf litter. Marattiaceous *Psaronius* tree ferns contributed from 16 to 18% of the peat. The trunk of *Psaronius* consisted of a polycyclic stele, an inner root mantle and an outer aerenchymatous root mantle which extended downward as the subterranean root mass of the plant

Table 1. Quantitative volumetric analysis of three cordaitean-dominated coal-ball peat deposits (percent volume).

	Williamson	Urbandale	Shuler
Total Cordaitean debris	56	51	71
Roots and aerial wood	30	36	40
Rootlets	8	2	11
Leaves	17	6	13
Inflorescences	4	0.6	5
Mitrospermum leeanum (seed)			
M. florini (seed)	0.1	0.4	0.1
Cardiocarpus spinatus (seed)	0.5	0.1	1
Nucellangium glabrum (seed)	3	0.1	1
Total *Medullosa* debris	18	21	5
Stem wood	1	2	0.7
Roots	2	0.8	0.4
Foliage	15	18	3
Pachytesta (seed)	—	0.5	1
Total *Psaronius* debris	16	18	17
Inner root zone	11	6	1
Outer root zone	5	12	16
Total *Arthropitys*	8	7	2
A. kansana	4	0.6	0.5
A. communis	—	6.0	—
A. sp A	4	—	0.3
Undescribed species	—	—	1.0
Roots	0.6	0.5	—
Sphenophyllum	0.1	0.3	0.1
Botryopteris	0.3	0.1	0.3
Anachoropteris	0.02	0.2	0.2
Etapteris	—		
Schopfiastrum	0.5	0.5	—
Callistophyton	0.3	0.2	—
Heterangium	0.4	0.04	—
Pteridosperm roots	0.1	0.1	—
Tyliosperma (seed)	0.03	0.1	—
Sporangia (undifferentiated)	1	1.0	2.0

(Ehret and Phillips, 1977). *Psaronius* stems and leaves appear to have decayed quickly; subterranean roots comprise the most common *Psaronius* fossil in these deposits. *Medullosa* was an unbranched seed fern with large fronds; some medullosas (*Medullosa noei*) have been reconstructed with a small tree habit (Stewart and Delevoryas, 1956). *Medullosa* contributed 5 to 21% of the peat. All the seeds attributed to *Medullosa* in these samples appear to belong to the same genus and species.

Two additional arborescent forms occur in these peats, *Arthropitys*, a calamites tree, and lycopods. *Arthropitys* contributed 2 to 8% of the peat. No arborescent lycopods were actually identified in the small samples used in this quantitative volumetric analysis of floral constituents; stigmarian roots belonging to arborescent lycopods do occur in a large sample of coal balls (approximately 90 coal balls) inspected from the Urbandale Mine. Lycopod roots comprise approximately 5% of this larger coal-ball sample.

Herbaceous or low scrambling forms contributed

less than 2% of the detritus. These forms belong to sphenopsids, coenopterid ferns and small seed ferns.

Reciprocal averaging analysis

Results of the reciprocal averaging analyses of the three samples appear in Fig. 1–3. Each figure shows the position of genera on reciprocal averaging axes 1 and 2. Peat constituents which plot close together often occur in the same coal-ball slabs. Peat constituents which are widely separated on the two axes seldom occur in the same coal-ball slabs.

In the reciprocal averaging analysis of the Williamson sample (Fig. 1) cordaitean trees, *Psaronius* and *Medullosa* occur along a single gradient. Cordaitean trees establish one end of the gradient. *Psaronius* falls in the middle of the gradient and *Medullosa* establishes the opposite end of the gradient.

In the reciprocal averaging analysis of the Urbandale sample (Fig. 2), cordaitean trees, *Medullosa* and *Psaronius* have a forked configuration. Cordaitean trees establish one end of the gradient. *Medullosa* ordinates in the upper right corner of the plot and forms a cordaitean – *Medullosa* gradient. *Psaronius* ordinates in the lower right corner of the plot and forms a cordaitean – *Psaronius* gradient.

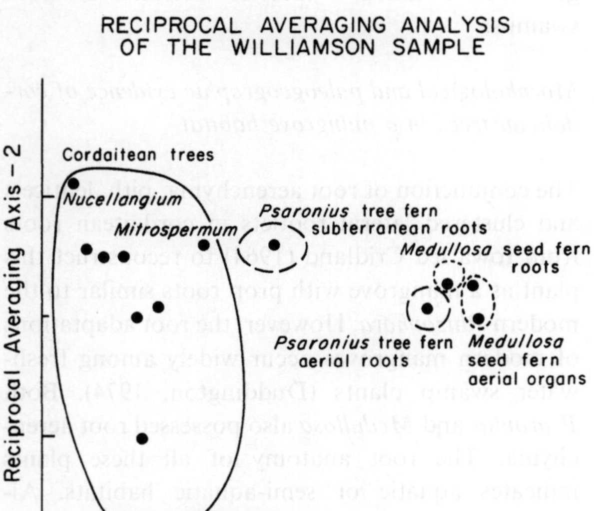

Fig. 1. Reciprocal averaging analysis of the Williamson Mine coal-ball vegetation.

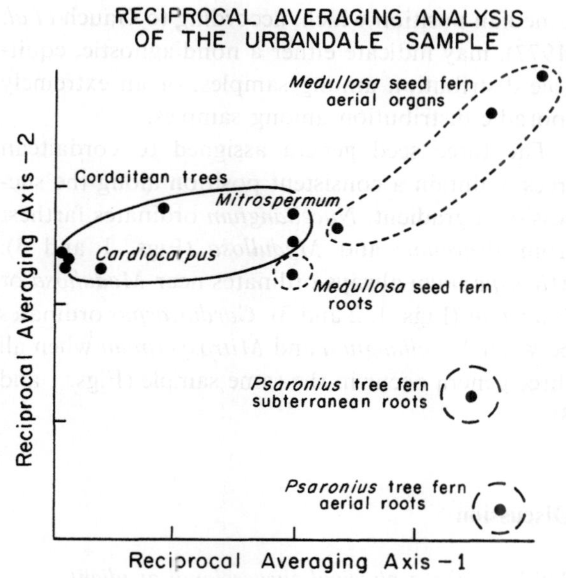

Fig. 2. Reciprocal averaging analysis of the Urbandale coal-ball vegetation.

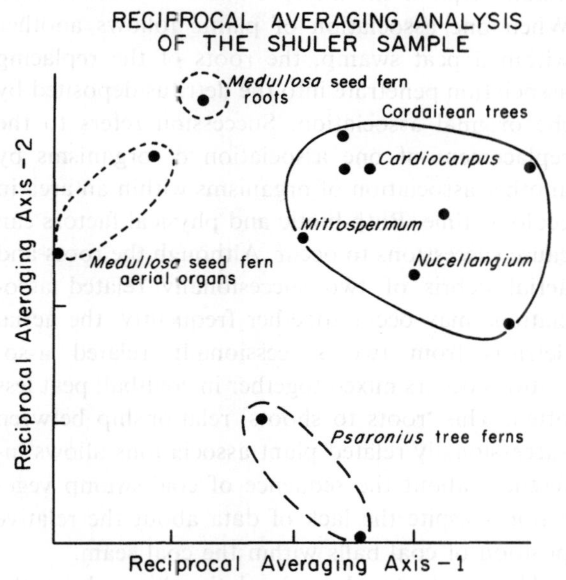

Fig. 3. Reciprocal averaging analysis of the Shuler Mine coal-ball vegetation.

In the reciprocal averaging analysis of the Shuler sample (Fig. 3) *Medullosa* establishes one end of the gradient and cordaitean trees establish the other. On this plot, *Psaronius* falls beneath the gradient in a median position which, according to Gauch *et al.* (1977), may indicate either a nondiagnostic, equitable distribution among samples, or an extremely sporadic distribution among samples.

The three seed genera assigned to cordaitean trees maintain a consistent position along the successional gradient. *Nucellangium* ordinates farthest from *Psaronius* and *Medullosa* (Figs. 1 and 3). *Mitrospermum* always ordinates near *Medullosa* or *Psaronius* (Figs. 1, 2 and 3). *Cardiocarpus* ordinates between *Nucellangium* and *Mitrospermum* when all three genera occur in the same sample (Figs. 1 and 3).

Discussion

Evidence for a physical successional gradient

An understanding of the process of peat accumulation helps in the interpretation of these results. When one association of plants follows another within a peat swamp, the roots of the replacing association penetrate into the detritus deposited by the original association. Succession refers to the replacement of one association of organisms by another association of organisms within an area in ecologic time. Both biotic and physical factors can cause successions to occur. Although the roots and aerial debris of two succesionally related associations may occur together frequently, the aerial detritus from two successionally related associations occurs mixed together in coal-ball peat less often. This 'roots to shoots' relationship between successionally related plant associations allows inferences about the sequence of coal-swamp vegetation, despite the lack of data about the relative position of coal balls within the coal seam.

The 'roots to shoots' relationships along the gradient in the Williamson sample (Fig. 1) indicate a successional gradient in which cordaitean trees are replaced by *Psaronius* which is subsequently replaced by *Medullosa*. In the Urbandale sample, *Medullosa* replaced cordaitean trees along one gradient and *Psaronius* replaced cordaitean trees along the other (Fig. 2). Although the order of *Medullosa* and cordaitean trees along the axis is reversed in the Shuler sample (Fig. 3), the 'roots and shoots' relationship of cordaitean trees and *Medullosa* in these samples indicate that *Medullosa* replaced the cordaitean tree association in this swamp. Because *Psaronius* falls below the gradient, its position along the successional gradient in the Shuler sample is unknown.

The three samples indicate a successional gradient in which either *Medullosa* or *Psaronius* replaces the cordaitean tree association. In one swamp *Medullosa* replaced *Psaronius*, but in the three samples, the relationship between *Medullosa* and *Psaronius* is not clear. In these samples, *Psaronius* and/or *Medullosa* roots penetrate the aerial debris of cordaitean trees. The spatial relations of *Psaronius* and *Medullosa* root systems with cordaitean aerial debris provide additional evidence that the interpreted succession took place on an ecological timescale. The reciprocal averaging analyses do not indicate whether this succession was driven by biotic or physical processes. In the absence of evidence which suggests a biotically driven succession, we assume that short-term environmental fluctuations caused the observed succession. Several lines of evidence permit a paleoecological interpretation of this successional gradient and the growth environment of these cordaitean-dominated swamps.

Morphological and paleogeographic evidence of cordaitean trees in a mangrove habitat

The conjunction of root aerenchyma, pith, lenticels and clustered lateral rootlets in cordaitean roots from Iowa led Cridland (1964) to reconstruct this plant as a mangrove with prop roots similar to the modern *Rhizophora*. However, the root adaptations of modern mangroves occur widely among freshwater swamp plants (Duddington, 1974). Both *Psaronius* and *Medullosa* also possessed root aerenchyma. The root anatomy of all these plants indicates aquatic or semi-aquatic habitats. Although the evidence of cordaitean root morphology is consistent with a mangrove growth habit, root morphology alone is not sufficient to establish that any of these plants grew in salt-water swamps.

Phillips (1980) defines the cordaitean interval of

the Late Carboniferous on the basis of abundant cordaitean trees (10% or more of the peat biomass) in many of the coal swamps of paralic coal basins (coal basins which border marine environments). It extends from the Westphalian A/B boundary (= 310 m.y.b.p.) into the early Westphalian D (= 295 m.y.b.p.). The cordaitean rich swamps of the early Westphalian D occur in portions of coal regions which also exhibit greater marine influence, as in the Western Coal Region, than does the Eastern Coal Region. The regional gradient of abundant and diverse cordaitean trees, diminishing west to east in the midcontinent area of the United States, also coincides with a suggested decrease in fresh water availability to swamps (Phillips *et al.*, 1980). Within the midcontinent region of North America, cordaitean trees attain their zenith in swamps with the first major marine transgression from the west. In turn, there is apparently a gradient of increasing abundance of cordaitean trees from a minimum of northeastern Oklahoma to a maximum in Iowa.

All these swamps occurred in paleotropical latitudes (Fig. 4) and the paleogeography of the Western Coal Region is consistent with brackish influences for the swamps. However, the exceptional dominance of cordaitean trees in Iowa suggests an environmental extreme compared to other areas in the Western Coal Region. Less freshwater availability from rainfall or runoff, or the preservation of a swamp environment in Iowa that is seldom preserved or seldom exposed elsewhere in the Western Coal Region could account for the differences between the Iowa swamps and those of the rest of the Western Coal Region. The dominance of cordaitean gymnosperms in the peat deposits and the dearth of aquatically reproductive lycopods such as *Lepidodendron* and *Lepidophloios* would be consistent with strong brackish influences and a mangrove habitat.

The most direct evidence of marine influence on coal swamps known from coal-ball peats is the mixture of marine invertebrate fossils with the plant debris resulting in a coal-ball concretion referred to as mixed, in contrast to a normal coal ball composed entirely of peat, or a faunal coal ball without peat. The relevance of this phenomenon to inferences about the swamp environments of the Iowa swamps is limited but consistent with the suggestion of other important factors relating to salinity. Most coal-ball deposits are found in coal seams with marine sediment covering them and coal swamps in the midcontinent region of the United States were usually terminated by marine transgressions. Thus, the presence of marine invertebrates in coal balls at the tops of the coal beds is not an uncommon phenomenon in these swamps. However, it is more common in the Western Region than in the Eastern Region. In relatively few cases has it been demonstrated that the marine influence recurred several times during the course of the swamp history (Mamay and Yochelson, 1964) and actually altered the vegetational composition prior to termination of the swamp (Eggert and Phillips, 1979; DiMichele, 1978). So far as we know, there have been no mixed or faunal coal balls from the Iowa mines reported on in this study. The lack of invertebrates in any coal balls from Iowa may be due to insufficient sampling, or it may indicate that the habitats were sufficiently removed from high-energy environments, or protected from the ocean in such a way as to exclude such shell hash.

The habit of Psaronius and Medullosa

Medullosa and *Psaronius* occur in other Upper Carboniferous coal-ball deposits as well as compression-flora deposits (leaves, seeds and other aerial debris preserved as compressions in clastic sediments). In his survey of Upper Carboniferous compression floras from England, Davies (1929) placed the foliage genera assigned to *Medullosa* (*Alethopteris*) and *Psaronius* (*Pecopteris*) in 'dry' freshwater environments. Scott (1977) also placed the foliage genus assigned to *Medullosa* in a levee-flood plain or lake shore environment. In their survey of the Upper Carboniferous flora of the Francis Creek Shale and Colchester (No. 2) Coal, Peppers and Pfefferkorn (1970) placed ferns, including *Psaronius*, in wet and dry freshwater swamp environments, as well as drier stream bank and levee environments. They placed seed ferns, including *Alethopteris*, in stream bank and levee bank deposits. No evidence from other Upper Carboni-

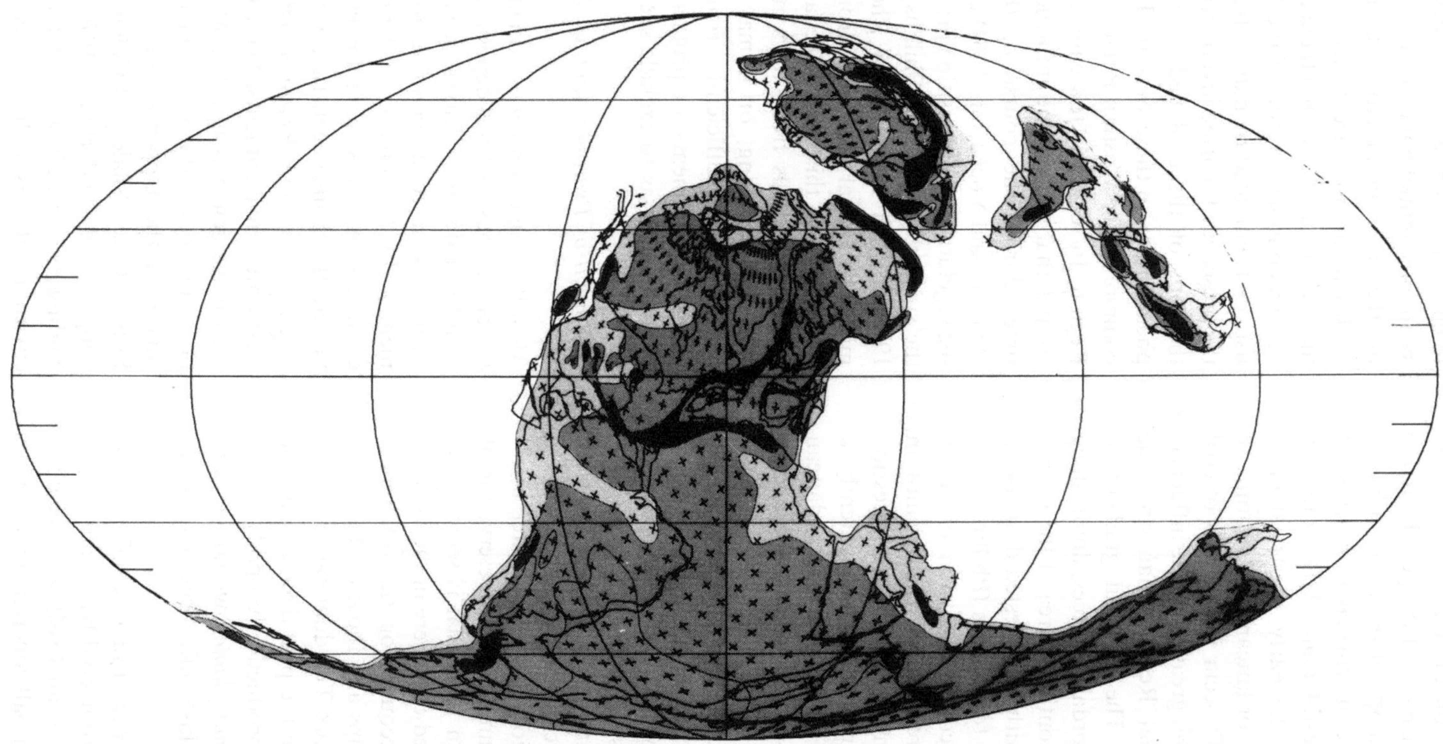

Fig. 4. Upper Carboniferous paleogeographic map showing the equatorial location of the cordaitean-dominated swamps from Iowa.

ferous floral deposits places *Psaronius* and *Medullosa* in salt water or brackish water environments.

Paleoenvironmental interpretation of Iowa cordaitean swamps

The results of these studies suggest that the successional gradient between cordaitean trees and *Psaronius* or *Medullosa* displayed in the reciprocal averaging analysis of the Iowa coals can also be interpreted as an environmental gradient. This environmental gradient links freshwater environments dominated by *Medullosa* and *Psaronius* with environments dominated by cordaitean trees. The root morphology of some cordaitean trees as well as the paleogeographic position of these Iowa swamps is consistent with the interpretation of the cordaitean-dominated portion of the swamp as a mangrove environment. The differing positions of seeds assigned to cordaitean trees along the interpreted physical successional gradient suggests that not all cordaitean trees shared a similar habitat. *Mitrospermum* consistently ordinates near *Medullosa* or *Psaronius* and may have come from a freshwater cordaitean tree. Cordaitean trees growing in a mangrove habitat may have produced *Nucellangium* and *Cardiocarpus spinatus*.

Non-morphological criteria for establishing compression floras as mangrove deposits

The methods used to establish Late Carboniferous swamp communities as possible mangrove associations can be applied to putative mangrove communities from structurally preserved peats. However, most putative mangrove communities occur as compression-flora deposits. The following non-morphological criteria, derived from a compilation of described Mesozoic and Cenozoic mangrove deposits, may permit evaluation of putative mangroves preserved as compression floras.

Rooted horizons in sediments that contain marine or brackish water faunas provide one of the most reliable indicators of fossil mangroves (Retallack and Dilcher, 1981; Retallack, 1975; Daber, 1968). Harris (1966) and also Krasilov and Zakharov (1975) describe fossil mangroves based on the association of aerial debris in sediments which contain marine faunas. Such an assessment requires caution because rivers can transport fresh water plant debris into near-shore marine environments. The freshwater coastal forms, *Barringtonia* and *Cocos*, are found associated with marine invertebrate fauna and mangrove pollen in near-shore marine sediments (Lakhanpal, 1974). Fossil plants which occur exclusively in near-shore marine sediments are more likely to be mangroves than fossil plants which occur in both freshwater and marine deposits. However, at least some modern mangroves are facultative halophytes which will grow in nearly freshwater environment (Tansley and Fritsch, 1905) and could presumably become preserved in both freshwater and near-shore marine sediments.

In the absence of co-occurring marine faunas, authigenic marine minerals provide the clearest evidence of marine depositional environments. Wesley (1966) has described a mangrove association preserved in fine-grained limestone. Sedimentary diagenesis refers to all biological and geochemical changes which occur in sediments after deposition. Although the same authigenic minerals can form during both early and late diagenesis, only early diagenetic minerals provide clear evidence for marine depositional environments. For minerals which form during both early and late diagenesis, crystal form and ion ratios may help specify environment or time of formation. Pyrite and pyrite precursors usually form in interstitial pore waters of marine sediments, mainly as a result of the metabolism of sulfate-reducing bacteria which are rare in freshwater environments (Berner, 1971). Framboidal pyrite forms rapidly in sediments with reducing pore water (Rikard, 1969). Calcites formed interstitially during early diagnesis of marine sediments have high Mg to Ca ratios; siderite and Fe-dolomites precipitate after Mg-calcites in this system. Early diagenetic siderites indicate reducing freshwater sediments (Matsumoto and Iijima, 1981; Berner, 1971). Evaporitic minerals such as gypsum, anhydrite and halite may precipitate within sediments or at the sediment water interface; they require hypersaline conditions (Berner, 1971). Gypsum can also form as a weathering product of sulfides in shales and thus does not always indicate hypersaline conditions (Retallack and Dilcher, 1981).

In desert latitudes (10°–20° N and 10°–20° S) high net evaporation can cause marine evaporites such as gypsum, anhydrite and dolomite to form within sediments or supratidal flats known as salt pans or sabhkas (Kinsman, 1969). Mangrove vegetation can grow on the seaward side of salt pans, but few mangroves can tolerate the extreme fluctuations between hypersaline and freshwater conditions that occur in salt pan interiors. In some salt pans, sandy hummocks trap rain water and provide an environment in which freshwater desert plants can grow. While these plants share some morphological characteristics with mangroves, such as xeromorphic morphology and thick cuticles, they lack physiological mechanisms for growth in salt water (Walter, 1971). The presence of root aerenchyma, which indicates growth in standing water, and the occasional co-occurrence of a fossil plant with normal marine or brackish water faunas can distinguish mangroves from drought-tolerant freshwater plants growing on salt pans.

Modern mangrove swamps form only in tropical and subtropical climates. Most of the more northern and southern recent mangrove swamps occur on islands with mild oceanic climates (Walter, 1973). These mangroves still have not adapted to the triple stress of hard winters, salt tolerance and standing water. For continental configurations similar to the recent (those with land masses over one or both poles) the distribution of putative mangroves should fall within tropical and subtropical climatic belts. Retallack (1977) and Krasilov and Zakharov (1975) have described a Triassic mangrove with a nearly bipolar paleogeographic distribution. During geologic periods such as the Triassic without polar land masses, the distribution of tropical and subtropical climate zones may have been quite different (Robinson, 1973). The presence of tropical terrestrial faunas from Triassic polar regions (Parrish et al., in press) provide paleoclimatic evidence consistent with the presence of Triassic mangroves near the poles.

Oldham (1976) identifies a possible mangrove on the basis of its preservation in monospecific fossil beds. Increased understanding of taphonomic processes suggests that plants growing in many environments, such as lake shores and flood plains, can produce low diversity or monospecific plant deposits in the absence of current transportation. Fossil plant deposits with high diversity may form as the result of current transportation which mixes debris from many plant communities (Black, 1929; Scott and Higgs, 1979). Although zonation among mangroves often leads to low diversity and monospecific stands, the occurrence of fossil plants in a low diversity accumulation is, by itself, a poor criterion for identifying fossil mangroves.

Conclusions

Paleoecological and paleogeographical analysis of one kind of a well-preserved swamp assemblage from the Upper Carboniferous suggests that some members of an extinct group of gymnosperms, the cordaites, grew in salt-water swamps. The fossil deposits in which these cordaitean trees occur also contain freshwater swamp plants. The results of an ordination analysis suggest that freshwater swamp associations dominated by *Medullosa* and *Psaronius* replaced salt-water swamp associations dominated by cordaitean trees on an ecologic time scale. Short term environmental fluctuations may have caused this successional replacement of associations. Ordination analysis also suggests that the cordaitean trees within these swamps had different ecological amplitudes. One cordaitean seed, *Mitrospermum*, associates primarily with freshwater swamp plants. Two other cordaitean seeds, *Cardiocarpus* and *Nucellangium*, appear to have been produced by mangroves.

Acknowledgements

We thank Elso Barghoorn, Harvard University, for providing the specimens and James Runkel and Robert C. Aller, University of Chicago, for reviewing an earlier draft of the manuscript.

Drawings were prepared by Patricia P. Phillips, University of Illinois, Urbana, and the paleogeographic map was provided by A.M. Ziegler and C.R. Scotese, University of Chicago.

This study was supported in part by a DuPont-Radcliffe research grant for women in science and by an NSF Fellowship.

Literature cited

Andrews, H.N. 1951. American coal-ball floras. Bot. Rev. 17:431–469.

Berner, R.A. 1971. Principles of Chemical Sedimentology. McGraw-Hill, New York, 240 pp.

Black, M.C. 1929. Drifted plant beds of the Upper Estuarine Series of Yorkshire. Quart. Jour. Geo. Soc. London 85:389–439.

Churchill, D.M. 1973. The ecological significance of tropical mangroves in the early Tertiary floras of southern Australia. Spec. Publs. Geol. Soc. Aust. 4:79–86.

Cridland, A.A. 1964. *Amyelon* in American coal-balls. Palaeontology 7:186–209.

Daber, R. 1969. A *Weichselia-Stiehleria* Matoniaceae community within the Quedlinburg Estuary of Lower Cretaceous age. Jour. Linn. Soc. (Bot) 61:75–85.

Darrah, W.C. 1939. The fossil flora of Iowa coal balls. I. Discovery and occurrence. Bot. Leaflets Harvard Univ. 7:125–136.

Darrah, W.C. 1941. Studies in American coal balls. Am. Jour. Sci. 239:33–53.

Davies, D. 1929. Correlation and palaeontology of the Coal Measures in East Glamorganshire. Phil. Trans. R. Soc. (B) 217:91–154.

DiMichele, W.A. 1978. Ecotypic variation in *Lepidodendron* Felix. Abstr. Bot. Soc. Amer. Misc. Ser. Pub. 156, p. 32.

Duddington, C.L. 1974. Evolution and Design in the Plant Kingdom. Thomas Y. Crowell Co., New York. 259 pp.

Eggert, D.L. and T.L. Phillips. 1979. Guidebook to Environments of Plant Deposition – Coal Balls, Paper Coals, and Gray Shale Floras in Fountain and Parke Counties, Indiana. 9th Int. Cong. Carb. Strat. Geol. Urbana. 74 pp.

Ehret, D.L. and T.L. Phillips. 1977. *Psaronius* root systems morphology and development. Palaeontogr. Bd. 161. Abt. B. 147–164.

Gauch, H.G., R.H. Whittaker and T.R. Wentworth. 1977. A comparative study of reciprocal averaging and other ordination techniques. Jour. Ecol. 65:157–174.

Harris, T.M. 1966. Dispersed cuticles. Palaeobot. 14:102–105.

Hill, O.M. 1973. Reciprocal averaging: an eigenvector method of ordinations. Jour. Ecol. 61:237–249.

Kinsman, D.J.J. 1969. Modes of formation, sedimentary associations, and diagnostic features of shallow-water and supratidal evaporites. Bul. Am. Assoc. Pet. Geo. 53:830–840.

Krassilov, V.A. and Yu. D. Zakharov. 1975. *Pleuromeia* from the Lower Triassic of the far east of the U.S.S.R. Rev. Palaeobot. Paly. 19:221–232.

Lakhanpal, R.N. 1974. Physical conditions of the Indian Tertiary in the light of palaeobotanical evidence. In: Aspects and Appraisal of Indian Palaeobotany, pp. 516–524, Birbal Sahni Institute of Palaeobotany, Lucknow.

Mamay, S.H. and E.L. Yochelson. 1962. Occurrence and significance of marine animal remains in American coal balls. U.S.G.S. Prof. Pap. 354:193–233.

Matsumoto, R. and A. Iijima. 1981. Origin and diagenetic evolution of Ca-Mg-Fe carbonates in some coal fields of Japan. Sedimentology 28:239–259.

Muller, J. 1964. A palynological contribution to the history of the mangrove vegetation in Borneo. In: Ancient Pacific Floras (ed. L.M. Cranwell), pp. 33–42, University of Hawaii Press, Honolulu.

Oldham, T.C.B. 1976. Flora of the Wealden plant debris beds of England. Palaeontology 19:437–502.

Parrish, J.M., J.T. Parrish and A.M. Ziegler. In press. Permo-Triassic paleogeography and paleoclimatology and implications for therapsid distributions. In: The Biology and Ecology of Mammal-like Reptiles. Smithsonian Press, Washington, DC.

Peppers, R.A. and H.W. Pfefferkorn. 1970. A comparison of the floras of the Colchester (No. 2) Coal and Francis Creek Shale. In: Depositional Environments in parts of the Carbondale Formation – Western and Northern Illinois. Illinois State Geological Survey Guidebook Series No. 8:61–74.

Phillips, T.L. 1979. Paleobotanical studies in the Pennsylvanian System of the Illinois Basin. In: Depositional and Structural History of the Pennsylvanian System of the Illinois Basin Part 2: Invited Papers pp. 4–7, Illinois State Geological Survey, Urbana.

Phillips, T.L. 1980. Stratigraphic and geographic occurrences of permineralized coal-swamp plants – Upper Carboniferous of North America and Europe. In: Biostratigraphy of Fossil plants, pp. 25–92, Dowden, Hutchinson & Ross, Inc., Stroudsburg, Pennsylvania.

Phillips, T.L., M.J. Avcin and D. Berggren. 1976. Fossil peat of the Illinois Basin – a guide to the study of coal balls of Pennsylvanian age. Illinois Geol. Surv. Ed. Series 11. 39 pp.

Phillips, T.L., A.B. Kunz and D.J. Mickish. 1977. Paleobotany of permineralized peat (coal balls) from the Herrin (No. 6) Coal Member of the Illinois Basin. In: Interdisciplinary Studies of Peat and Coal Origins. pp. 18–49, Geol. Soc. America Microform Pub. 7.

Phillips, T.L., R.A. Peppers, M.J. Avcin and P.F. Laughan. 1974. Fossil plants and coal: patterns of change in Pennsylvanian coal swamps of the Illinois Basin. Science 184:1367–1369.

Phillips, T.L., J.L. Shepard and P.J. DeMaris. 1980. Stratigraphic patterns of Pennsylvanian coal swamps in Midcontinent and Appalachian regions of the United States. GSA Abst. with Progs. Vo. 12:498–499.

Ravn, R.L. In press. Stratigraphic Ranges of Miospores in Coals of the Des Moines Series of Southern Iowa. Iowa Geological Survey Technical Paper No. 7.

Retallack, G. 1975. The life and times of a Triassic lycopod. Alcheringa 1:3–29.

Retallack, G. and D.L. Dilcher. 1981. A coastal hypothesis for the dispersal and rise to dominance of flowering plants. In: Paleobotany, Paleoecology and Evolution Vol. 2 pp. 27–77. Praeger Publishers, New York.

Rickard, D.T. 1969. The chemistry of iron sulphide formation at low temperatures. Stockholm Contrib. to Geol. 20:67–95.

Robinson, P.L. 1973. Paleoclimatology and continental drift. In:

Implications of Continental Drift to the Earth Sciences Vol. 1 pp. 449–476, Academic Press, London.

Scott, A.C. 1977. A review of the ecology of Upper Carboniferous plant assemblages, with new data from Strathclyde. Palaeontology 20:449–473.

Scott, A.C. and K. Higgs. 1979. The distribution of Upper Carboniferous plants from the Sanquhar Coal field, Southern Scotland. Abst. of Papers, Ninth Int. Cong. Carb. Strat. Geol. Urbana. p. 199.

Stewart, W.N. and T. Delevoryas. 1956. The medullosan pteridosperms. Bot. Rev. 22:45–80.

Stopes, M.C. and D.M.S. Watson. 1908. On the present distribution and origin of the calcareous concretions in coal seams, known as coal balls. Phil. Trans. Roy. Soc. London 200B:167–218.

Tansley, A.G. and F.E. Fritsch. 1905. The flora of the Ceylon littoral. New Phytol. 4:27–55.

Walter, H. 1973. Vegetation of the Earth. Springer-Verlag, New York. 237 pp.

Wesley, A. 1966. The fossil flora of the Grey Limestones of Veneto, northern Italy, and its relationships to the other European floras of similar age. Palaeobotanist 14:124–130.

CHAPTER 3

Ecological notes on the mangroves of Fujian, China

LIN PENG* and WEI XIN-MEN**

* *Department of Biology, Xiamen University, Xiamen, Fujian, China*
** *Third Institute of Oceanography, National Bureau of Oceanography, Xiamen, Fujian, China*

Abstract. Funjian Province, in the subtropical zone of China, has six mangrove species: *Bruguiera gymnorrhiza, Kandelia candel, Excoecaria agallocha, Acanthus ilicifolius, Aegiceras corniculatum*, and *Avicennia marina*. Along the southern coastal portion of Fujian all six species and several associations occur; in the middle region four species and fewer associations are found; and in the northern region only *Kandelia candel* occurs. The northern region at 27° 20′ is the northern most extent of mangroves on the China coast.

The height of *Kandelia candel* was correlated with soil salinity and subsurface nitrogen content of the soil. Heavy growth of barnacles on *Kandelia candel* trunks was associated with poor growth of the trees.

Introduction

Fujian is a coastal province in the subtropical zone of China where mangroves are an important ecosystem of the seashore and the estuaries of the rivers and streams. The mangrove swamps are discontinuous and vary in width from a few to several hundred meters, depending on the availability of muddy substrates.

Mangrove swamps cover a large area along the seashore of Fujian Province. The largest areas occur along the southeastern coast of Fujian. They are found on the deltas on the Jinjiang, Jiulongjiang and Zhangjiang rivers, and smaller patches are scattered along the coast and at smaller river mouths.

Environmental condition along the coast in Fujian

The seashore of Fujian is situated 23° 46′–27° 20′ N, the northern extent of which is the northern most limit of mangroves in China. Variation of the meteorological factors in different coastal counties of this province are presented in Table 1.

The days are usually windy and the weather is affected by typhoons during the summer months. The mean annual wind velocity is 4–8 m/sec. It is a southeastern wind from June to August and a northeastern wind from September to May, so the mangroves are found most often along the southwestern coasts.

Mangrove species and communities

The subtropical mangroves of Fujian consist of six species belonging to five families. They are *Bruguiera gymnorrhiza* (L.) Lamk. and *Kandelia candel* (L.) Druce of the Rhizophoraceae; *Excoecaria agallocha* L. of the Euphorbiaceae; *Acanthus ilicifolius* L. of Acanthaceae; *Aegiceras corniculatum* Blanco of Myrsinaceae; and *Avicennia marina* (Forsk.) Vierh of the Verbenaceae. A description of these species and their habitats and associates follows.

Tasks for vegetation science, Vol. 8 ed. by H.J. Teas
 ISBN 978-90-481-8526-9

Table 1. The chief meteorological factors along the coast of Fujian.

Metereological factor	County									
	Zhao-an	Yunxiao	Longhai	Xiamen	Jinjiang	Putian	Fuqing	Fuzhou	Ningde	Fuding
Latitude (° N)	23° 46′	23° 59′	24° 27′	24° 31′	24° 49′	25° 26′	25° 43′	26° 05′	26° 32′	27° 20′
Mean annual rainfall (mm)	1306.7	1642.3	1284.4	1036.0	1091.0	1264.4	1276.0	1280.8	2061.7	1631.7
Mean annual temperature (° C)	21.3	21.1	20.9	20.8	20.4	20.2	19.7	19.7	19.0	18.5
Monthly mean lowest temperature (° C)	12.9	12.9	12.2	12.3	11.6	11.2	10.4	10.3	9.3	8.4
Mean annual range (° C)	15.4	15.4	16.5	16.0	17.8	17.9	17.9	18.5	19.3	19.8

Acanthaceae

Acanthus ilicifolius L. (Laoshulei, local name)

Shrublet to 50–120 cm tall. Leaves opposite, oblong or oblong-lanceolate, margins, nodes and apices armed with sharp thorns; flower white with or without purplish tinge; 2 bractlets, 3–4 cm long; fruit oblong, capsular, 2–3 cm long, slightly compressed, splitting lengthwise in the middle to reveal 2–4 flattened seeds, 10–13 mm, yellowish (see Fig. 1).

On stiff or heavy loam along inland tidal creek banks or sometimes in *Kandelia candel* forest with the same soil type.

Euphorbiaceae

Excoecaria agallocha L. (Haiqi, local name)

Small to medium-sized, unbuttressed tree to 5 m tall;characteristic by its white latex and coppicing habit. Bark surface grey or pale brown. Leaves alternate, elliptic or obovate-elliptic with tapering, shortly pointed apices; petioles small, slender. Inflorescences in axillary racemoses, males and females on different trees; petals absent. Fruits trilobed, to 10 mm diameter splitting open when dry.

On heavy and stiff sandy loam near the shore *Hibiscus tiliaceus* L. is the main associate.

Myrsinaceae

Aegiceras corniculatum (L.) Blanco (Tonghuashu, local name)

Small bushy shrub to 1–2 m tall. Bark dark chocolate brown. Leaves opposite, elliptic or obovate, salt excretion from leaf surface. Inflorescences in axillary or terminal spikes, petal white, fruits

Fig. 1. Acanthus ilicifolius.

curved, bean-like, green.

On the newly formed mud flats at river mouths or along low banks; may also occur inland on the same soil type.

Rhizophoraceae

Bruguiera gymnorrhiza (L.) Lamk. (Mulan, local name)

Small to large buttressed tree to 3–6 m tall with stout knee-like pneumatophores. Bark surface pinkish-brown. Leaves green, opposite; stipules white-green, lanceolate. Flowers pinkish-yellow, calyx frequently red, sepals 10–14, stamen is two times of calyx. Extruded hypocotyl to 10–14 cm long, stout dark green, flush deep purple (see Fig. 2).

Is found on deep muddy shore, in association with *Kandelia candel*; scattered along shore inland of river mouth.

Kandelia candel (L.) Druce

Small stilt-root or buttressed tree to 4–8 m tall, base fluted or swollen (see Fig. 3). Bark surface reddish-brown to chocolate brown. Leaves opposite, oblong or obovate-oblong, with rounded or

Fig. 2. Bruguiera gymnorrhisa.

Fig. 3. The buttress-like root of *Kandelia candel.*

blunt apices, to 5–10 cm. Flowers white sepals 5, reflexed in fruit; extruded hypocotyl to 15–25 cm long, slender with suddenly tapering, pointed tip.

On the soft mud along in-land river banks, associated with *Aegiceras corniculatum* as gregarious undergrowth on the same habitat.

Verbenaceae

Avicennia marina (Forsk.) Vierh (Beigurang, local name)

Shrub to medium-sized tree to 2–5 m tall. No buttresses but thin or gregarious pneumatophores, finger-like (see Fig. 4) and short aerial roots may project from the trunk. Bark grey, white or brownish. Leaves opposite, elliptic, entire, and the lower surface often bears minute salt crystals and densely clothed with grey, white hair. Fruit glaucous-green more or less heart-shaped, slightly flattened.

A pioneer species on newly formed mud with a high proportion of sand, but does not seem to grow on pure mud. It is found at the mouth of the river or lower tidal area.

Distribution of mangroves on the Fujian Coast

The distribution of mangroves on the Fujian coast is summarized in Table 2. The main features with respect to geography are:

(a) In the southern region from Zhao-an to Xiamen all six species are present, and their com-

Fig. 4. Avicennia marina and its pneumatophores.

Fig. 5. Aegiceras corniculatum association.

munities are composed of a *Kandelia candel-Bruguiera gymnorrhiza – Aegiceras corniculatum* association; a *Kandelia candel – Aegiceras coniculatum* association; an *Aegiceras corniculatum* association (see Fig. 5); an *Avicennia marina* association; and a *Kandelia candel* association. *Kandelia candel* is the dominant species, the tallest trees reaching a height of eight m (see Fig. 6).

(b) In the middle region from Xiamen to Putian, there are four species (*Kandelia candel, Aegiceras corniculatum, Acanthus ilicifolius* and *Avicennia marina*) and the last four types of communities mentioned in (a).

(c) In the northeast region from Putian to Fuding, there is only one species, *Kandelia candel.*

This summary of species and community distributions suggests that the number of species and community diversity decrease from the south to north as a function of the decreasing mean annual temperature (Table 1).

Table 2. Distribution of species of mangroves in Fujian compared with other parts of the world.

Family	Species	Fujian Seashore			Guangdong Province	Taiwan Province	Japan	Asian coasts	Australia coast	Eastern African coast	Atlantic coasts
		Southern part	Middle part	Northeast part							
Acanthaceae	Acanthus ilicifolius	+	+	0	+	0	0	+	+	0	0
Euphorbiaceae	Excoecaria agallocha	+	0	0	+	+	0	+	+	0	0
Myrsinaceae	Aegiceras corniculatum	+	+	0	+	0	0	+	+	0	0
Rhizophoraceae	Bruguiera gymnorrhiza	+	0	0	+	+	?	+	+	+	0
	Kandelia candel	+	+	+	+	+	+	+	0	0	0
Verbenaceae	Avicennia marina	+	+	0	+	+	0	+	+	+	0

Fig. 6. Kandelia candel–Aegiceras corniculatum association.

Mangrove growth in relation to ecological factors

The salinity of sea water may greatly affect the morphological and physiological characteristics of mangrove plants. It is usually found that as the osmotic pressure of the water increases the leaves become thicker and coriaceous and their stomata sink into the epidermis. Salinity is also related to the growth and development of mangrove plant and coexisting aquatic animals. Our observations indicate that when the salinity of water is low, 7.5–21.2‰, *Kandelia candel* grows normally and flowers and produces fruit. The five-year-old trees may reach 1.6–2 m in height; *Balanus* (barnacles) adhere to them in small numbers or are absent. However, when the salinity is high, i.e. 25.6–37.4‰, *Kandelia candel* grows poorly so that the five-year-old trees are only about one m in height, their leaves are small and succulent and they do not bloom normally. Some plants growing at high salinity die, e.g. about 30% were dead in the area surveyed. Under high salinity conditions there are frequently large numbers of *Balanus* attached to the trunks. The relationships of the factors affecting *Kandelia candel* are detailed in Table 3. The correlation between the height of *Kandelia candel* and the soil salinity appears to be a parabolic curve (see Fig. 7).

The growth in height of *Kandelia candel*, where salinity is favorable, can be correlated with the nitrogen content (%) of the soil. The mud of the first layer (0–20 cm) is unstable and is readily influenced by the waves, so the nitrogen content of the second layer (20–40 cm) was measured. It was found that there was a semi-logarithmic linear correlation between tree height and soil nitrogen. The relationship was: $\log y = 0.9397 + 3.124\ x$

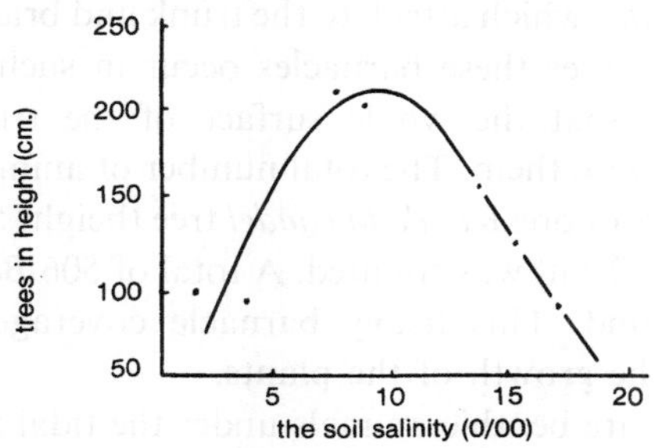

Fig. 7. The relationship between the soil salinity and the growth in height of *Kandelia candel.*

Table 3. The relationship between the salinity of sea water and the characteristics of five-year-old *Kandelia candel* trees.

County (village)	The salinity of sea water (‰)	Tree height (cm)	Reproductive characteristics	Relative frequency of *Balanus*
Fuqing (Donghan)	1.04	130	Few flowers and fruits	Absent
Hui-an* (Longcun)	5.3	75	Few flowers and fruits	Absent
Fuqing (Limei)	7.5	200	Blossoms, fruits and seedling normally	Absent
Fuqing (Shanghu)	21.2	160	Blossoms, fruits and seedling normally	Absent
Fuqing (Qianhua)	25.6	100	Flowers and fruits absent	Moderate
Putian* (Duoxia)	26.2	90	Flowers and fruits absent	Moderate
Putian* (Zhuangdou)	28.5	120	Few flowers and fruits	Moderate
Hui-an* (Tiandong)	32.7	65	Flowers and fruits absent	Frequent
Hui-an* (Sucuo)	33.0	80	Flowers and fruits absent	Frequent
Hui-an* (Dongyuan)	33.8	57	Flowers and fruits absent	Frequent
Hui-an* (Xiyuan)	37.3	60	Flowers and fruits absent	Frequent

* Hui-an county, located near Jinjian county (see Table 1).

where y is the tree height (m), x is the edaphic nitrogen content (%). The correlation coefficient (r) was 0.8875; the degrees of freedom (df) was 7. The determinate region: trees height 50–220 cm tall, the edaphic nitrogen content 3–43‰.

Animal associates of the mangroves

In the mangrove ecosystem the ecological function of associated animals is also significant. Insects and birds are the common inhabitants. At some times more than 30% of mangrove plants may be consumed by insects. In the community of *Kandelia candel* in Sanjiankou of Putian, it was found that 95% of the leaves showed feeding by a species of tortrix moth.

There are two kinds of animals that attach to the trunk and leaves of tidal forest, i.e. *Chithamalus* sp., which adhere to leaf blades and *Balanus amphitrite albicostatus* which attach to the trunk and branches. In some cases these barnacles occur in such large numbers that the whole surface of the trunk is covered with them. The total number of animals on the trunk of one *Kandelia candel* tree (height 53 cm, diameter 2 cm) was counted. A total of 506 *Balanus* were found. This heavy barnacle coverage may reduce the growth of the plants.

There are benthic animals under the tidal forest, such as *Sesarma* sp., *Pachygrapsus crassipies* and *Littorinopsis melanostoma* and many others that dig holes in the soil. In one case there were 122 holes, each of 0.5–1.2 cm diameter, in a 70 × 70 cm quadrat. Such porous soil would probably have increased aeration which would probably favor growth of the mangrove plants.

Usefulness of mangroves to man

The mangroves from a natural shelterbelt which decreases the tidal force. Mangroves are also of great economic value because of utilization of hypocotyl starch, cortical tannin, trunk wood etc. Therefore, care should be taken to protect mangrove communities.

Literature cited

Guangdong institute of Botany. 1976. Guandong vegetation, Science Publishing House, pp. 106–121.

Hou, K.Z. and C.L. Ho. 1953. Mangroves in China, Shengwuxue Tongbao, 10:366–369.

Ho, G. 1957. The ecology of mangroves, Shengwuxue Tongbao 8:1–5.

Li, H.L. 1963. Woody Flora of Taiwan, Livingston Publishing Company, pp. 632–635.

Macnae, W. 1968. A general account of the fauna and flora of mangroves swamps and forests in Indo-West-Pacific region. Advances in Marine Biology 6:74–241.

Percival, M. and J.S. Womersley. 1975. Floristics and Ecology of the Mangrove Vegetation of Papue New Guinea, Lae, Papua New Guinea, pp. 1–29.

Schimper, A.F.W. 1903. Plant-Geography upon a Physiological Basis. Englemenn and Wheldon and Wesley. Ltd. (Reprinted 1960), pp. 395–410.

Zhang, H.D., C.C. Zhang and B.S. Wang, 1957. Act. Scien. Natural Univ., Sunyatsen. 1:122–145.

CHAPTER 4

A general account of the mangroves of Princess Charlotte Bay with particular reference to zonation of the open shoreline

J.A. ELSOL* and P. SAENGER**

* *Queensland Herbarium, Meiers Road, Indooroopilly, Qld 4068, Australia*
** *c/- Scientific Services, Queensland Electricity Generating Board, G.P.O. Box 1424, Brisbane, Qld 4001, Australia*

Abstract. The mangrove vegetation of Princess Charlotte Bay is described. Three land groupings are distinguished – supra-tidal land (predominantly supporting a mixed grassland), high tidal flats (vegetation free), and lower intertidal areas (supporting mangrove vegetation). The mangrove zonation pattern associated with the open shoreline is distinct and comprises a landward *Avicennia-Aegialitis* zone, a *Ceriops-Avicennia* zone, a mixed zone of *Aegialitis, Bruguiera, Ceriops, Avicennia, Xylocarpus* and *Rhizophora*, a *Rhizophora* zone and occasionally a narrow *Sonneratia* zone at the lowest tidal level. Interspecific competition between *Avicennia* and *Ceriops* is examined and shown to occur in the two landward zones. The zonation pattern described is similar to that occurring in other areas of northern Australia.

Introduction

In contrast with other areas of Australia, there exist few descriptive accounts of the mangrove communities occurring on Cape York Peninsula (that area of Queensland extending north of 16° S). Brief accounts are contained in Jones (1971), Bunt (1978), Hynes and Tracey (1980), Stoddart (1980) and Bunt and Williams (1981). Specht *et al.* (1977) list mangrove species occurring at Weipa on the west coastline, and Byrnes *et al.* (1977) list those species present on Lizard Island in the Great Barrier Reef Province adjoining eastern Cape York.

With the aid of black and white aerial photography flown on 4 July 1970 at 25,000′ (7,700 metres) the findings of a survey in May 1979 to the Marrett River, Princess Charlotte Bay, are presented and discussed in the following account.

The study area

Setting

Princess Charlotte Bay lies at approximately 14° S, 144° E on eastern Cape York Peninsula (Fig. 1).

The Marrett River lies to the east of the three other main rivers draining into the southern end of the bay; the North Kennedy, Bizant and Normanby Rivers. Of the catchments present on eastern Cape York Peninsula, i.e. east of the main dividing ranges, the Normanby River drainage basin which empties into Princess Charlotte Bay is the largest (ca. 25,000 km²). It also experiences the lowest rainfall. Relief is generally low and much of the catchment consists of gently undulating plains (Isbell, 1978).

The plains consist predominantly of terrigenous material carried by the large river systems entering the bay (Frankel, 1974), though some marine deposits are evident within close proximity to the coast. These consist of linear ridges of marine-shell

Tasks for vegetation science, Vol. 8 ed. by H.J. Teas

ISBN 978-90-481-8526-9

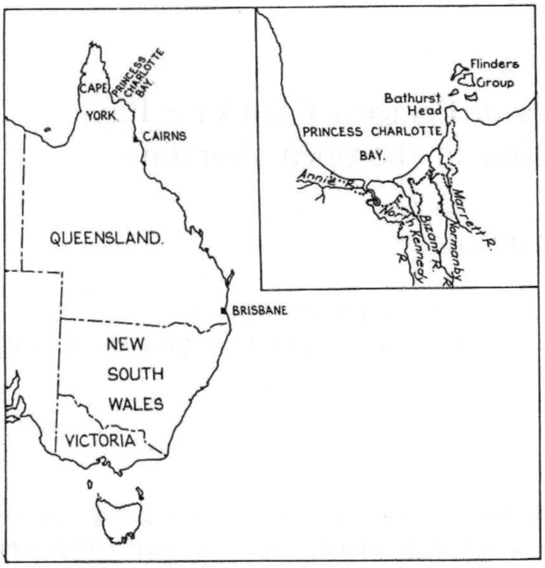

Fig. 1. Locality plan, Princess Charlotte Bay, North Queensland (14° S, 144° E), Australia.

material deposited in the mangrove fringe of the open-shoreline. Advancement of the coastline has left these ridges stranded within the extensive system of flats formed behind the current fringe. These ridges are particularly evident radiating away from the first meander of the Marrett River's west bank (Fig. 2).

Former Quaternary coastlines are apparent in the region indicating an advancing, or at least prograded coastline. This is evidenced by the linear shell ridges mentioned above, areas of closely spaced river meanders, and in western areas of the bay, parallel beach ridges of Pleistocene age.

Due to the presence of offshore barrier reefs and especially the protective influence of the headlands to the east of the bay, the south Pacific swell and Trade wind induced wave action are well dissipated. As a result the southern coastline of the bay is exposed only to wave action of low energy, thus allowing mangrove development to occur along the shoreline in addition to that occurring along the rivers.

The area is in a near natural condition, since European habitation in the region is very limited. Wild pigs (*Sus scrofa* L.) are seasonal inhabitants and crocodile (*Crocodylus porosus* Schneider) numbers have been reduced by past hunting activities. The study area is within the lands of the Mitchell-Normanby area documented by Galloway (1972) and Galloway *et al.* (1972). Floristic information and vegetation accounts of areas which include Princess Charlotte Bay are found in Pedley and Isbell (1971), Jones (1971), Storey (1972), Bunt (1978) and Bunt and Williams (1981). The soils of the area are described by Gunn (1972a). Additional features of the physical environment are contained in Lucas and De Keyser (1965), De Keyser and Lucas (1968) and Whitaker and Gibson (1977).

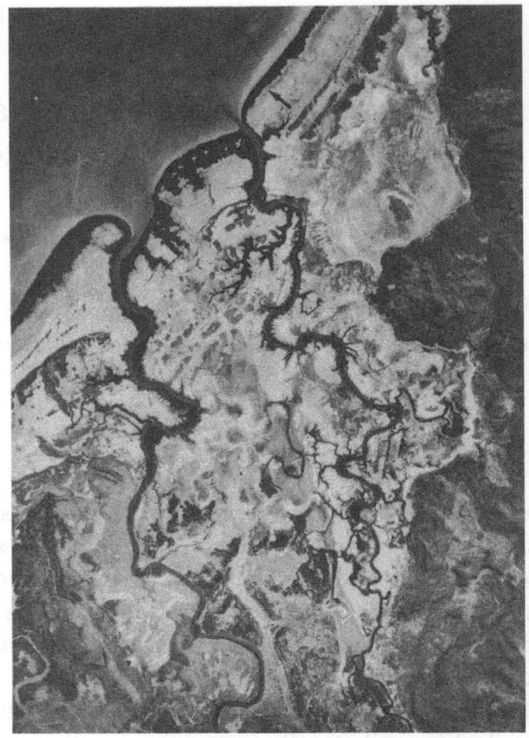

Fig. 2. Aerial view of study area taken during 1970 dry season showing Marrett and Normanby Rivers. The two river mouths are separated by approximately 3.5 km. Arrow indicates established transect adjacent to the mouth of the Marrett River. (This photograph is Crown Copyright and has been reproduced by permission of the Director, Division of National Mapping, Department of National Development and Energy, Australia.)

Climate

This area of Cape York Peninsula has a subhumid to humid tropical climate with marked with wet and dry seasons.

According to Nix and Kalma (1972), shifts of the intertropical convergence zone dominate the climate of the region. During the dry season (May to October) whilst southeasterly Trade winds prevail, the intertropical convergence zone is well to the north of the Australian continent and dry relatively cool weather and only very light rainfalls occur. Local convectional thunderstorms occur at the beginning of the wet season in November. In December–January, the intertropical convergence zone moves south and the influx of moist equatorial and tropical maritime air masses brings heavy rain to the area. In the December–March period, tropical cyclones originating in the Coral Sea are an additional source of often torrential rain.

Of the near 1250 mm average annual rainfall, approximately 95% is expected between November and April. The average number of rainy days during the dry season (for the Mitchell-Normanby region) is about seven (7) and rainfall in this period is generally less than 50 mm (Gunn, 1972b).

Air temperatures for the area are fairly high throughout the year; the extreme weekly mean maxima lie in the range of 30–35° C and the extreme weekly mean minima, 15–20° C (Nix and Kalma, 1972). Sea surface temperatures are about 4° C higher in summer than in winter and lie in the approximate range of 24–28° C (Webster and Stretton, 1972).

Landform, soils and vegetation

Galloway *et al.* (1972) described two land systems occurring in the Princess Charlotte Bay area. These are the Inkerman and Battersea land systems. The former includes the upper reaches of the rivers and adjacent lands, whilst the latter comprises the immediate coastline incorporating the coastal and much of the riverine vegetation.

The study was conducted mainly in the latter area. Relative to tide levels, three groupings as outlined below are distinguished in the vicinity of the Marrett River.

1. Supra-tidal lands

This occurs on relatively flat areas of land up to several metres higher than the high tidal flats (see below). Many of these areas are interspersed by extensive formations of high tidal flats. In many places their alignment within these flats follows apparent former coastlines. Less fragmented and more extensive areas occur upstream.

The soils are cracking, strongly alkaline clays and often contain calcareous shell fragments. The linear ridges of marine deposits are included within this grouping.

The vegetation contains a large number of herbaceous species and is a mixed (herbland) grassland (low shrubland), with occasional shrubs, low trees and trees.

The most common grasses include *Rottboellia exaltata*, *Sorghum laxiflora*, *Themeda quadrivalvis* and *Xerochloa imberbis*. Some areas contain a mixture of shrubs from 1 to 4 m high. These include *Breynia* sp., *Cleistanthus apodus*, *Securinega melanthesoides*, *Terminalia subacropta* and *Thespesia* sp.. The monocarpic palm *Corypha elata* occurs in scattered localities. Shrubs of *Myoporum acuminatum* often occur on the lower slopes adjacent the high tidal flats, forming an (open-heath) low shrubland.

The lowest limit of this unit corresponds with the upper limit of the high tidal flats. This area has its own distinctive vegetation. A fringe of chenopods, dominated by *Halosarcia* spp., with occasional *Sporobolus virginicus* var. *minor* is often present. The herbaceous *Batis argillicola* is rarely present. Scattered mangroves to 4 m are present in some areas; these include *Aegialitis annulata* (usually in moister areas), *Avicennia marina*, *Ceriops decandra*, *Excoecaria agallocha*, *Lumnitzera racemosa* and *Osbornia octodonta*. Seedlings, and plant litter comprising leaves, Rhizophoraceae propagules and *Heritiera littoralis* fruits may also be present.

Some vegetated areas often only several metres across are unprotected by such a fringe, and appear to be eroding. As a result, soil, shell material and plant litter are being deposited on the high tidal flats.

2. *High tidal flats*

These correspond to the contemporary flats of long term development of Spenceley (1976) and the high tidal flat zone of Thom *et al.* (1975) and are similar to those described by Saenger and Hopkins (1975). Large logs may occur on the surface, as was also noted by Thom *et al.* (1975), predominantly of *Ceriops*.

The soils are saline-alkaline cracking clays and uniform fine textured soils with a sometimes soft flocculated salty layer at the surface when dry. These flats are characteristically bare of macroscopic vegetation.

The occurrence of high tidal saline mud flats has been known for some time and is further discussed in Fosberg (1961), Saenger and Hopkins (1975), Coleman *et al.* (1966), Spenceley (1966) and Cintron *et al.* (1978). Whilst the factors leading to the formation of the vegetation free zones are still unclear, these zones are characterized by generally low soil moisture content and by high soil salt concentrations. Samples from the study area showed soil chlorinities ranging from 9–15 times that of seawater, and it seems likely that the osmotic equivalent of such soil concentrations are sufficiently high to inhibit the permanent colonization by higher plants. Total soil chlorophyll determinations indicate however, that a sparse algal standing crop was present.

3. *Lower intertidal lands*

These are present at the land-water interface and the soils consist predominantly of wet saline clays.

The vegetation consists of mangroves; a list of intertidal species occurring along the Marrett River is given in Appendix A. The mangroves occur at (a) the open-shoreline, and along (b) the river system.

(a) The open shoreline

To investigate the mangrove vegetation of the open shoreline a permanent transect was laid out on the shoreline immediately to the east of the mouth of the Marrett River (Fig. 2.). The transect is 5 m wide and extends throughout the entire mangrove zone, a distance of approximately 130 m.

Within the transect, all trees and seedlings were given co-ordinates, the species recorded, their heights determined and all specimens were numbered. Girth-at-breast-heights were also determined on all specimens greater than 1.8 m in height. Canopy cover percentages at 10-m intervals were measured by photographing the canopy from about 1 m height and screening the slides onto a 100 point grid. A mean value was determined after four readings of each slide. The average canopy cover % for any zone given in Table 1 represents the average of the mean values recorded over the specified distance. The values in parentheses represent the upper and lower extremes of the mean values.

The transect was divided into a series of 5 m × 5 m quadrats so as to determine the Relative Importance of each species per quadrat. This was done by summing the three indices below, based on the observations of adult individuals of each species. Adults were given the operational definition of greater than 0.5 m in height, except for *Aegialitis*, where 0.2 m was taken as the minimum height. The indices are defined.

Table 1. Open shoreline zonation sequence based on data from permanent transect.

Zone	Description	Lateral extent	Width in transect (m)	Canopy cover (%)
1	*Avicennia-Aegialitis* zone fringing bare high tidal flats	Continuous	15 (0–15)	40
2	*Ceriops-Avicennia* zone	Continuous	45 (15–60)	(72)–75–(78)
3	Mixed zone consisting of *Aegialitis*, *Bruguiera*, *Ceriops*, *Avicennia*, *Xylocarpus* and *Rhizophora*	Continuous	20 (60–80)	(58)–71–(79)
4	*Rhizophora* zone	Continuous	45 (80–125)	(74)–75–(75)
5	*Sonneratia* zone at lowest level	Sporadic	7 (125–132)	40

Relative Transect Density (%) =

$$\frac{\text{Number of individuals of one species in quadrat} \times 100}{\text{Number of individuals of all species in densest quadrat}}$$

This parameter gives a measure of density of a species in any quadrat compared with the densest one, and indicates which species contributes most to the total density of that quadrat.

Relative Quadrat Density (%) =

$$\frac{\text{Number of individuals of one species} \times 100}{\text{Number of individuals of all species in that quadrat}}$$

For each quadrat, this parameter equals 100% and it measures the density of each species in relation to the other species in the same quadrat.

Biomass Index (%) =

$$\frac{100 \sum (\text{G.B.H. in dm} + \text{Height in m for each species in quadrat})}{\text{Largest quadrat value of above in transect}}$$

(G.B.H. = girth at breast height).

This parameter estimates the biomass contribution of each species in each quadrat.

The Relative Importance values of the species are diagramatically shown in Fig. 3. This Figure quantifies the intuitive zonation sequence arrived at during the field work and which is summarized in Table 1. The vegetation is now described.

Between the high tidal flats and the first zone of mangroves, low chenopod vegetation occurs. This is similar to that described occurring between the high tidal flats and the supra-tidal vegetation.

The mangrove canopy rises at about a 30° angle from the bare flats. The *Avicennia-Aegialitis* zone is narrow and is quickly replaced by a *Ceriops–Avicennia* zone, numerically dominated by *Ceriops*. An inverse relationship exists between density and height. Where the height is less than 6 m multiple stems are common. Stands of *Ceriops* have an even canopy and constitute (low) closed-forests. They may reach 10 m height. *Avicennia*, and *Osbornia* when present, are often emergent above the *Ceriops* canopy by several metres. *Bruguiera exaristata*, *Lumnitzera* and *Xylocarpus australasicus* are occasionally present, mostly reaching a height below that of the *Ceriops*. *Rhizophora stylosa* is occasionally present along lines of drainage or other minor depressions.

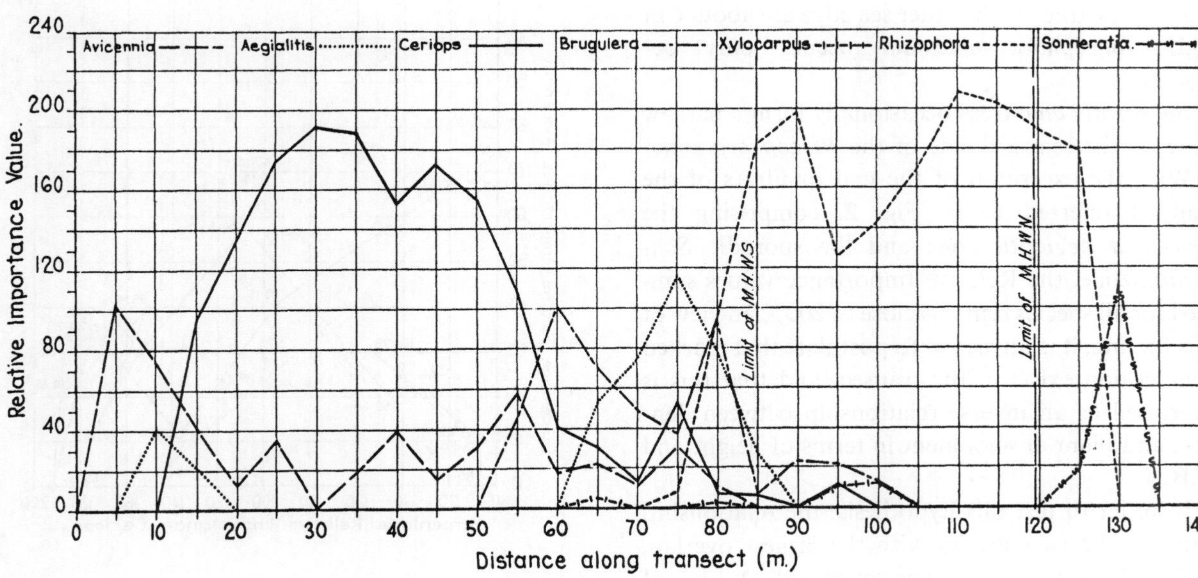

Fig. 3. Open shoreline zonation, Princess Charlotte Bay. Distribution of Relative Importance values across permanent transect.

In some areas, death of mangroves on the landward edge has occurred with no regeneration, resulting in an extension of the high tidal flats. On the west bank of the Marrett River where the *Ceriops* dominated zone is wider, this same phenomenon has also occurred actually within the zone, forming bare areas (often oblong and up to 130 m long) surrounded by mangroves. In such closed situations, an *Avicennia* (sometimes with *Aegialitis*) zone may establish around the perimeter.

To the seaward side of the *Ceriops–Avicennia* zone lies a mixed zone of *Aegialitis*, *Bruguiera exaristata*, *Ceriops*, *Avicennia*, *Xylocarpus australasicus* with occasionally *Rhizophora* in wet depressions. In this zone the community structure varies markedly, from (low) closed-forest, to open-scrub where it adjoins the *Rhizophora* zone. *Avicennia* characteristically is emergent above the other species.

The next zone consists of *Rhizophora stylosa* and adjoins the mixed zone at a height of about 2 to 6 m. The height increases up to 10–12 (15) m forming a closed-forest. Where the canopy is highest, the tree density is reduced and the trees are often multi-trunked. Generally no understorey is present, though concentrations or ridges of shell material support occasional *Aegialitis*, *Bruguiera exaristata*, *B. gymnorhiza*, *Osbornia* and *Xylocarpus australasicus*. The trees of the outer sea edge are about 6 m high. *Rhizophora apiculata* occurs only occasionally.

Sonneratia alba trees occasionally form a narrow zone on the seaward side of the *Rhizophora* zone.

With the exception of the end quadrats of the transect referred to in Fig. 2, comprising the *Avicennia–Aegialitis* zone and the sporadic *Sonneratia* zone, the Relative Importance values summed for all species remains close to 200%. In view of this finding, it is attractive to postulate that marked competition exists in the transect and that this is expressed as an inverse relationship between density, and plant development in terms of height and G.B.H..

In order to test this hypothesis, the relationship between the two species with the largest overlap, *Avicennia* and *Ceriops*, was examined. A plot of their Relative Importance values (Fig. 4) indicates that two broad bands are recognizable; one from 0–55 m in which their importance estimates lie on a negatively sloping line, i.e. one varies inversely with the other, and another 60 m onwards with a positive slope, where both vary in proportion to the other. This suggests that from 0–55 m the environmental conditions are optimal for both species and they compete with each other. From 60 m onwards, the conditions are no longer as favourable and both species decline in importance.

(b) The river system

The rivers extend considerable distances inland, often in excess of 25 km. The position of the rivers is not static; with time the size, shape and position of the meanders change. The radii of curvature of the river meanders vary considerably. The meanderine nature of the rivers is similar to that described for the Gulf of Carpentaria (Whitehouse, 1944). Oxbow lakes have formed on some of the rivers. The actual drainage pattern present in any particular area is dependent upon the topography and substrate of the land adjacent the river. Irregularities or departures from the generalized meanderine drainage pattern do occur.

The mangrove vegetation of the rivers varies

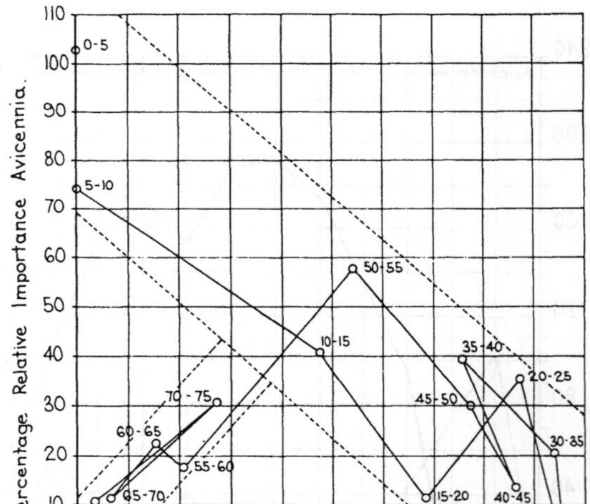

Fig. 4. Relationships between the Relative Importance values (%) of *Avicennia* and those of *Ceriops*.

considerably in composition and areal extent. The mangroves form a zone between the river and the sparsely vegetated supra-tidal plains or bare high tidal flats. Large areas of mangroves also occur on the dendritic drainage patterns which frequently enter the concave banks of the river meanders.

An important feature of the system is the marked seasonal variation in river flow. Erosion of the concave banks and accretion of the convex lobes are largely associated with the intermittent seasonal flow of flood waters. The continuing movement of the rivers is suggested by two features: species zonation of the convex lobes is associated with accreting conditions (e.g. the occurence of *Rhizophora stylosa* at the rivers' edge), and the depletion of the number of zones present on concave banks due to erosion.

Surface salinities of the Marrett River (recorded 10 May 1979) ranged from 32‰ at the mouth to 0‰ at ca. 30 km upstream (about 20 km in a straight line from the mouth). Accordingly mangrove species zonation occurs lengthwise along the river in response to changing salinities. However this is confounded by the influences erosion and accretion have on lateral zonation of communities associated with meanders. *Rhizophora stylosa* for example usually occurs on the accreting convex banks in the downstream areas, but not on the eroding banks nor between the meanders where conditions are more stable.

The mangrove communities near the mouth of the Marrett River are dominated by *Ceriops* and *Rhizophora stylosa. Aegialitis, Avicennia, Bruguiera* spp., *Excoecaria, Lumnitzera, Maytenus emarginata, Osbornia* and *Xylocarpus* are also present to varying degrees. *Aegiceras corniculatum* reaches its seaward limit at about 5 km upstream.

Further upstream there is a general narrowing of the river and its associated littoral vegetation. The influence of seawater is reduced still further and *Aegiceras* occurs more frequently. After 8 km, large trees of *Heritiera littoralis* are occasionally present. *Hibiscus tiliaceus* occurs in scattered locations after 16 km. Trees of *Melaleuca* spp., particulary *M. leucadendron*, are present on the banks after 19 km and *Diospyros littorea* is locally common after 21 km.

In the upper reaches, *Aegiceras* and *Avicennia* are the only common mangrove species present that are also present downstream. *Maytenus*, though present along the total length of the river, is present only in small numbers. Because of the reduced marine influence, mixtures of species are common. *Diospyros* and *Melaleuca* spp. (*M. leucadendron, M. stenostachya* and occasionally *M. acacioides* and *M. viridiflora*) are locally common. Occasional other species on the banks include *Canthium latifolium, Clerodendrum inerme* (common to abundant in uppermost reaches), *Cynometra iripa, Derris trifoliata* (vine), *Premna acuminata, Strychnos lucida* and *Terminalia subacropta. Flagellaria indica* is locally common in the upper reaches. Some of these areas approach what could be termed gallery rainforest.

Two common low growing species that were also recorded along the middle and upper reaches of the river are *Acanthus ilicifolius* and *Acrostichum speciosum*.

Discussion

There are many ecological questions that can be investigated and answered by studies in naturally zoned communities, for these can be treated as the outcome of a natural experiment (Pielou, 1977). Especially where uniformly sloping environmental gradients are involved, the pattern of zonation can be used to express the ecological requirements of the constituent species. Where irregular environmental gradients are involved, the resultant patterns of zonation may be so complex as to be indiscernible without detailed experimental investigations. Attempts to unravel such zonal associations by dividing transects into arbitrary zones which ignore existing environmental gradients (e.g. Bunt and Williams, 1980, 1981), are likely to lead to inappropriate data which even sophisticated pattern analysis will not overcome (Harrington *et al.*, 1981).

In the mangroves of Princess Charlotte Bay, the formations of the open shoreline can be interpreted as a zonation pattern reflecting a more or less uniformly sloping environmental gradient, in this instance a relatively uniform change in tidal levels. With the qualifications that the seaward fringe is dominated by *Rhizophora* (with perhaps a sporadic, narrow *Sonneratia* zone) rather than by *Avicennia*, and that the landward zones are numerically dominated by *Ceriops decandra* rather than *C. tagal*, the basic zonation pattern corresponds well with that described for other uniformly sloping shorelines (Macnae, 1966; 1967; Saenger and Hopkins, 1975; Thom *et al.*, 1975; Saenger *et al.*, 1977). In the situations described by these authors, the different mangrove zones can be viewed as steady state responses to their respective tidal levels (Thom, 1967). In other words, each constituent species finds, within its tolerance range to salinity, its place in the environmental gradient resulting from the substrate level and the tidal range. This environmental sorting selects the species which are capable of growing in particular zones, while intra- and inter-specific competition, as described for *Avicennia* and *Ceriops*, largely determines the fine structure of that zone.

As already mentioned, some minor irregularities occur along the shoreline of the study area, including occasional large deposits of shells forming ridges in the *Rhizophora* zone, and localized mixing of shell material in the substrate. While these irregularities may modify the edaphic conditions locally, they barely alter the substrate level in relation to tides. Consequently it is not surprising that these have little or no discernible effects on the overall zonation pattern.

In contrast to those of the open shoreline, the riverine mangrove formations are more complex, due to the non-uniform environmental gradients. For example river meanders may affect species presence/absence in addition to the effects of longitudinal salinity gradients along the river. Such effects may be more pronounced with large diameter meanders where salinity-induced limits may be obscured for several kilometres along the river by the geomorphological restraints imposed by erosion and accretion. Nevertheless it is interesting to note that the limited observations of longitudinal river zonation reported for the Gulf of Carpentaria (Saenger and Hopkins, 1975) are similar to those described here.

It follows that zonation can be recognized at two scales – in a particular transect at 90° to the shoreline, and longitudinally along rivers. The interaction of these two scales is expressed in a number of diverse floristic assemblages with no overall clear pattern, as was found by Bunt and Williams (1980). For regional comparisons, therefore, greatest emphasis should be placed in the open shoreline rather than upriver sites since, in the former, easily interpretable floristic patterns are not obscured by these two scales of zonation interacting with one another.

Acknowledgements

In the field the assistance of Mr. Trevor Stanley, Queensland Herbarium, Messrs. Frank Olsen, Dan Mayer and Peter Stephens, Queensland Fisheries Service, and Mr. Jim Davie, Botany Department, University of Queensland, was gratefully appreciated. Many of the plant specimens were identified by Mr. Stanley and all illustrations were prepared by Mr. Mayer.

In particular we thank Dr. R.W. Johnson, Queensland Herbarium, for critically reading the first draft, and Queensland Fisheries Service for providing us the opportunity to visit and study Princess Charlotte Bay.

Literature cited

Bunt, J.S. 1978. The mangroves of the eastern coast of Cape York Peninsula north of Cooktown. Great Barrier Reef Mar. Park Auth. Workshop series 1:253–269.

Bunt, J.S., and W.T. Williams. 1980. Studies in the analysis of data from Australian tidal forests ('mangroves'). I. Vegetational sequences and their graphic representation. Aust. Jour. Ecol. 5:385–390.

Bunt, J.S. and W.T. Williams. 1981. Vegetational relationships in the mangroves of tropical Australia. Mar. Ecol. Prog. Ser. 4:349–359.

Byrnes, N.B., S.L. Everist, S.T. Reynolds, A. Specht and R.L. Specht. 1977. The vegetation of Lizard Island, north Queensland. Proc. Roy. Soc. Qd 88:1–15.

Cintron, G., A.E. Lugo, D.J. Pool and G. Morris. 1978. Mangroves of arid environments in Puerto Rico and adjacent islands. Biotropica 10:110–121.

Coleman, J.M., S.M. Gagliano and W.G. Smith. 1966. Chemical and physical weathering on saline high tidal flats, northern Queensland Australia. Bull. Geol. Soc. Amer. 77:205–206.

De Keyser, F. and K.G. Lucas. 1968. Geology of the Hodgkinson and Laura Basins, north Queensland. Bur. Miner. Resour. Aust. Bull. 84.

Fosberg, F.R. 1961. Vegetation – free zone on dry mangrove

coasts. U.S. Geol. Surv., Prof. Pap., 424D:216–218.
Frankel, E. 1974. Recent sedimentation in the Princess Charlotte Bay area, Great Barrier Reef Province. Proceedings of the Second International Coral Reef Symposium 2. Great Barrier Reef Committee, Brisbane December 1974, pp. 335–369.
Galloway, R.W. 1972. Introduction and summary description of the Mitchell-Normanby area. In: Lands of the Mitchell-Normanby area, Queensland. Land Res. Ser. CSIRO Aust. 26:7–14.
Galloway, R.W., R.H. Gunn and R. Story. 1972. Land Systems of the Mitchell-Normanby area. In: Lands of the Mitchell-Normanby area, Queensland. Land Res. Ser. CSIRO Aust. 26:15–38.
Gunn, R.H. 1972a. Soils of the Mitchell-Normanby area. In: Lands of the Mitchell-Normanby area. Queensland. Land Res. Ser. CSIRO Aust. 26:53–74.
Gunn, R.H. 1972b. Climate of the Mitchell-Normanby area. In: Lands of the Mitchell-Normanby area, Queensland. Land Res. Ser. CSIRO Aust. 26:38–44.
Harrington, G.N., G.T. Dawes and J.A. Ludwig. 1981. An analysis of the vegetation pattern in a semi-arid *Eucalyptus populnea* woodland in north-west New South Wales. Aust. Jour Ecol. 6:279–287.
Hynes, R.A., and J.G. Tracey. 1980. Vegetation of the Iron Range area, Cape York Peninsula, In: Contemporary Cape York Peninsula. Proceedings Royal Society of Queensland Symposium, Brisbane 29 October 1979, pp. 11–30.
Isbell, R.F. 1978. Land and river systems influencing the Great Barrier Reef in Lizard Island – Cape York area, north Queensland. Great Barrier Reef Mar. Park Auth., Workshop series 1:72–89.
Jones, N.T. 1971. Mangroves. In: L. Pedley and R.F. Isbell (eds). Plant Communities of Cape York Peninsula. Proc. Roy. Soc. Qd 82:67.
Lucas, K.G. and F. De Keyser. 1965. Cape Melville, Qd – 1:250,000 Geological Series. Bur. Miner. Resour. Aust. explanatory notes D/55–9.
Macnae, W. 1966. Mangroves in eastern and southern Australia. Aust. Jour. Bot. 14:67–104.
Macnae, W. 1967. Zonation within mangroves associated with estuaries in north Queensland. In: Estuaries. Am. Assoc. Adv. Sci. 83:432–441.
Nix, H.A. and J.D. Kalma. 1972. Climate as a dominant control in the biogeography of northern Australia and New Guinea, In: Bridge and Barrier: the natural and cultural history of Torres Strait, pp. 61–91.
Pedley, L. and R.F. Isbell. 1971. Plant Communities of Cape York Peninsula. Proc. Roy. Soc. Qd 82:51–74.
Pielou, E.C. 1977. Mathematical Ecology. Wiley Interscience Publication, John Wiley & Sons, pp. 385.
Saenger, P. and M.S. Hopkins. 1975. Observations on the mangroves of the southeastern Gulf of Carpentaria, Australia. In: Proc. Int. Symp. Biol. Manag. Mangr. Mangroves, Inst. Food and Agr. Sci., Univ. Florida, Gainesville, Forida 1:126–136.
Saenger, P., M.M. Specht, R.L. Specht and V.J. Chapman. 1977. Mangal and coastal salt-marsh communities in Australasia. In: Ecosystems of the world: Wet Coastal Ecosystems, Elsevier, Amsterdam, pp. 293–345.
Spenceley, A.P. 1976. Unvegetated saline tidal flats in north Queensland. Jour. Trop. Geography 42:78–85.
Stoddart, D.R. 1980. Mangroves as successional stages, inner reefs of the northern Great Barrier Reef. Jour. Biogeogr. 7:269–284.
Story, R. 1972. Vegetation of the Mitchell-Normanby area. In: Lands of the Mitchell-Normanby area, Queensland. Land Res. Ser. CSIRO, Aust. 26:75–81.
Thom, B.G. 1967. Mangrove ecology and deltaic geomorphology, Tabasco, Mexico. Jour. Ecol. 55:301–343.
Thom, B.G., L.D. Wright and J.M. Coleman. 1975. Mangrove ecology and deltic-estuarine geomorphology: Cambridge Gulf-Ord River, Western Australia. Jour. Ecol. 63:203–232.
Webster, P.J. and N.A. Streton. 1972. Aspects of late Quaternary Climate in tropical Australasia. In: Bridge and Barrier: The natural and cultural history of Torres Strait pp. 39–60.
Whitaker, W.G. and D.L. Gibson. 1977. Ebagoola, Queensland – 1:250,000 Geological series, Bur. Miner. Resour. Aust. and Geol. Surv. Qd. Explanatory notes D/54–12.
Whitehouse, F.W. 1944. The natural drainage of some very flat monsoonal lands. Aust. Geogr. 4:183–196.

Appendix A

Species growing within tidal limits, Marrett River

Acanthus ilicifolius L.
Acrostichum speciosum Willd.
Aegialitis annulata R.Br.
Aegiceras corniculatum (L.) Blanco
Amyema conspicuum (F.M. Bailey) Danser*
Halosarcia pergranulata (J.M. Black) P.G. Wilson ssp. *queenslandica* P.G. Wilson
H. indica (Willd.) P.G. Wilson ssp. *leiostachya* (Benth.) P.G. Wilson
Avicennia marina (Forsk.) Vierh.
Batis argillicola van Royen
Bruguiera cylindrica (L.) Blume
B. exaristata Ding Hou
B. gymnorhiza (L.) Lam.
Canthium latifolium F. Muell.
Ceriops decandra (Griff.) Ding Hou
Clerodendrum inerme (L.) Gaertn.
Cynometra iripa Kostel.
Dendropthoe glabrescens (Blakely) Barlow†
Derris trifoliata Lour.
Diospyros littorea (R.Br.) Kosterm.
Excoecaria agallocha L.
Flagellaria indica L.
Heritiera littoralis Aiton ex Dryander

Hibiscus tiliaceus L.
Lumnitzera racemosa Willd.
Maytenus emarginata (Willd.) Ding Hou
Melaleuca acacioides F. Muell.
M. leucadendron (L.) L.
M. stenostachya S.T. Blake
M. viridiflora Solander ex Gaertn.
Osbornia octodonta F. Muell.
Premna acuminata R.Br.
Rhizophora apiculata Blume
R. stylosa Griff.
Sonneratia alba J. Smith
Sporobolus virginicus (L.) Kunth var. *minor* F.M. Bailey
Strychnos lucida R.Br.
Terminalia subacropta Domin
Xylocarpus australasicus Ridley
X. granatum Koenig

Footnotes

* Mistletoe, parasitic on *Xylocarpus australasicus*, one only seen.

† Mistletoe, parasitic on *Excoecaria agallocha*, numerous.

CHAPTER 5

Aspects of the development of mangals in the Townsville Region, North Queensland, Australia

A.P. SPENCELEY

Department of Geography, University of Papua New Guinea, Box 320 University, Papua New Guinea

The development of mangals and the underlying factors which control the zonation of mangroves have been considered for some time. As a rule these two aspects of mangrove swamps have been considered in isolation from each other. The idea has been suggested that contemporary factors such as salinity, frequency of inundation and duration of tidal emergence and height of the land surface are the most important controlling influences of the zonation of mangroves (Watson 1928; Davids 1940; Macnae 1966, 1967, 1968; Baltzer 1969; Clarke and Hannon 1969). The relationship between geomorphological and soil textural factors has also been noted (Gledhill 1963; Giglioli and Thornton 1965; Thom 1967; Hervieu 1968; Vieillefon 1969). More recently a temporal element has been introduced into such a discussion (Thom 1975; Thom et al. 1975). The relationship of the soil chemical status of the underlying substratum to the zonation of mangroves has also been commented upon (Navalkar and Bharucha 1950; Thornton and Giglioli 1965; Giglioli and King 1966; Vieillefon 1969; Bunning, cited in Ding Hou 1958; Kartawinata and Walujo 1977). However, the latter group of studies has tended to consider one type of coastline only and it is unclear as to how widespread the relationship is. The object of this paper is to consider the vegetation/soil relationships on both an open accreting coastline and an estuarine coastline (Diemont and Wijngaarden 1975) in the Townsville region, north Queensland.

Site characteristics

Two sites were chosen for detailed analysis. The west coast of Magnetic Island, located 8 km offshore from Townsville, is an example of an open accreting coastline. A second site, at Saunders Beach, 25 km north of Townsville (Fig. 1) is an example of an estuarine coastline. Geologically the area is young. The mangal on Magnetic Island grows on Recent sediment. Saunders Beach is situated on a Pleistocene and Quaternary infilled coastal plain with the coastal fringe comprising interdigitating marine and terrestrial deposits.

Climate

Townsville experiences a seasonally dry climate with its hot wet season in December to April. This is followed by a cooler dry season. Mean annual temperature is 24.4° C with a mean maximum temperature of 28.2° C. Temperatures are highest in January (mean 27.6° C) and lowest in July (mean 19.9° C) (Oliver 1978). In spite of this trend the mean receipt of solar radiation is greatest between September and December when there is the least cloud cover and relatively longer days.

Rainfall is highly seasonal, averaging 1163 mm per annum. On average 89% of the mean annual total falls between November and April. Because of high evaporation rates, January, February and March are on average the only months with a

Tasks for vegetation science, Vol. 8 ed. by H.J. Teas

 ISBN 978-90-481-8526-9

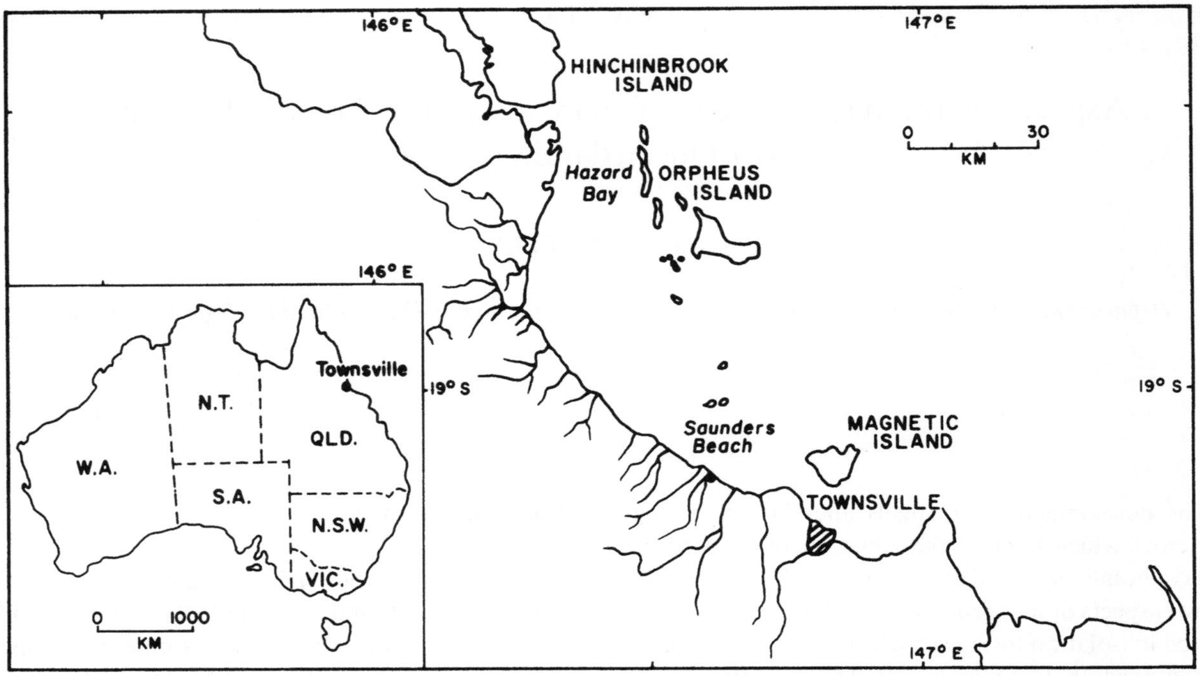

Fig. 1. Location of study sites.

moisture surplus. The basic features of the rainfall are a low reliability and high variability with much of the rainfall occurring as falls of high intensity and short duration (Murtha and Reid 1976). Much of the rain is associated with rain depressions which result from the development of tropical cyclones. On average 3–4 cyclones per year affect the Queensland coastline (Oliver 1973).

Few climatic data are available for Magnetic Island which probably experiences a similar climatic regime to Townsville. Average rainfall is higher than Townsville's by between 30%–50%. No accurate data are available for Saunders Beach.

Vegetation – Magnetic Island

Three basic vegetation zones are found on the island. These are from the land to the sea (Figs. 2 and 3):

(i) *Ceriops* zone.
(ii) *Rhizophora* zone.
(iii) *Avicennia/Sonneratia* fringe.

(i) The Ceriops zone is dominated by *Ceriops tagal* var. *australis*, height 1–5 m. At MHWST this zone is interrupted by a salt flat which is surrounded by samphire vegetation. On the landward fringe of the salt flat is a zone of *Arthrocnemum leiostachyum* with the occasional *Avicennia eucalyptifolia.* On the seaward side of the salt flat two samphire vegetation zones are found. The first of them is dominated by *A. leiostachyum* with the occasional *A. eucalyptifolia* and *Suaeda maritima*, whilst the second zone is dominated by *A. eucalyptifolia* and *S. maritima*.

(ii) The *Rhizophora* zone is the widest zone being approximately 200 m wide along the transect line. The zone is dominated by *Rhizophora stylosa.* In the areas devastated by Cyclone Althea, December 1971, seedlings of *Bruguiera spp.* along with those of *Rhizophora* spp. are present. Trees range in height from 3–8 m depending on their age.

(iii) The narrow *Avicennia/Sonneratia* fringe is dominated by *A. eucalyptifolia* and *Sonneratia alba.* They are approximately 2–4 m high.

A total of 11 sampling stations were set up across the intertidal zone. They were set up within the

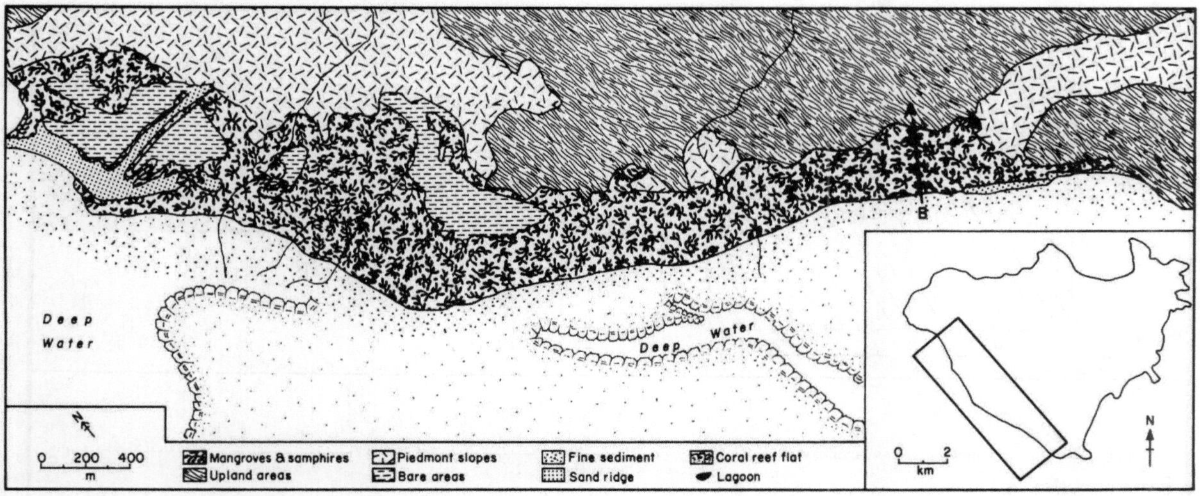

Fig. 2. Geomorphological map of Magnetic Island.

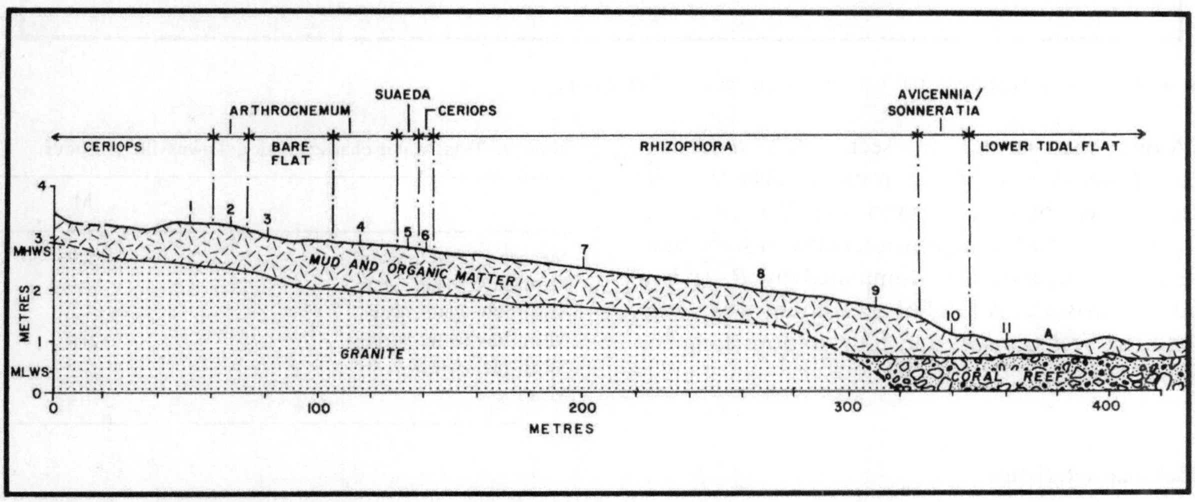

Fig. 3. Zonation of vegetation through the mangal on Magnetic Island.

vegetation zones and on the bare lower intertidal slope (Fig. 3).

Vegetation – Saunders Beach

Two transects were laid down at Saunders Beach (Fig. 4). The generalized zonational pattern from the land to the sea is:

(i) A zone of *Ceriops tagal*, trees 2–4 m high.

(ii) A mixed creek bank community dominated by *Rhizophora stylosa*. Other species present include *Aegialitis annulata*, *Avicennia eucalyptifolia*, *Bruguiera exaristata*, *Ceriops tagal*, *Lumnitzera racemosa* and *Xylocarpus australasicum*. Trees range in height from 4–5 m. Some variations to this pattern are displayed along the two transects, SB1, the *Ceriops* zone is interrupted by a narrow *Sporobolus virginicus* situated on top of the creek bank. Further inland, but at a lower elevation, a salt flat has developed with a bordering *Arthrocnemum leiostachyum* zone.

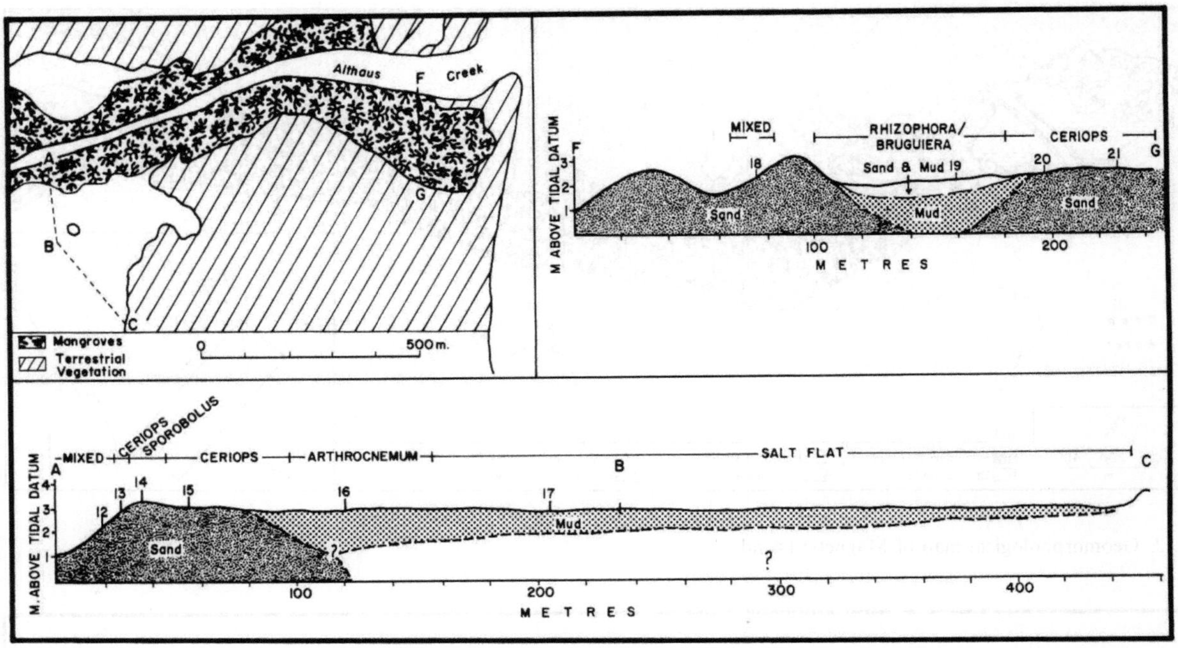

Fig. 4. Zonation of vegetation through the mangal at Saunders Beach.

Along the second transect, SB2, a *Rhizophora/Bruguiera* zone is present between the mixed creek bank zone and the *Ceriops* zone. Situated in a shallow depression between some sand banks, the vegetation is dominated by *R. stylosa* and *B. exaristata*. A total of six stations was set up along the first transect and four stations long the second transect.

Table 1. Tidal water characteristics, Townsville Harbour.

	M
Mean high water springs	2.9
Mean high water neaps	2.0
Mean low water neaps	1.2
Mean low water springs	0.4
Mean high water	2.4
Mean level	1.59

Tidal characteristics

The tidal levels for Townsville Harbour are given in Table 1. For the period October 1973 to December 1975, inclusive, the actual tidal records were extracted from the tidal gauge records maintained by the Townsville Harbour Board. High tide levels were compared with stations' heights within the zones as well as the height of vegetation zones (Fig. 5). It can be seen that, although the distribution of the vegetation zones are in the same relative order at the various sites, their location on the intertidal slope is at quite different heights. Consequently, similar vegetation zones have different inundation characteristics within the Townsville area.

Methods

Stations on Magnetic Island were sampled every month during 1975 on the spring tides. Stations at Saunders Beach were sampled every three months starting in January 1975 and ending in January 1976. Ground water samples were taken, where possible, and soil samples were collected from the surface, 10 cm and 30 cm depth. These were analyzed for pH; water soluble chloride by titration with silver nitrate; water soluble sulphate using a spectrophotometer method; and soluble and ex-

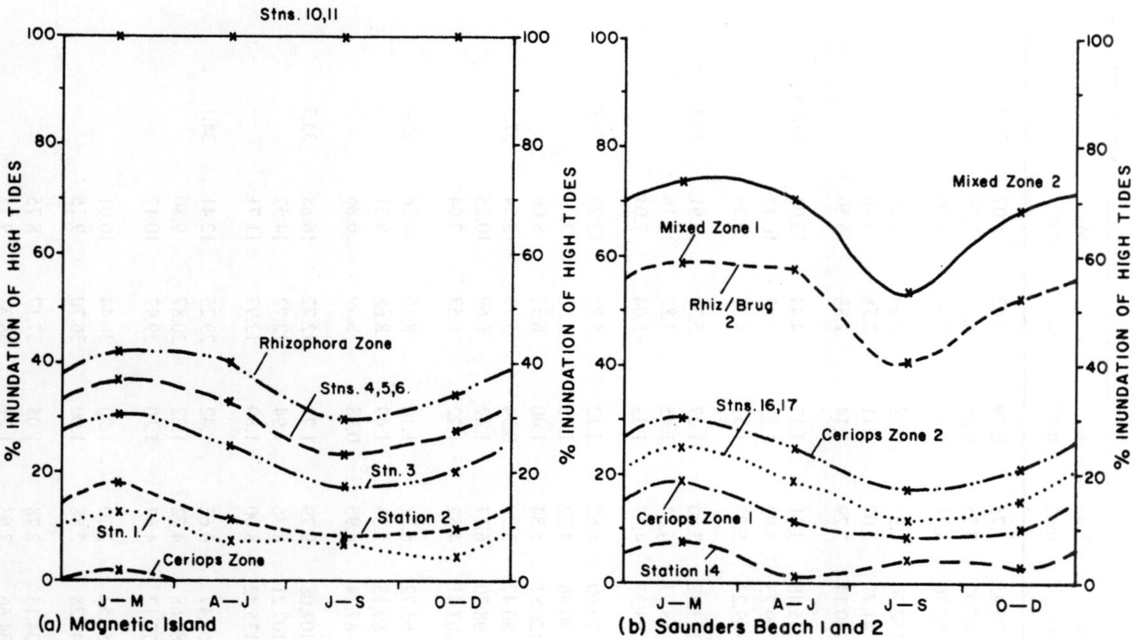

Fig. 5. Tidal inundation curves for stations and zones on Magnetic Island and Saunders Beach.

changeable sodium, potassium, calcium and magnesium using an atomic absorption spectrophotometer.

Results

The average data for each variable at each sampling site are given in Tables 2 and 3. The general trends which are highlighted by the results are given in Tables 4 and 5. The spatial trend in the data is slightly different along each transect. On Magnetic Island the general trend for soluble ions is a maximum value on the salt flat decreasing landward and seaward. The exception is water soluble sulphate which has a maximum value in the *Rhizophora* zone. For exchangeable ions maxima values occur in the *Rhizophora* zone decreasing landward and seaward.

A spatial trend in the data can be identified also at Saunders Beach. Along SB1 minima values occur on top of the creek bank in the *Sporobolus* zone increasing creekward and to the land. Maximum values occur on the salt flat. Along SB2 the highest concentrations are present in the second zone, *Rhizophora*/*Bruguiera* zone, decreasing towards the creek and inland. Values in this zone are often higher than those in other zones even though it is located at a lower elevation.

The major difference between the sites is found when considering the variations in elemental concentrations on a seasonal basis. On Magnetic Island most of the variables that were measured showed little if any identifiable seasonal variation. At Saunders Beach, however, there are marked seasonal variations in almost all variables. Low concentrations are present in the wet season and high concentrations occur in the dry season.

Discussion

This trends that are found in the data augment observations discussed by Diemont and Wijngaarden (1975). They note gross differences which exist between an estuarine coastline and an open accreting coastline. For the vaiables considered in this study, it has been shown that not only are there differences in the elemental concentrations between the two types of coastline in the Townsville

Table 2. Soil chemical data for Magnetic Island[a].

Station	Depth (cm)	pHKCl	pHH_2O	Water soluble		Soluble				Exchangeable				Ground water salinity ppt
				Cl me%	SO_4 me%	Na me%	K me%	Ca me%	Mg me%	Na me%	K me%	Ca me%	Mg me%	
1	0	6.43	6.93	26.54	1.66	172.28	5.27	2.52	33.45	1.75	0.39	3.62	5.07	31.2
	10	6.16	6.35	44.45	2.91	297.54	8.14	4.08	47.07	3.75	0.68	4.77	7.29	
	30	5.29	6.29	46.83	3.46	304.53	9.86	5.24	52.70	4.71	0.67	5.67	8.06	
2	0	6.94	7.70	82.80	8.38	596.04	18.48	4.55	152.70	2.9	0.98	4.97	9.55	53.7
	10	6.32	7.17	39.02	3.82	264.85	7.79	3.63	41.77	3.07	1.11	2.78	5.70	
	30	5.82	7.10	38.02	2.63	257.14	7.10	3.70	40.14	3.29	1.14	2.88	5.50	
3	0	8.19	8.67	89.92	7.87	696.71	18.87	5.90	145.48	3.61	1.17	4.44	15.04	118.0
	10	7.75	8.26	71.34	5.53	501.75	14.31	4.48	99.25	3.92	1.32	2.83	10.12	
	30	6.56	6.92	105.14	11.22	689.27	28.86	4.78	169.23	5.23	1.31	4.37	6.94	
4	0	7.77	8.49	70.24	6.62	498.32	13.75	4.99	103.57	3.80	1.27	5.39	9.91	75.1
	10	7.10	7.58	50.98	3.86	413.22	8.70	4.55	68.97	4.34	1.45	3.82	7.18	
	30	6.20	6.55	64.34	6.36	484.17	11.39	4.88	93.37	4.87	1.23	5.04	7.04	
5	0	7.52	7.90	57.37	5.06	458.04	10.88	5.80	79.60	6.62	2.12	5.81	12.27	73.9
	10	7.47	7.82	44.40	2.91	320.23	7.86	4.21	50.86	4.32	1.43	3.52	7.09	
	30	4.95	5.72	72.28	10.24	518.98	15.49	5.54	126.81	5.84	1.40	6.52	8.06	
6	0	6.78	7.50	61.77	5.79	453.96	12.86	4.84	80.47	3.98	0.94	5.73	8.34	57.5
	10	6.04	6.27	69.54	6.82	459.01	16.39	6.31	98.70	6.17	1.42	6.69	10.25	
	30	5.08	5.52	68.59	8.31	469.91	15.79	5.75	107.41	4.82	1.22	6.58	7.04	
7	0	5.73	6.02	47.11	7.51	337.34	10.38	5.03	63.72	6.27	1.10	8.13	10.56	33.4
	10	4.66	5.08	48.60	11.20	356.22	13.48	5.86	89.13	5.33	1.49	8.68	8.21	
	30	4.65	5.01	50.32	11.21	411.59	10.96	5.52	87.24	4.95	0.98	8.39	6.86	
8	0	5.55	6.05	64.47	10.02	475.91	15.21	9.04	107.08	9.79	1.78	12.72	16.88	33.7
	10	5.07	5.55	59.44	12.23	445.53	15.18	8.09	107.21	8.86	1.94	12.72	14.52	
	30	4.49	5.20	71.12	17.24	524.43	19.64	8.60	135.42	8.00	1.80	12.97	11.71	
9	0	6.88	7.49	49.99	6.64	375.41	9.51	10.72	69.47	6.62	1.42	29.25	12.41	34.1
	10	6.48	6.60	62.34	12.68	400.70	13.44	10.52	80.68	4.72	1.22	20.87	9.80	
	30	6.70	6.81	58.27	10.86	412.50	14.84	11.19	72.12	4.67	1.10	29.97	10.17	
10	0	7.58	7.71	41.50	5.91	290.44	15.03	11.97	51.53	3.70	1.20	36.44	10.01	
	10	7.41	7.71	37.34	5.84	237.22	8.89	10.19	41.28	2.88	1.00	36.70	9.15	
11	0	7.69	8.43	35.83	4.34	241.34	7.72	10.08	34.15	2.78	1.05	37.17	8.75	
	10	7.79	8.53	37.68	5.59	244.14	8.03	9.58	34.66	2.63	1.21	38.31	9.75	

[a] pH to Ex Mg inclusive, average of 12 montly values; groundwater Salinity, average of 26 monthly values.

Table 3. Soil chemical data for Saunders Beach.

Station	Depth (cm)	pHKCl	pHH_2O	Water soluble		Soluble				Exchangeable				Ground water salinity ppt
				Cl me%	SO_4 me%	Na me%	K me%	Ca me%	Mg me%	Na me%	K me%	Ca me%	Mg me%	
12	0	7.52	9.08	21.92	1.64	121.53	1.93	75.16	22.23	2.32	0.76	263.03	5.34	33.98
	10	7.36	8.46	14.78	1.03	100.50	1.61	63.72	13.29	2.42	0.82	135.20	4.04	
	30	6.88	7.80	16.56	0.80	129.20	2.03	68.67	19.82	2.70	0.87	71.01	4.13	
13	0	7.16	8.62	24.82	1.85	141.94	2.53	62.39	30.35	1.90	0.65	139.74	4.12	46.43
	10	6.88	8.52	11.28	0.70	71.84	1.13	41.10	9.85	1.96	0.64	59.44	3.14	
	30	6.70	8.12	14.38	0.88	97.14	1.37	48.87	13.34	2.26	0.70	56.70	3.56	
14	0	6.26	8.04	2.14	0.07	7.26	0.32	17.25	1.44	0.68	0.34	68.30	2.25	nd
	10	6.14	8.56	3.00	0.05	8.83	0.20	13.41	1.64	1.05	0.31	35.42	2.82	
	30	6.12	8.38	4.26	0.13	14.58	0.33	25.36	1.80	1.14	0.40	39.47	2.09	
15	0	7.08	8.10	25.02	2.19	148.43	4.04	52.08	37.98	1.42	0.70	113.62	3.58	44.78
	10	6.92	8.20	16.70	0.84	110.39	1.88	51.15	16.45	2.94	0.81	71.10	4.67	
	30	6.78	8.14	16.52	0.89	117.61	1.51	52.21	17.23	3.12	0.89	74.63	4.69	
16	0	8.38	9.38	100.00	6.92	496.54	12.02	103.63	198.76	2.90	0.93	300.67	8.61	48.86
	10	8.04	8.64	35.30	1.89	245.72	4.22	76.70	38.52	4.29	1.39	61.44	5.70	
	30	7.82	9.12	27.00	1.40	200.04	2.62	72.70	28.32	4.61	1.42	56.94	5.36	
17	0	7.52	8.84	96.48	6.45	579.05	11.75	116.95	210.38	5.15	1.28	159.30	8.24	85.94
	10	7.40	8.72	52.76	3.21	386.45	5.29	94.11	79.26	6.17	1.69	44.92	7.19	
	30	7.04	8.34	56.34	3.67	417.43	5.48	83.64	88.21	7.52	1.98	52.46	6.33	
18	0	7.52	8.74	17.06	1.11	114.22	2.38	66.24	21.11	1.36	0.67	136.02	3.54	31.63
	10	7.68	8.58	13.68	1.00	91.71	1.96	62.72	14.00	1.29	0.58	86.61	3.33	
	30	7.74	8.38	18.14	1.19	106.19	2.07	57.35	16.27	2.10	0.71	122.31	3.72	
19	0	6.78	7.46	36.34	2.68	220.27	3.12	105.06	43.71	13.66	2.55	168.22	13.76	35.97
	10	6.42	7.04	36.28	3.06	261.60	4.32	122.35	47.84	15.25	3.21	163.40	13.11	
	30	6.30	6.96	42.50	4.02	292.58	5.47	116.11	59.43	14.60	3.36	173.89	13.11	
20	0	6.96	7.90	19.42	1.39	117.21	2.13	55.11	27.50	2.41	0.76	68.61	4.65	42.05
	10	6.74	7.38	18.76	1.35	131.36	1.76	50.09	17.22	4.06	1.12	76.71	5.43	
	30	6.74	7.60	25.58	1.55	147.51	2.03	60.55	26.15	4.46	1.23	65.49	5.73	
21	0	7.06	7.84	23.08	1.69	142.03	2.63	58.23	31.94	1.92	0.67	75.90	3.83	46.2
	10	7.08	7.90	18.88	1.27	113.17	1.80	45.77	19.13	2.15	0.68	57.13	3.68	
	30	7.00	7.96	21.78	1.50	149.16	2.23	56.15	24.93	2.62	0.80	60.23	4.34	

[a] pH to Ex Mg inclusive, average of 5 values; groundwater salinity, stations 12, 16, 17 average of 5 values; station 13 average of 3 values; station 15 average of 4 values; stations 18, 19, 20, 21 average of 6 values.

[b] SB1 – stations 12 – 17 inclusive; SB2 – stations 18 –21 inclusive.

Table 4. Trends in the soil chemical data, Magnetic Island.

pHKCl	Spatial trend, highest values on the salt flat and in the *Rhizophora* zone. Lack of seasonal trend.
pHH_2O	Similar to pHKCl but with a slightly larger range in values.
WSCl	Highest values on the salt flat decreasing to land and sea. Seasonal trend more developed in landward than seaward zones.
$WSSO_4$	Values increase with depth with greatest values in the *Rhizophora* zone.
SolNa	Highest values on salt flat decreasing to land and sea. Greater variation in surface layers than at depth. Concentration least in wet season and greatest end of the dry season.
SolK	Lack of spatial trend but landward and seaward two zones have lower concentrations than other zones. Higher values at end of dry season lower in wet season, with highest values on upper portion of intertidal slope.
SolCa	Concentrations increase from the land to the sea. Lack of reliable seasonal trends.
SolMg	Landward *Arthrocnemum* zone and salt flat have highest concentrations decreasing to land and sea. Surface values show greatest variation.
ExNa	Increase from land to the *Rhizophora* zone. Lack of a seasonal trend.
ExK	Highest values in the *Saeda/Avicennia* zone and station 8 in the *Rhizophora* zone. Lack of a seasonal trend.
ExCa	Increase from the land to the sea, sharply within *Rhizophora* zone. No clear seasonal pattern.
ExMg	Highest concentration on salt flat and in *Rhizophora* zone. Lower values tend to occur in the wet season.
ExNa/ExK	Lowest values on salt flat and highest in *Rhizophora* zone.
ExCa/ExMg	Similar to ExNa/ExK.
Salinity	Highest values on salt flat decreasing to sea and land. Marked seasonal variations, lowest in the wet season. Variation greatest on salt flat least in *Rhizophora* zone.

region, but the behaviour of the elements also differs. On an open accreting coastline, although there is a spatial variation, most variables have little or no regular variation with time. However, on an estuarine coastline there is both a spatial and a temporal variation in elemental concentration.

Another difference also exists between the two types of coastline. It can be demonstrated, statistically, that on Magnetic Island, the open accreting coastal site, the multivariate soil chemical profile of stations located within the same vegetation zone, either the *Rhizophora* or *Ceriops* zones, are different from each other. However, stations located on the upper intertidal slope, in zones with a low discontinuous or absent plant cover, are similar to each other. In contrast to this, at Saunders Beach, the estuarine coastal site, the reverse situation is present. On SB1 the four zones closest to the creek bank are similar to each other but the *Arthrocnemum* zone and the salt flat are different from each other and from every other zone. On SB2 all stations are similar to each other except for the *Rhizophora/Bruguiera* zone.

The tidal inundation information has shown that similar vegetation zones are located on different portions of the intertidal slope. Hence they have a different frequency of tidal inundation and emergence. Also it has been shown (Spenceley 1976) that salt flats have a variety of modes of origin. Hypersalinity appears to be a function of the type and density of plant cover more than solely a function of distance from the sea or creek.

Consequently, contrary to findings from other studies, for the soil chemical variables that were measured no predictable soil/vegetation relationship could be recognized in the Townsville mangals. This is not to say that soil chemical variables do not play a significant role in influencing the distribution of plant species across the intertidal slope.

Thom and his co-workers (Thom 1967, 1975; Thom *et al.* 1975) have demonstrated the role of the geomorphological development of an area on the distribution of mangroves. However, concomittant with changes in the geomorphological environment are changes in the oxidation/reduction conditions in the soil. The expected changes that take place have been outlined previously (cf van Beers 1962; Pons and Zonneveld 1965; Ponnamperuma 1972; Moorman and Pons 1975). Attention to this aspect of the soil environment has been directed primarily to the agricultural potential of the land (cf Dost (ed.) 1973; Moorman and Pons 1975; Metson *et al.* 1977) and not with respect to the effect on the distribution of plants within the mangals. The more

Table 5. Trends in the soil chemical data, Saunders Beach.

pHKCl	SB1 – lowest under the *Sporobolus* zone increasing 1 creekward and landward.
	SB2 – lowest in the *Rhizophora/Bruguiera* zone increasing creekward and landward. Lack of a seasonal trend.
pHH_2O	Similar trend to above but with higher values.
WSC1	SB1 – decrease from the creek to the *Sporobolus* zone then increasing to the salt flat. Marked seasonal trend especially in the surface layers.
	SB2 – greatest concentration in the *Rhizophora/Bruguiera* zone decreasing to land and the creek. Lack of a seasonal trend.
$WSSO_4$	SB1 – similar to WSCl.
	SB2 – similar to WSC1 but with a seasonal trend.
SolNa	SB1 – lowest value in the *Sporobolus* zone and highest on the salt flat. Marked seasonal trends with surface concentrations varying the most.
	SB2 – highest values in the *Rhizophora/Bruguiera* zone but with 30 cm layer having the greatest concentration. Unclear seasonal pattern.
SolK	SB1 – increase in concentrations to the salt flat with marked seasonal variation especially in the surface layer.
	SB2 – greatest concentrations in the *Rhizophora/Bruguiera* zone especially in the sub-surface layers. Seasonal variation present.
SolCa	SB1 – lowest concentrations on the *Sporobolus* zone and highest on the salt flat. Seasonal variation present.
	SB2 – highest concentration in the *Rhizophora/Bruguiera* zone, seasonal variation present.
SolMg	SB1 – increase in concentration to the salt flat. Seasonal variation greatest in the surface layer.
	SB2 – highest values in the *Rhizophora/Bruguiera* zone especially in the sub-surface layers. A seasonal trend is present.
ExNa	SB1 – lowest values in the *Sporobolus* zone and highest on the salt flat. No seasonal trend apparent.
	SB2 – highest values in the *Rhizophora/Bruiguiera* zone with no seasonal trend apparent.
ExK	SB1 – similar trend to ExNa.
	SB2 – similar trend to EXNa.
ExCa	SB1 – lowest value in the *Sporobolus* zone increasing to land and the creek. Seasonal variation greatest in the surface layer.
	SB2 – highest values in the *Rhizophora/Bruguiera* zone with marked seasonal variations.
ExMg	SB1 – lowest values in the *Sporobolus* zone, highest on the salt flat. No appreciable seasonal variation.
	SB2 – highest value in the *Rhizophora/Bruguiera* zone. No appreciable seasonal variation.
ExNa/ExK	SB1 – marked variation along the transect. Possible seasonal variation.
	SB2 – lack of a seasonal trend with great variation in the ratio except in the *Rhizophora/Bruguiera* zone.
ExCa/ExMg	SB1 – general decrease landward, seasonal trend present.
	SB2 – lowest ratio in the *Rhizophora/Bruguiera* zone. Seasonal trend present.
Salinity	SB1 – general increase landward, seasonal pattern greatest on the salt flat.
	SB2 – increase in a landward direction, with a seasonal pattern present.

labile elements are known to affect the viability of plants (Rorison 1973). Iron and manganese have been shown to influence the distribution of plants in dune swales (Jones 1972a, b). Thus the relationship between the geomorphological controls and the oxidation/reduction factor may provide a more useful line of inquiry to follow when considering the causes behind the distribution of plants in mangals.

Literature cited

Baltzer, F. 1969. Les formations végétales associées au delta de la Dumbéa (Nouvelle Caledonie) et leurs indications écologiques, géomorphologiques et sédimentologiques, mises én évidence par la cartographie. Cah. Office de la Recherche Scientifique et Technique Outre Mer, sér Géol. 1: 59–84.

van Beers. 1962. Acid Sulphate Soils IILRI Bull No. 3.

Clarke, L.D. and N.I. Hannon. 1969. The mangrove swamp and salt marsh communities of the Sydney District. II. The holocoenotic complex with particular reference to physiography. Jour. Ecol. 57: 213–234.

Davis, J.H. 1940. The ecology and geologic role of mangroves in Florida. Pap. Tortugas Lab. 32: 303–412.

Diemont, W.H. and W. van Wijngaarden. 1975. Sedimentation patterns, soils, mangrove vegetation and land use in the tidal areas of West Malaysia. In: Int. Symp. Biol. and Management of Mangroves. II: 513–528.

Ding Hou. 1958. Rhizophoraceae, Flora Malesiana ed. van Steenis 5: 429–493.

Dost, H. (ed.). 1973. Acid Sulphate Soils IILRI Pub No. 18, 2 vols.

Giglioli, M.E.C. and I. Thornton. 1965. The mangrove swamps of Keneba, Lower Gambia river basin. I. Descriptive notes on the climate, the mangrove swamps and the physical composition of their soils. Jour. Appl. Ecol. 2: 81–103.

Giglioli, M.E.C. and D.F. King. 1966. II. Seasonal variations in the chloride and water content of swamp soils with observations on the water content of swamp soils with observations on the water levels and chloride concentration of free soil water under a barren mud flat during the dry season. Jour. Appl. Ecol. 3: 1–19.

Gledhill, D. 1963. The ecology of the Aberdeen creek mangrove swamp. Jour. Ecol. 51: 693–703.

Hervieu, J. 1968. Contribution à l'étude de l'alluvionnement en milieu tropical. Mem Office de la Recherche Scientifique et Technique Outre Mer No. 24.

Jones, R. 1972a. Comparative studies of plant growth and distribution in relation to waterlogging. V. The uptake of iron and manganese by dune and dune slack plants. Jour. Ecol. 60: 131–140.

Jones, R. 1972b. VI. The effect of manganese on the growth of dune and dune slack plants. Jour. Ecol. 60: 141–145.

Kartawinata, K. and E.B. Walujo. 1977. A preliminary study of the mangrove forest on Pulau Rambut, Jakarta Bay. Marine Research in Indonesia 18: 119–129.

Macnae, W. 1966. Mangroves in eastern and southern Australie. Aust. Jour. Bot. 14: 67–104.

Macnae, W. 1967. Zonation within mangroves associated with estuaries in north Queensland. In: Estuaries ed. Lauff, 432–441.

Macnae, W. 1968. A general account of the fauna and flora of mangrove swamps and forests in the Indo-West Pacific Region. Adv. Mar. Biol. 6: 73–270.

Metson, A.J., E.J. Gibson, J.E. Cox and D.B. Gibbs. 1977. The problem of acid sulphate soils, with examples from North Auckland, New Zealand. N.Z.J. Sci. 20: 371–394.

Moorman, F.R. and L.J. Pons. 1975. Characteristics of mangrove soils in relation to their agricultural land use and potential. Int. Symp. Biol. and Management of Mangroves pp 529–547.

Murtha, G.G. and R. Reid. 1976. Soils of the Townsville area in relation to urban development. Commonwealth Scientific and Industrial Research Organization (Austr.), Div. Soils Div. Rept. No. 11.

Navalkar, B.S. and F.R. Bharucha. 1950. Studies in the ecology of mangroves VI. Exchangeable bases of mangrove soils. Jour. Univ. Bombay 19: 7–16.

Oliver, J. 1973. Australian weather example 1. Tropical cyclone. Aust. Geogr. 12: 257–263.

Oliver, J. 1978. The climatic environment of the Townsville area. In: Geographical Studies of the Townsville. Occ Pap., James Cook University of North Queensland 2: 3–17.

Ponnamperuma, F.N. 1972. The chemistry of submerged soils. Adv. Agron. 29–96.

Pons, L.J. and I.S. Zonneveld. 1965. Soil ripening and soil classification. IIRI Pub. No. 13.

Rorison, I.H. 1973. The effect of extreme soil acidity on the nutrient uptake and physiology of plants. In: Acid Sulphate Soils 1: 223–254.

Spencely, A.P. 1976. Unvegetated saline tidal flats in North Queensland. Trop. Geogr. 42: 78–85.

Thom, B.G. 1967. Mangrove ecology and deltaic geomorphology Tabasco, Mexico. Jour. Ecol. 55: 301–343.

Thom, B.G. 1975. Mangrove ecology from a geomorphic view point. Int. Symp. Biology and Management of Mangroves II: 469–481.

Thom, B.G., L.D. Wright and J.M. Coleman. 1975. Mangrove ecology and deltaic-estuarine geomorphology. Cambridge Gulf-Ord River, Western Australia. Jour. Ecol. 63: 203–232.

Thornton, I. and M.E.C. Giglioli. 1965. The mangrove swamps of Keneba, Lower Gambia river basin. II. Sulphur and pH in the profiles of swamp soils. Jour. Appl. Ecol. 2: 257–269.

Vieillefon, J. 1969. La pédogenese dans les mangroves tropicales. Un example de chronsequence. Sci. Sol. 2: 115–152.

Watson, J.G. 1978. Mangrove forests of the Malaya Peninsula. Mal. For. Rec., No. 6, 275 pp.

CHAPTER 6

Distribution of mangrove species in Australia

A.G. WELLS

Department of Environmental Physics, School of Physics, University of Sydney, NSW, Australia 2006

Abstract. In Australia, mangroves occur in small pockets around the coastline. They are most prominent around the northern coastline and have only isolated occurrences in Victoria, South Australia and south of Carnarvon in Western Australia. No mangrove species is recorded for Tasmania.

The geographical location of the continent provides scope to look at present distribution of the various species through an extremely wide latitudinal range (11°–38° S). Many species show restricted distributions which may be attributed to past and present historical and environmental conditions. The number of mangrove species in Australia (33) decreases with increasing latitude with most distributions extending further down the eastern than the western coastline and at the southernmost limit of mangroves in the world – Corner Inlet, Victoria (38°45′ S, 146°30′ E) – only the species *Avicennia marina* var. *australasica* remains.

Introduction

Within Australia, studies by various workers, overviewed by Gill (1975) and Saenger et al. (1977) and additional researches by George and Kenneally (1975); Lear and Turner (1977); Messel et al. (1977); Saenger and Robson (1977); Specht et al. (1977); Semeniuk et al. (1978); Bunt (1978); Dowling (1979); Duke and Bunt (1979); Messel et al. (1979, a–g), (1980, a–f), (1981, a–c); Bunt and Williams (1980); Olsen et al. (1980); Semeniuk et al. (1980); Dowling and McDonald (1981); Kenneally (1981) and Wells (1981a, b), now provide considerable information on occurrences of the various mangrove tree, shrub and herb species around this continent.

With a mainland coastline estimated at 19,650 km and extending from 11° to 39° S, mangrove communities change in floristics and structure under various climatic, physiographic and edaphic conditions. The area occupied by mangrove vegetation is estimated by Galloway (1981) at 11,617 km^2, of which 10,827 km^2 occurs north of the Tropic of Capricorn (23° S). His estimate for the Queensland coastline (4,629 km^2) is closely followed by the Northern Territory (4,130 km^2) and Western Australia (2,517 km^2), with the southern mainland states of New South Wales, Victoria and South Australia accounting for only 341 km^2, or 3%, of mangrove swamp in Australia (Galloway, 1981).

Floristically and structurally, mangrove vegetation shows greatest development in north-eastern Queensland where at least thirty-three 'mangrove' species belonging to sixteen families of angiosperms have been recorded. Within the Northern Territory occur twenty-six species from thirteen plant families, whilst only seventeen species from twelve families are presently recorded from Western Australia (Table 2).

From the point of view of species distributions, the only portion of the coastline that now requires

Tasks for vegetation science, Vol. 8 ed. by H.J. Teas

 ISBN 978-90-481-8526-9

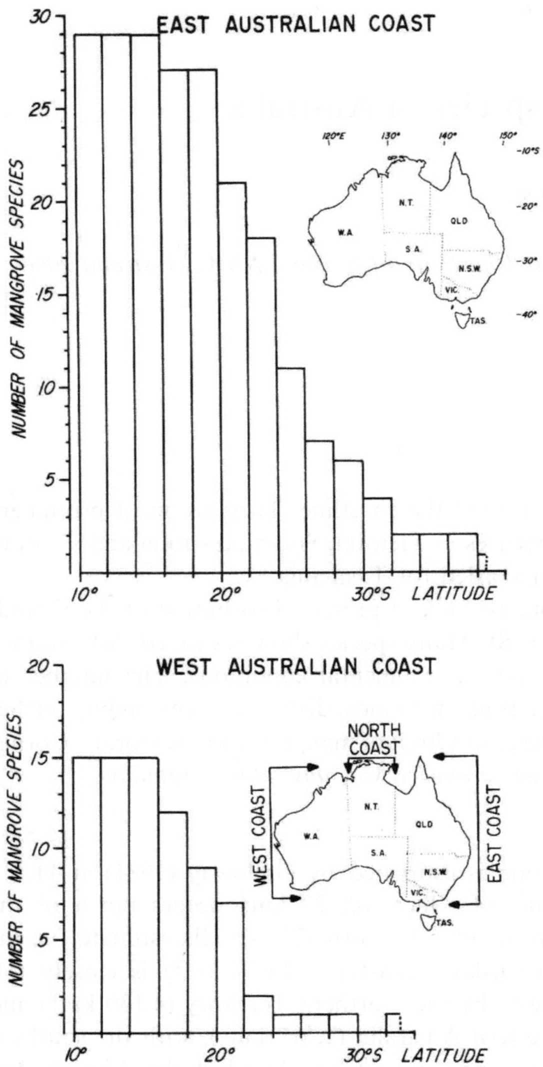

Fig. 1. Number of Mangrove species recorded in 2° latitudinal subdivisions down the east and west coasts of Australia.

Table 1. Mangrove species recorded in Australia.

Acanthaceae	*Acanthus ebracteatus* Vahl.
	Acanthus ilicifolius L.
Avicenniaceae	*Avicennia marina* (Forsk.) Vierh.
	Avicennia officinalis L.
Bignoniaceae	*Dolichandrone spathacea* (L.f.) K. Sch.
Bombacaceae	*Camptostemon schultzii* Mas.
Combretaceae	*Lumnitzera littorea* (Jack.) Voigt.
	Lumnitzera racemosa Willd.
	Lumnitzera rosea (Gaud.) Presl.
Euphorbiaceae	*Excoecaria agallocha* L.
Leguminoseae	*Cynometra ramiflora* var. *Bijuga* (Spanoghe) Benth.
Meliaceae	*Xylocarpus australasicus* Ridl.
	Xylocarpus granatum Koen.
Myrsinaceae	*Aegiceras corniculatum* (L.) Blanco
Myrtaceae	*Osbornia octodonta* F. Muell.
Nypaceae	*Nypa fruticans* Wurmb.
Plumbaginaceae	*Aegialitis annulata* R. Br.
Rhizophoraceae	*Bruguiera cylindrica* (L.) Bl.
	Bruguiera exaristata Ding Hou.
	Bruguiera gymnorhiza (L.) Lamk.
	Bruguiera parviflora (Roxb.) W & A ex Griff.
	Bruguiera sexangula (Lour.) Poir.
	Ceriops decandra (Griff.) Ding Hou.
	Ceriops tagal (Perr.) C.B. Rob. var. *australis* C.T. White
	Ceriops tagal (Perr.) C.B. Rob
	Rhizophora apiculata Blume
	Rhizophora lamarckii Montr.
	Rhizophora mucronata Lamk.
	Rhizophora stylosa Griff.
Rubiaceae	*Scyphiphora hydrophyllacea* Gaertn. f.
Sterculiaceae	*Heritiera littoralis* (Dryand) Ait.
Sonneratiaceae	*Sonneratia alba* J.E. Sm.
	Sonneratia caseolaris (L.) Engl.
	Sonneratia spp.

intensive study lies between Weipa and Karumba on the eastern shores of the Gulf of Carpentaria (Queensland). Work is currently underway at the Australian Institute of Marine Science to describe several new *Sonneratia* species and their respective distributions.

Methods

This study involved a detailed survey of species present on most tidal waterways across northern Australia where previously distributions of mangrove species were very imperfectly known. The extreme distributional limits of many species on the east coast of Queensland draw largely on the findings of Jones (1971), Bunt (1978), Duke and Bunt (1979), the latter authors' current research at the Australian Institute of Marine Science and of Dowling at the Queensland Herbarium. For much of the west Australian coastline not surveyed by the author of this article, the findings of Semeniuk et al. (1978) have also proved invaluable.

Table 2. Distribution of mangrove species in Australia in latitudinal subdivisions.

Species	Northern coastline (N.T.)	East Australian coastline															West Australian coastline											
LAT	10–12° S	10–12° S	12–14°	14–16°	16–18°	18–20°	20–22°	22–24°	24–26°	26–28°	28–30°	30–32°	32–34°	34°–36°	36° + 38°	–38° 45′	10–12° S	12–14°	14–16°	16–18°	18–20°	20–22°	22–24°	24–26°	26–28°	28–30°	30–32°	33° 16′
Acanthus ebracteatus		×																	×									
Acanthus ilicifolius	×	×	×	×	×	×	×	×									×	×										
Aegialitis annulata	×	×	×	×	×	×	×	×	×								×	×	×	×	×	×						
Aegiceras corniculatum	×	×	×	×	×	×	×	×	×	×	×	×	×	×	×		×	×	×	×	×	×						
Avicennia marina	×	×	×	×	×	×	×	×	×	×	×	×	×	×	×	×	×	×	×	×	×	×	×	×	×	×		×
Avicennia officinalis	×																											
Bruguiera cylindrica		×	×	×																								
Bruguiera exaristata	×	×	×	×	×	×	×	×									×	×	×	×	×	×						
Bruguiera gymnorhiza	×	×	×	×	×	×	×	×	×	×	×																	
Bruguiera parviflora	×	×	×	×	×	×	×										×	×	×									
Bruguiera sexangula	×	×	×	×	×	×																						
Camptostemon schultzii	×	×	×	×													×	×	×	×								
Ceriops decandra	×	×	×	×	×	×																						
Ceriops tagal var. *australis*	×	×	×	×	×	×	×	×	×	×	×						×	×	×	×	×	×						
Ceriops tagal	×	×	×	×	×	×	×	×	×																			
Cynometra ramiflora var. *bijuga*		×	×	×	×	×	×																					
Dolichandrone spathacea		×																										
Excoecaria agallocha	×	×	×	×	×	×	×	×	×	×	×	×					×	×	×	×	×							
Heritiera littoralis		×	×	×	×	×	×	×																				
Lumnitzera littorea	×	×	×	×	×	×																						
Lumnitzera racemosa	×	×	×	×	×	×	×	×	×	×							×	×	×	×								
Lumnitzera rosea							×																					
Osbornia octodonta	×	×	×	×	×	×	×	×	×								×	×	×	×	×	×						
Nypa fruticans	×		×	×	×	×																						
Rhizophora apiculata	×	×	×	×	×	×	×	×																				
Rhizophora lamarckii	×	×	×	×	×	×	×	×																				
Rhizophora mucronata		×	×	×	×	×																						
Rhizophora stylosa	×	×	×	×	×	×	×	×	×	×	×	×					×	×	×	×	×	×	×					
Scyphiphora hydrophyllacea	×	×	×	×	×	×											×	×										
Sonneratia alba	×	×	×	×	×	×	×	×									×	×	×	×	×							
Sonneratia caseolaris	×		×	×	×	×																						
Xylocarpus australasicus	×	×	×	×	×	×	×	×									×	×	×	×								
Xylocarpus granatum	×	×	×	×	×	×	×	×	×										×									
Totals:	26	29	29	29	27	27	21	18	11	7	6	4	2	2	2	1	15	15	15	12	9	7	2	1	1	1	0	1

Species distributions shown as dot patterns in the ensuing figures are not indicative of a particular locality recorded with a herbarium specimen, although this may be the case, but rather the specific area of the coastline over which the species has been recorded.

Voucher specimens of mangroves from most portions of the coastline have been deposited in the various herbaria throughout Australia.

Results

Mangrove species currently recorded in Australia are shown in Table 1 with distributions along the eastern, northern and western coastlines (in latitudinal subdivisions) being provided in Figure 1 and Table 2.

Distributional information for individual species is presented under each species.

ACANTHACEAE

Acanthus ebracteatus

This species has only been recorded from a few localities in Australia – King River near Wyndham (15° 30′S, 128° 05′E) in Western Australia (Semeniuk et al., 1978) and in the upper reaches of the Escape River (10°59′S) on Cape York Peninsula (Duke, pers. comm.).

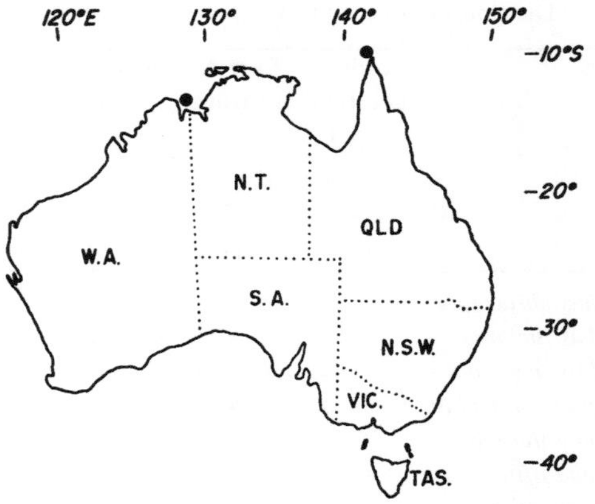

Acanthus ilicifolius

The southernmost limits of distribution of this species are the Daly River (13°10′S, 130°14′E) on the west coast of the Northern Territory (Wells, 1981b) and St Lawrence, Qld. (22°15′S) on the east Australian coastline (Jones, 1971).

Acanthus ilicifolius, although often locally abundant, is discontinuous in occurrence on tidal waterways of the Northern Territory eastwards of the Daly River. The species is particularly uncommon on tidal waterways entering the Gulf of Carpentaria but is more abundant (although still discontinuous in occurrence) along the north-eastern Australian coastline (Bunt, 1978; Duke and Bunt, Dowling, pers. comm.).

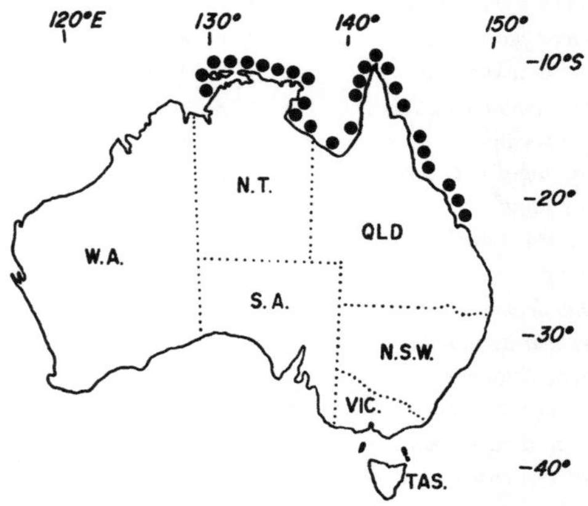

AVICENNIACEAE

Avicennia marina

The southernmost limits of distribution of this species are Leschenault Inlet, Bunbury (33°16′S, 115°42′E) on the west Australian coast and at Corner Inlet, Victoria (38°45′S, 146°42′E) on the south-east Australian coast.

Avicennia marina at Corner Inlet is the southernmost occurrence of any mangrove species in the world.

The occurrence of *Avicennia marina* at Bunbury is considered by Semeniuk et al., 1978, to be an outlier as the species does not reappear to the north along the west Australian coastline until the Abrolhos Islands are reached. Northwards from this point *Avicennia marina* is common at suitable sites around the entire coastline to Corner Inlet in Victoria. Isolated pockets of this species also occur in protected coastal pockets throughout South Australia (Butler et al., 1977).

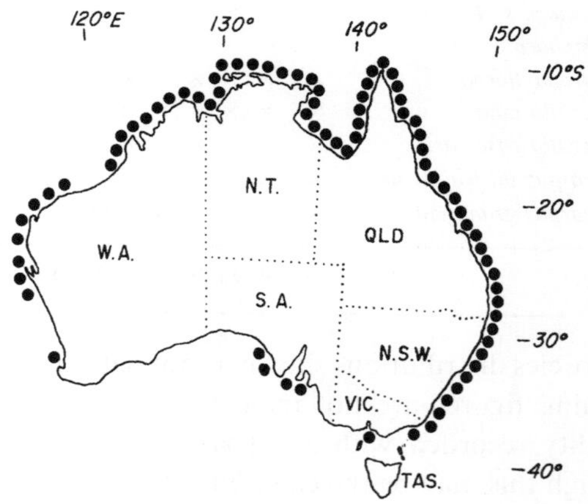

Avicennia officinalis

The restricted distribution of this species could be a result of its being at the edge of its geographical range in northern Australia.

Avicennia officinalis has not been recorded west of Buffalo Creek, Darwin, NT (12°20′s, 130°57′E) or east of the Glyde River (12°16′S, 135°03′E) in eastern Arnhem Land (Wells, 1981b). That *Avicennia officinalis* does not occur in rivers entering the north coast of Melville Island is most unusual as this species is quite widely distributed throughout southeast Asia (Chai, 1973; Chapman, 1975).

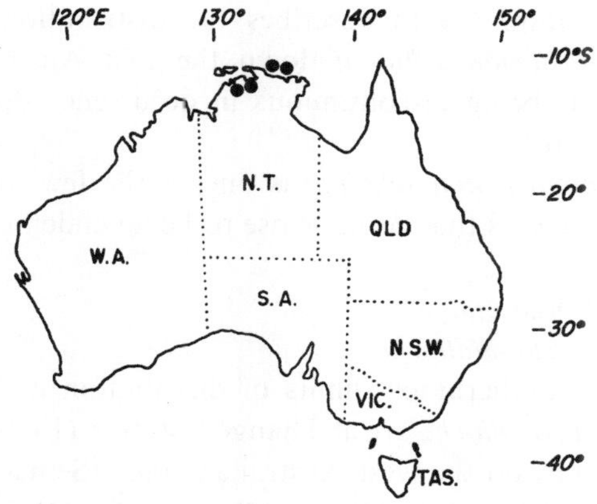

BIGNONIACEAE

Dolichandrone spathacea

This species is currently recorded from only two sites in north-east Queensland – Cowal Creek (10°54′S) and the Olive River (12°10′S) on Cape York Peninsula (Duke, pers comm.).

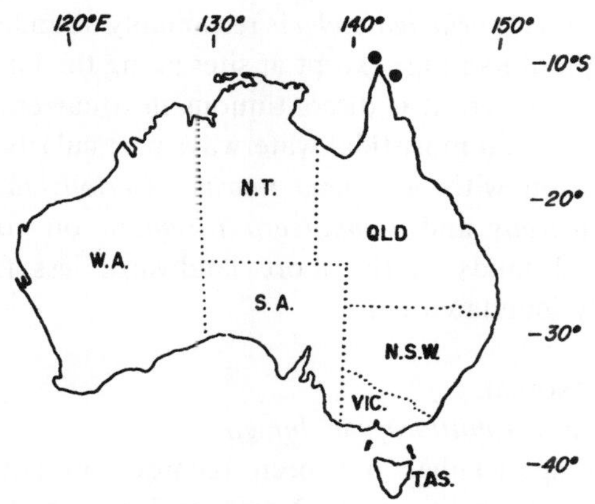

BOMBACACEAE

Camptostemon schultzii

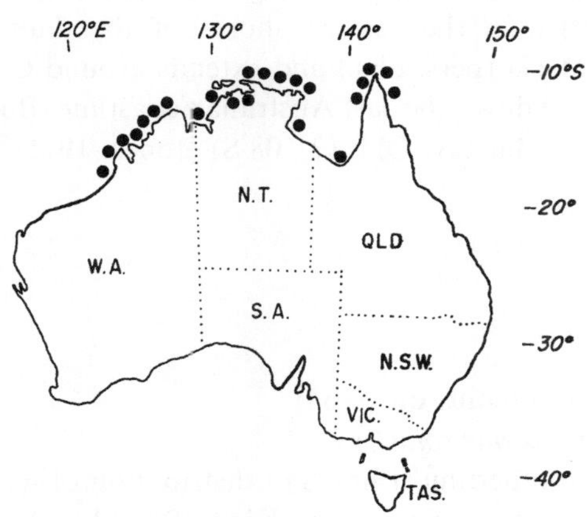

The southernmost limits of distribution of *Camptostemon schultzii* are Cape Bossut (18°43′S, 121°38′E) on the west Australian coastline (Semeniuk et al., 1978) and the Pascoe River, Qld (12°31′S) on the east Australian coastline (Duke, pers. comm.). The species is common from Cape Bossut to Gove (12°13′S, 136°42′E) in the Northern Territory, excluding the Joseph Bonaparte Gulf region where it is uncommon and the north coast of Melville Island, where it is unrecorded (Wells, 1981b). *Camptostemon schultzii* has not been recorded on any tidal waterway entering the western shores of the Gulf of Carpentaria although isolated trees of this species were collected on the Limmen Bight River (15°06′S, 135°42′E), which enters into the southern portion of the Gulf of Carpentaria. The species is unknown on any other tidal waterway entering the southern shoreline of the Gulf of Carpentaria across to Karumba (Saenger and Hopkins, 1975; Wells, 1981b).

Along the eastern shores of the Gulf of Carpentaria this species occurs at least as far south as Weipa Qld. (12°40′S, 141°50′E) (Specht et al.,

1977). Bunt (1978) describes the distribution of *Camptostemon schultzii* down the east Australia coast as being discontinuous in occurrence down Cape York.

Camptostemon schultzii is one of the few mangrove species that comes close to being endemic to Australia. As well as its wide distribution across northern Australia, the species also occurs in tidal waterways entering the Gulf of Papua (Percival and Womersley, 1975; Floyd, 1977) and Ambon Island in the Moluccas (Chapman, 1976).

EUPHORBIACEAE

Excoecaria agallocha

The southernmost limits of distribution of *Excoecaria agallocha* are at Thangoo Station (18°16′S, 122°10′E) on the west Australian coast (Semeniuk et al., 1978) and the Hastings River, NSW (31°26′S, 152°26′E) on the east Australian coast (Floyd, 1980). *Excoecaria agallocha* is reasonably abundant throughout its range except at sites along the NSW coastline, where it is discontinuous in occurrence.

It occurs on most tidal waterways particularly in association with *Avicennia marina*, *Ceriops tagal* var. *australis* and *Lumnitzera racemosa* on consolidated muds in the more landward, less frequently inundated zones.

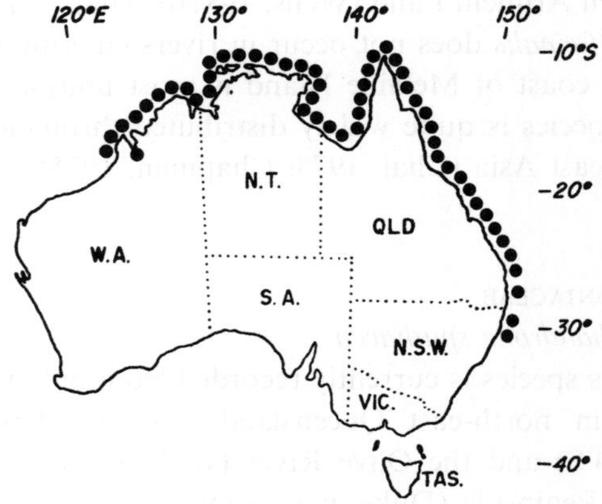

LEGUMINOSEAE

Cynometra ramiflora var. *bijuga*

This species has not been recorded in either Western Australia nor the Northern Territory. It is first recorded at Port Musgrave, Qld (12°00′S, 141°54′E) on the eastern shores of the Gulf of Carpentaria (pers. obs.) and extends around Cape York and down the east Australian coastline (Bunt, 1978) to Mackay, Qld. (21°08′S) (Jones, 1971).

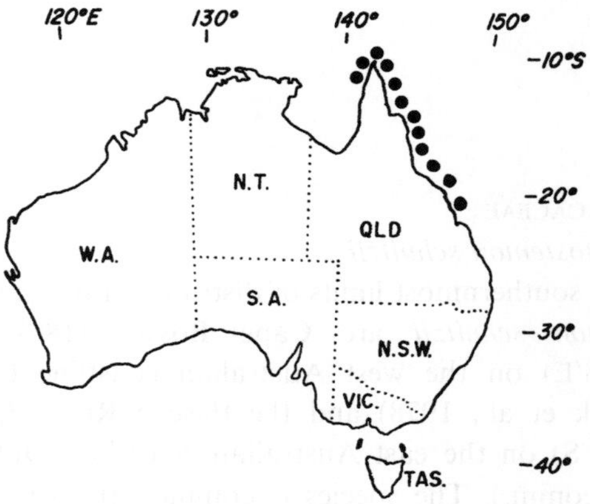

MELIACEAE (cont. on p. 64)

Xylocarpus australasicus

The southernmost limits of distribution of *Xylocarpus australasicus* are King Sound, Derby (17°19′S, 123°38′E) on the west Australian coast (Semeniuk et al., 1978) and Gladstone, Qld. (23°50′S) on the east Australian coast (Saenger and Robson, 1977).

Throughout this area *Xylocarpus australasicus* commonly occurs as isolated trees amongst the

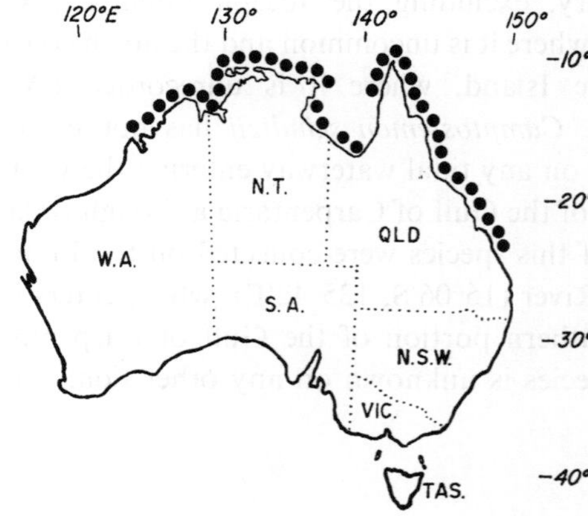

fringing vegetation along tidal waterways. In portions of tidal waterways of the Northern Territory that remain brackish for long periods, *Xylocarpus australasicus* may occur as a dominant tree species typically in association with *Diospyros ferrea* var. *humilis*, *Bruguiera parviflora* and *Rhizophora stylosa* on suitable substrates.

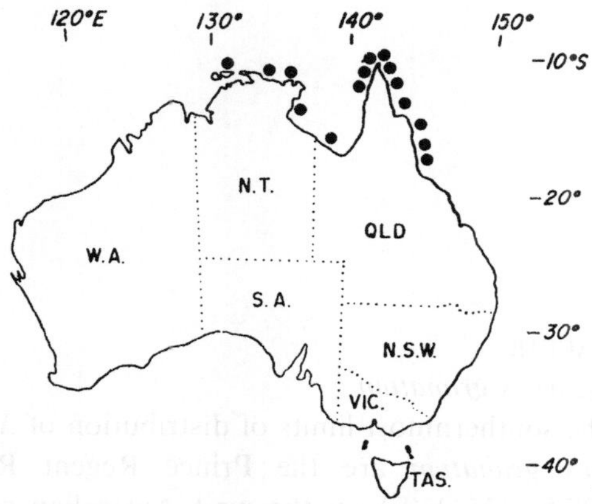

COMBRETACEAE

Lumnitzera littorea

Lumnitzera littorea has not been recorded west of the Johnston River (11°18′S, 131°10′E) on Melville Island. Distribution of the species across the northern portion of Arnhem Land is discontinuous and it has not been recorded south of the Rose River (14°17′S, 135°44′E) on the western shore of the Gulf of Carpentaria (Wells, 1981b).

Saenger and Hopkins (1975) record *Lumnitzera littorea* at Tarrant Point (17°23′S, 139°25′E) in the south-eastern portion of the Gulf of Carpentaria and Woolston (1973) also records the species at Mornington Island.

Along the eastern shore of the Gulf of Carpentaria, *Lumnitzera littorea* occurs at least as far south as Weipa (Love, pers. comm.). On the east Australian coast this species is recorded south to the Herbert River, Qld. (18°32′S) (Jones, 1971), although it has a discontinuous distribution throughout this area (Bunt, 1978; Duke and Bunt pers. comm.).

Lumnitzera littorea only occurs as a shrub to small tree (<6 m) in the Northern Territory (Wells, 1981b). On Cape York Peninsula this species often attains tree heights to 25 m (Bunt, 1978; pers. obs.).

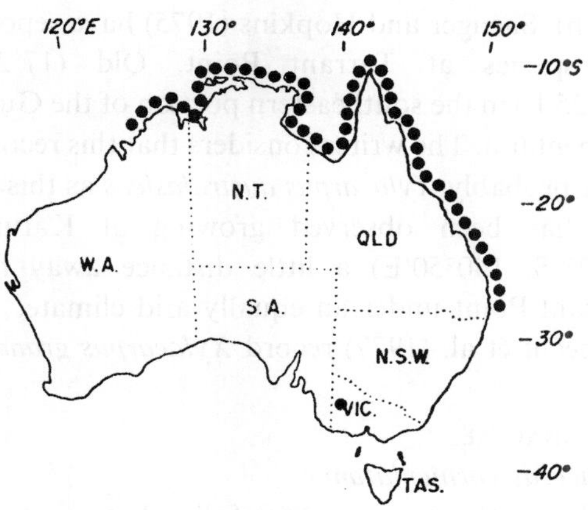

Lumnitzera racemosa

The southernmost limits of distribution of *Lumnitzera racemosa* are Beagle Bay (16°56′S, 122°32′E) on the west Australian coast (Semeniuk et al., 1978) and Moreton Bay, Qld. (27°30′S) on the east Australian coast (Dowling, 1979).

Excluding the Kimberley area of Western Australia, where the species is uncommon, *Lumnitzera racemosa* is widespread throughout the remainder of its range.

Lumnitzera rosea

This species is only known in Australia from Missionary Bay, Hinchinbrook Island. (18°15′S), (Tomlinson et al., 1978.)

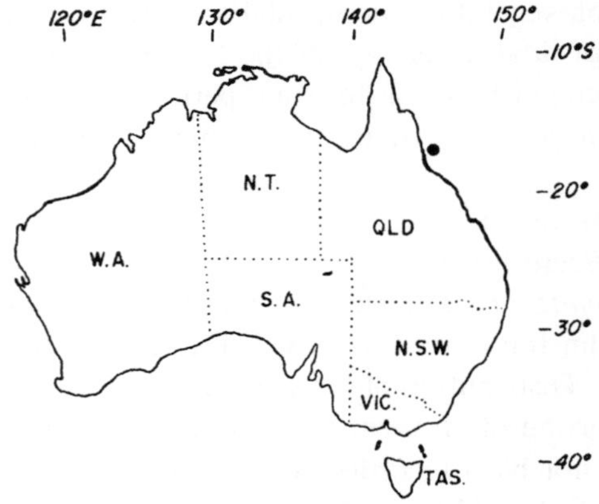

MELIACEAE

Xylocarpus granatum

The southernmost limits of distribution of *Xylocarpus granatum* are the Prince Regent River (15°27′S, 125°03′E) on the west Australian coast (Wells, 1981a) and Frazer Island, Qld. (25°30′S) on the east Australian coast (Dowling, pers. comm.).

In the Northern Territory *Xylocarpus granatum* occurs on Andranangoo Creek, Melville Island (11°21′S, 130°51′E) and on the Gove Peninsula (Yirrkala) (Specht, 1958).

The species has not been recorded on any other tidal waterway in the Northern Territory (Wells, 1981b). Saenger and Hopkins (1975) have reported the species at Tarrant Point, Qld (17°23′S, 139°25′E) in the southeastern portion of the Gulf of Carpentaria. The writer considers that this record is most probably *Xylocarpus australasicus* as this species has been observed growing at Karumba (17°28′S, 140°50′E) a little distance away from Tarrant Point under an equally arid climate.

Specht et al. (1977) record *Xylocarpus granatum* at Weipa and the species is reasonably abundant in the Port Musgrave area north of Weipa (pers. obs.).

Distribution of this species along the remainder of the western shores of Cape York Peninsula is unknown. *Xylocarpus granatum* is particularly common amongst fringing vegetation in tidal waterways down the east Australian coastline (Bunt, 1978; Duke and Bunt, pers. comm.).

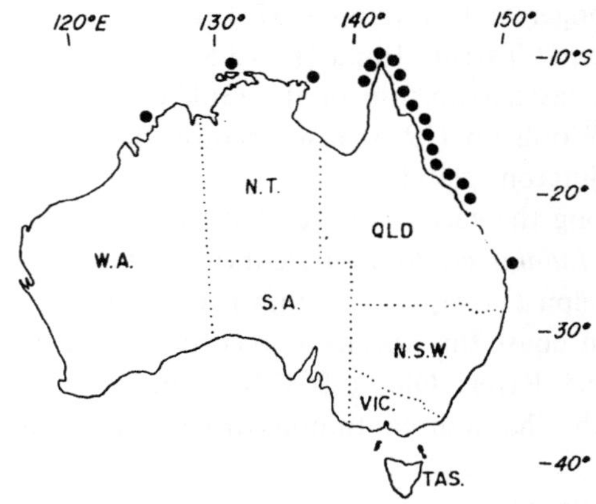

MYRSINACEAE

Aegiceras corniculatum

The southernmost limits of distribution of *Aegiceras corniculatum* are at Cossack (20°41′S, 117°12′E) on the west Australian coast (Semeniuk et al., 1978) and Merimbula, NSW (36°53′S, 149°55′E) on the south-east coast of NSW (Weate, pers. comm.).

Throughout this area *Aegiceras corniculatum* is reasonably abundant, particularly as a shrub understory to fringing vegetation along tidal waterways.

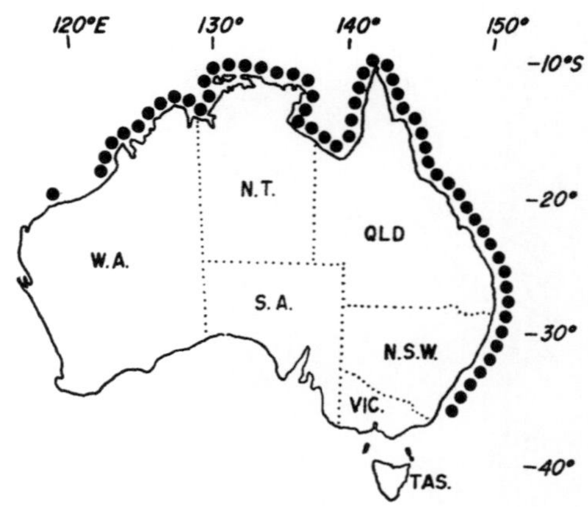

MYRTACEAE

Osbornia octodonta

The southernmost limits of distribution of *Osbornia octodonta* are at Cossack (20°41′S, 117°11′E) on the west Australian coast (Semeniuk et al., 1978) and Tin Can Bay, Qld. (25°30′S) on the east Australian coast (Dowling and McDonald, 1981).

Distribution of *Osbornia octodonta* is discontinuous around the coastline, particularly in the Gulf of Carpentaria. Specht et al. (1977) record this species at Weipa, Qld. (12°40′S, 141°50′E), although Saenger and Hopkins (1975) did not record it at Tarrant Point in the south-eastern portion of the Gulf of Carpentaria. Wells (1981b) records *Osbornia octodonta* from over half the tidal waterways of the Kimberley region of Western Australia and the Northern Territory and Bunt (1978) and Dowling (pers. comm.) record this species as being reasonably common at sites along the east coast of Queensland.

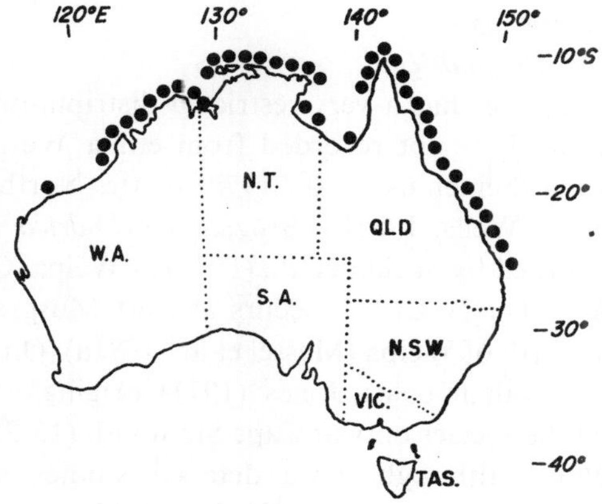

NYPACEAE

Nypa fruticans

Nypa fruticans has been recorded at only one locality, Maxwell Creek, Melville Island in the Northern Territory.

Nypa fruticans has been recorded by Messel et al. 1981a at Port Musgrave, Qld. (12°00′S, 141°54′E) on the eastern shores of the Gulf of Carpentaria. It is also known to occur on the Jardine river (10°55′S) and Cowal Creek (10°54′S). (Duke and Bunt, pers. comm.).

On the east Australian coast, *Nypa fruticans* occurs in a discontinuous distribution from the Olive River, Qld. (12°10′S) (Bunt, 1978) to the southern end of Hinchinbrook channel, Qld. (18°31′S) (Duke and Bunt, pers. comm.).

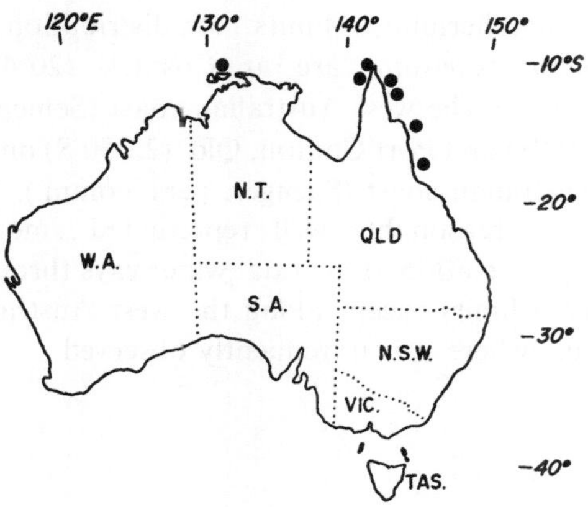

PLUMBAGINACEAE

Aegialitis annulata

The southernmost limits of distribution of *Aegialitis annulata* are at Exmouth Gulf (21°53′S, 114°22′E) on the west Australian coast (Semeniuk et al., 1978) and Frazer Island, Qld. (25°30′S) on the east Australian coast (Jones, 1971).

Throughout its range *Aegialitis annulata* is reasonably abundant.

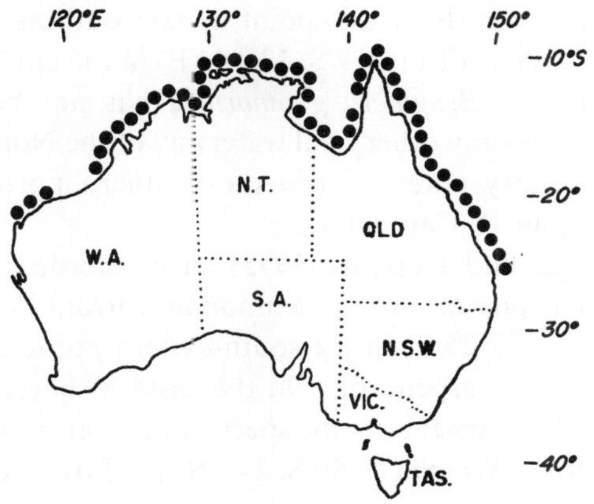

RHIZOPHORACEAE

Bruguiera cylindrica

This species has a very restricted distribution in Australia. It is not recorded from either Western Australia (Semeniuk et al., 1978) or the Northern Territory (Wells, 1981b). *Bruguiera cylindrica* was not recorded by Specht et al., (1977) at Weipa, Qld. (12°40′S, 141°50′E) but occurs at Port Musgrave, 65 km north of Weipa (Messel et al., 1981a). On the east Australian coast Jones (1971) originally recorded the species only at Cape Sidmouth (13°25′S, 143°36′E), although more detailed studies now show that although the distribution of this species is discontinuous; it occurs at several mainland and island localities north of the Normanby River (14°25′S) and at localities around the tip of Cape York (Bunt, 1978; Duke and Bunt, pers. comm.; Saenger, pers. comm.).

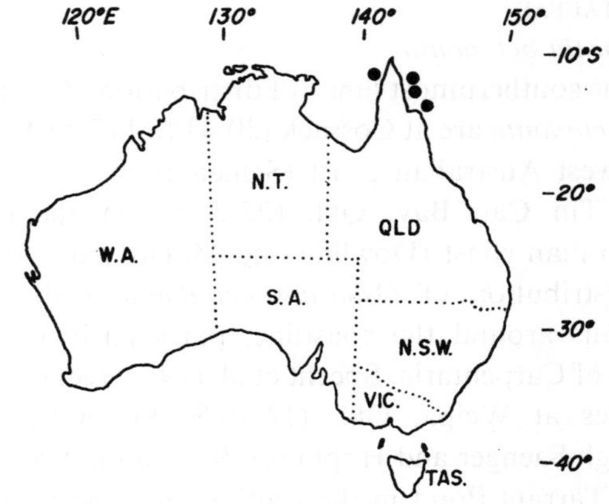

Bruguiera exaristata

The southernmost limits of distribution of *Bruguiera exaristata* are at Cossack (20°41′S, 117°11′E) on the west Australian coast (Semeniuk et al., 1978) and Port Clinton, Qld. (22°30′S) on the east Australian coast (Saenger, pers. comm.). The species is reasonably well represented amongst fringing vegetation along tidal waterways throughout these limits except along the west Australian coastline where it is infrequently observed.

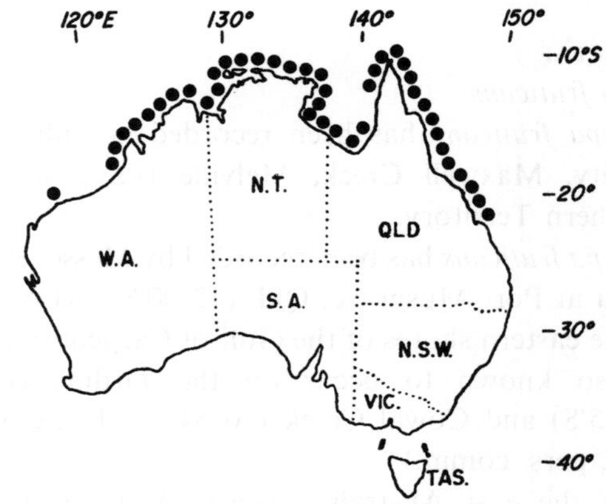

Bruguiera gymnorhiza

Bruguiera gymnorhiza has not been recorded west of Darwin, NT (12°25′S, 130°48′E) (Wells, 1981b).

Although discontinuous in occurrence the species occurs eastwards of this point as far south as the Roper River, NT (14°17′S, 135°44′E) in the Gulf of Carpentaria. *Bruguiera gymnorhiza* has not been observed on any other tidal waterway of the Northern Territory entering into the southern portion of the Gulf of Carpentaria.

Saenger and Hopkins (1975) have recorded the species as present but uncommon at Tarrant Point (17°23′S, 139°25′E) in the south-eastern portion of the Gulf of Carpentaria. On the eastern shores of the Gulf of Carpentaria the species occurs at least as far south as Weipa (12°40′S, 141°50′E) (Love, pers. comm.). Its distribution along the remainder of the tidal waterways entering the Gulf of Carpentaria

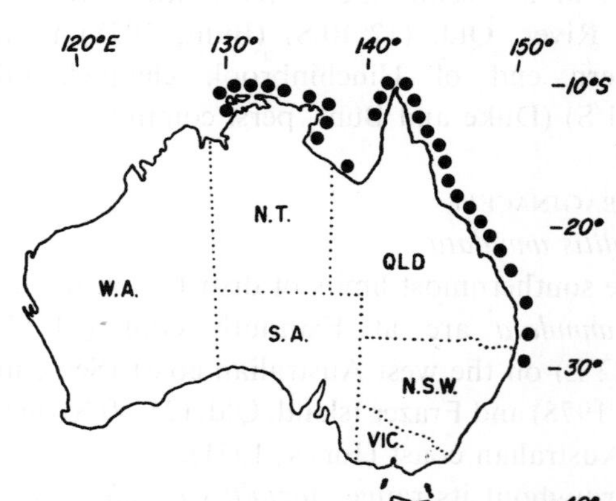

from Weipa to Karumba is unknown. On the east Australian coastline, *Bruguiera gymnorhiza* has been recorded south to the Clarence River, N.S.W.

Bruguiera parviflora

The southernmost limits of distribution of *Bruguiera parviflora* are the Sale River (15°57′S, 124°33′E) on the west Australian coast (Wells, 1981b) and Proserpine, Qld. (20°30′S) on the east Australian coast (Jones, 1971).

The species is uncommon throughout the Kimberleys and the Joseph Bonaparte Gulf region of Western Australia and the Northern Territory. Distribution of the species is continuous across the northern coastline of the Northern Territory.

Bruguiera parviflora is rare on tidal waterways entering the western shores of the Gulf of Carpentaria and has not been recorded eastwards of the Limmen Bight River (15°06′S, 135°42′E) in the Northern Territory. *Bruguiera parviflora* recently has been recorded at Port Musgrave Qld. (12°00′S, 141°54′E) (Messel et al., 1981a). The species is reasonably common along tidal waterways over the northern portion of the east Australian coast (Bunt, 1978) although it becomes discontinuous in occurrence towards the southern limit of distribution at Proserpine.

Bruguiera sexangula

Bruguiera sexangula has not been recorded previously from the Northern Territory. It is known from only the Cato and Peter John Rivers in Arnhem Bay, NT (12°15′S, 136°21′E) (Wells, 1981b).

On the eastern shores of the Gulf of Carpentaria, *Bruguiera sexangula* has been recorded south to Port Musgrave, Qld. (12°00′S, 141°54′E) (pers. obs.) while on the east Australian coastline, this species has been recorded south to Missionary Bay, Hinchinbrook Island, Qld. (18°15′S) (Duke and Bunt, pers. comm.). The distribution of *Bruguiera sexangula* down the east Australian coastline is discontinuous.

(29°25′S, 153°21′E) (McCormick, pers. comm.; Floyd, pers. comm.), although this species only occurs as isolated trees in northern N.S.W.

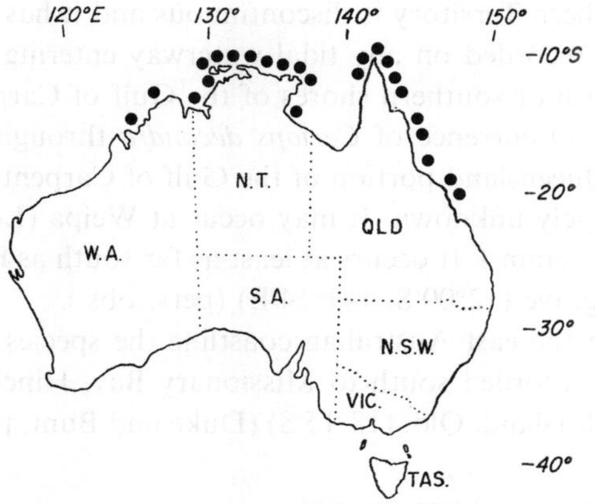

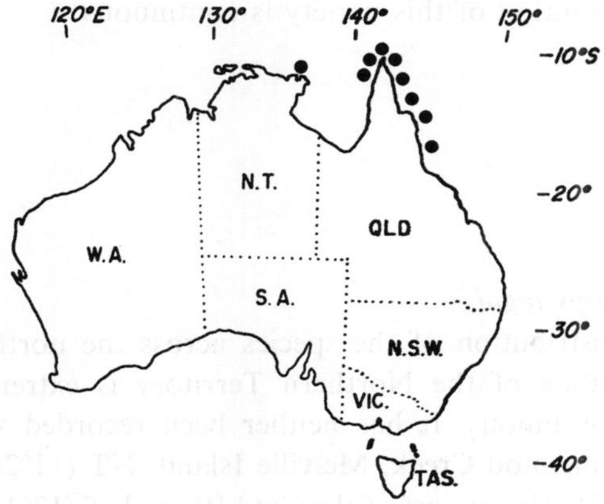

Ceriops decandra

Ceriops decandra has not been recorded west of Darwin, NT (12°25′S, 130°48′E) (Wells, 1981b). Its distribution across the northern coastline of the Northern Territory is discontinuous and it has not been recorded on any tidal waterway entering the western or southern shores of the Gulf of Carpentaria. Occurrence of *Ceriops decandra* throughout the Queensland portion of the Gulf of Carpentaria is largely unknown. It may occur at Weipa (Love, pers. comm.). It occurs at least as far south as Port Musgrave (12°00′S, 141°54′E) (pers. obs.).

On the east Australian coastline the species has been recorded south to Missionary Bay, Hinchinbrook Island, Qld. (18°15′S) (Duke and Bunt, pers. comm.) and is common between Cape York and Cooktown (Bunt, 1978), and south to Cairns, Qld. (16°55′S) (Dowling and McDonald 1981).

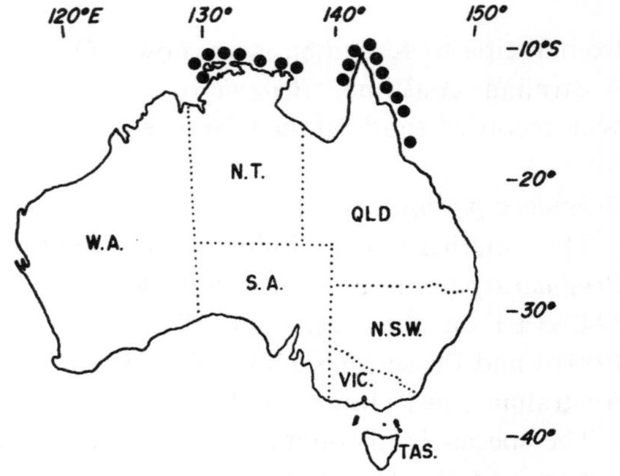

Ceriops tagal var. *australis*

The southernmost limits of distribution of *Ceriops tagal* var. *australis* are at Exmouth Gulf (21°53′S, 114°22′E) on the west Australian coast (Semeniuk et al., 1978) and the Tweed River, NSW (28°11′S, 153°32′E) on the east Australian coast (Lear and Turner, 1977). Throughout its range the distribution of this variety is continuous.

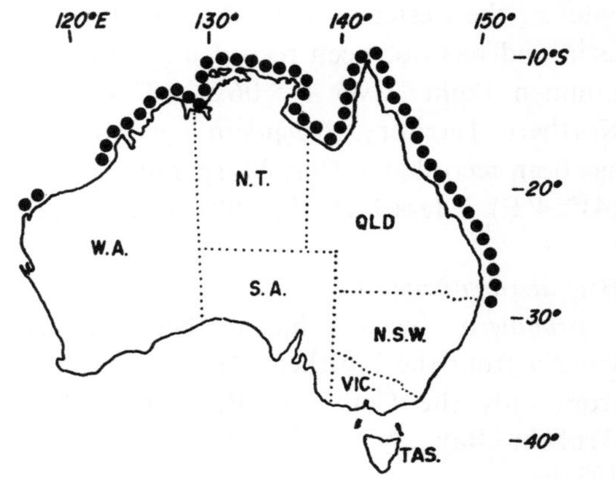

Ceriops tagal

Distribution of the species across the northern coastline of the Northern Territory is extremely discontinuous. It has neither been recorded west of Tinganoo Creek, Melville Island, NT (11°24′S, 131°29′E) nor east of the Cato River, NT (12°16′S, 136°21′E) (Wells, 1981b) nor on any tidal waterway within the Gulf of Carpentaria.

In eastern Australia, the species is quite abundant south from Cape York to Cooktown (Bunt, 1978) and becomes progressively rarer as it approaches the southernmost limit of its distribution at Bustard Head, Qld. (24°01′S) (Dowling and McDonald, 1981).

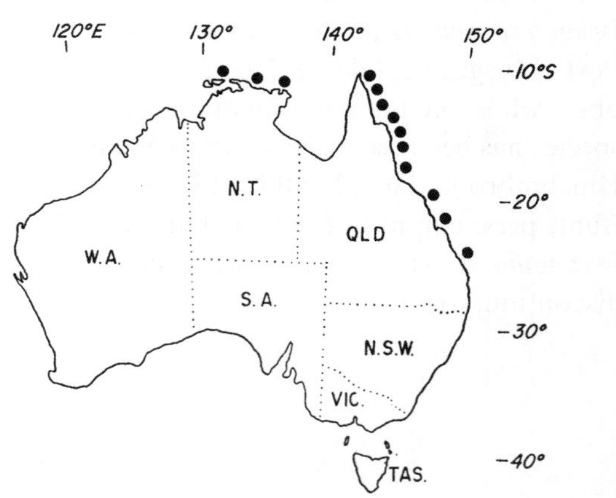

Rhizophora apiculata

Rhizophora apiculata has not been recorded west of Andranangoo Creek, Melville Island (11°21′S, 130°51′E) or east of Gove (12°13′S, 136°43′E) in the Northern Territory (Wells, 1981b). The species is absent from the greater portion of tidal waterways entering into the Gulf of Carpentaria. Specht et al. (1977) did not record *Rhizophora apiculata* at Weipa (12°40′S, 141°50′E) although this writer found this species to be common within Port Musgrave (12°00′S, 141°54′E) just 65 km north of Weipa.

On the east Australian coastline *Rhizophora apiculata* has been recorded south to Port Clinton, Qld (22°30′S) (Duke and Bunt, 1979); these authors suggest that this species may occur further south.

Rhizophora lamarckii

Rhizophora lamarckii has been recorded from the Alligator Rivers region of the Northern Territory (Hegerl, pers. comm.). It has not been recorded within any other tidal waterway of the Northern Territory although it could occur in Arnhem Bay.

This species has not been recorded in the Gulf of Carpentaria although it could occur in low numbers on the Jardine River (Duke and Bunt, 1979). Although not recorded by this writer on tidal rivers entering Port Musgrave, it may be present here also.

Along the east Australian coastline *Rhizophora lamarckii* shows a discontinuous distribution from Cape York to Port Clinton, Qld. (22°30′S) (Duke and Bunt, 1979). It is probable that this species occurs south of this given limit at Port Clinton.

Rhizophora mucronata

Rhizophora mucronata has not been recorded on any tidal waterway in Western Australia (Semeniuk et al., 1978) or the Northern Territory (Wells, 1981b).

Rhizophora mucronata has been recorded from Port Musgrave, Qld. (12°00′S, 141°54′E) (pers. obs.) and appears to be discontinuous in occurrence in tidal waterways from this point to Deluge Inlet, Qld (18°25′S) (Duke and Bunt, 1979).

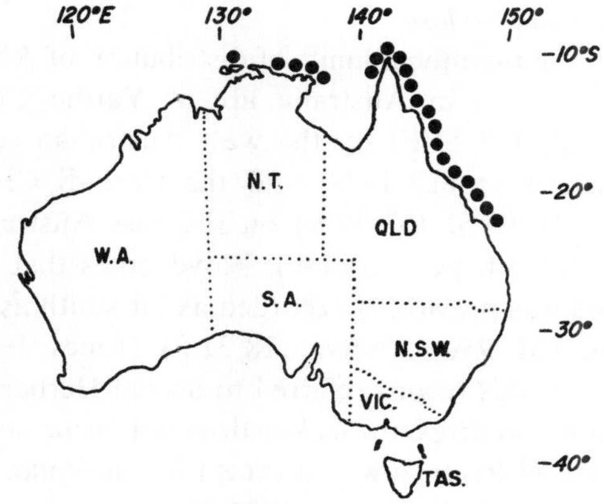

Duke and Bunt (1979) show that throughout its range along the east Australian coast, *Rhizophora apiculata* is continuous in occurrence.

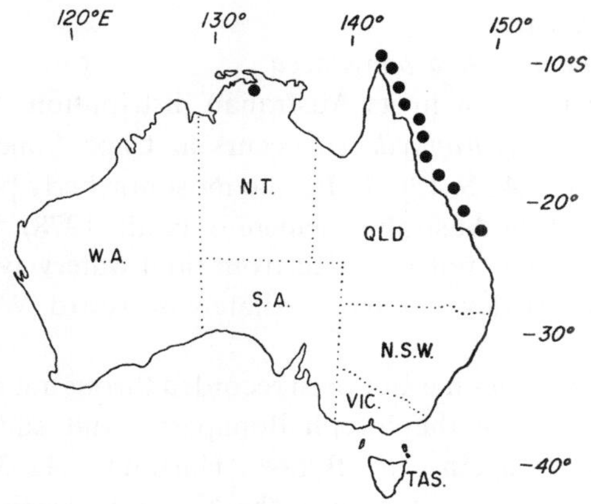

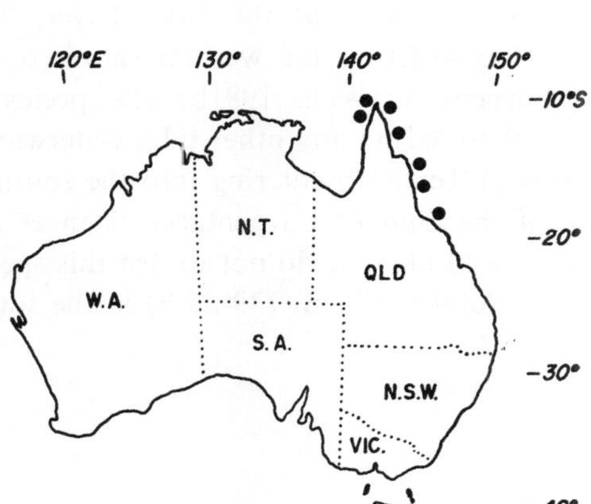

Rhizophora stylosa

The southernmost limits of distribution of *Rhizophora stylosa* in Australia are at Yardie Creek (22°20′S, 113°51′E) on the west Australian coast (Semeniuk et al., 1978) and the Corindi Creek, NSW (30°03′S, 153°12′E) on the east Australian coast (Floyd, pers. comm.). Floyd notes that this species was previously recorded as far south as the Richmond River, NSW (28°31′S) (Jones 1971); however, this record referred to an old Herbarium specimen. *Rhizophora stylosa* does not occur on the Richmond River now and except for the collection made at Corindi Creek in 1979, the species was not known to have occurred in recent times south of Cudgera Creek on the Tweed River, NSW (Floyd, pers. comm.).

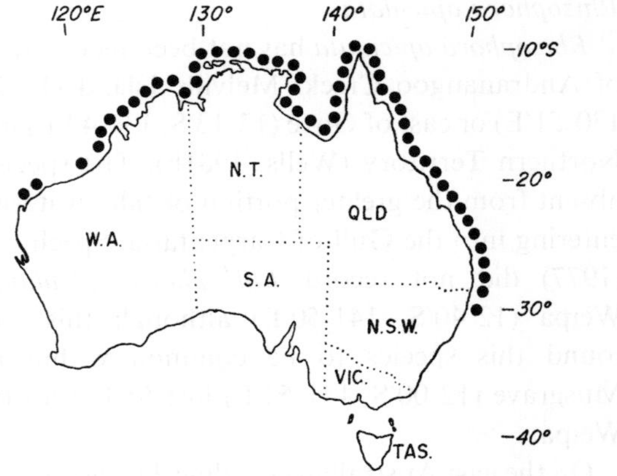

RUBIACEAE

Scyphiphora hydrophyllacea

To the west in its Australian distribution, *Scyphiphora hydrophyllacea* occurs at Cape Londonderry (13°48′S, 126°46′E), the most northerly point of Western Australia (Semeniuk et al., 1978). It is surprisingly, not recorded from tidal waterways of the Kimberley region immediately westward (Wells, 1981a).

The species has not been recorded throughout the remainder of the Joseph Bonaparte Gulf and reappears again in Bynoe Harbour (12°34′S, 130°31′E) near Darwin in the Northern Territory. *Scyphiphora hydrophyllacea* has a discontinuous distribution across the remainder of the northern coastline of the Northern Territory and has been recorded as far south as the Rose River, N.T. (14°17′S, 135°44′E) on the western shores of the Gulf of Carpentaria (Wells, 1981b). The species has not been observed on any other tidal waterway of the Northern Territory entering into the southern portion of the Gulf of Carpentaria. Saenger and Hopkins (1975) likewise do not record this species at Tarrant Point (17°23′S, 139°25′E) in the south-eastern portion of the Gulf of Carpentaria.

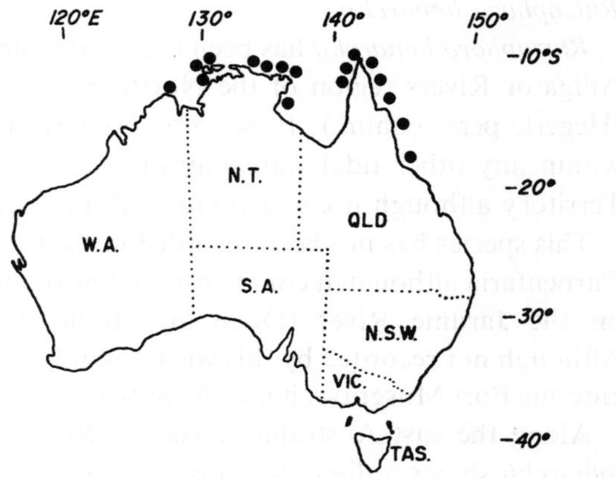

Specht et al., (1977) record the species as far south as Weipa, Qld. (12°40′S, 141°50′E) on the eastern shore of the Gulf of Carpentaria. However, its occurrence down the remainder of this shore to Karumba is unknown. On the east Australian coast this species occurs as far south as Townsville, Qld. (19°12′S) (Jones, 1971), although again it is discontinuous throughout this region.

STERCULIACEAE

Heritiera littoralis

Heritiera littoralis has not been recorded from Western Australia nor from the Northern Territory.

The species is first recorded at Port Musgrave, Qld. (Messel et al., 1981a) and thence forms a fairly continuous distribution around Cape York and down the east Australian coastline to Cooktown (Bunt, 1978) and Cairns Qld. (16°55′S) (Dowling and McDonald, 1981). the species is common along the upstream portions of many tidal waterways down the east coast to Hinchinbrook Island. The southernmost limit of distribution of *Heritiera littoralis* is St Lawrence, Qld. (22°15′S) (Jones, 1971).

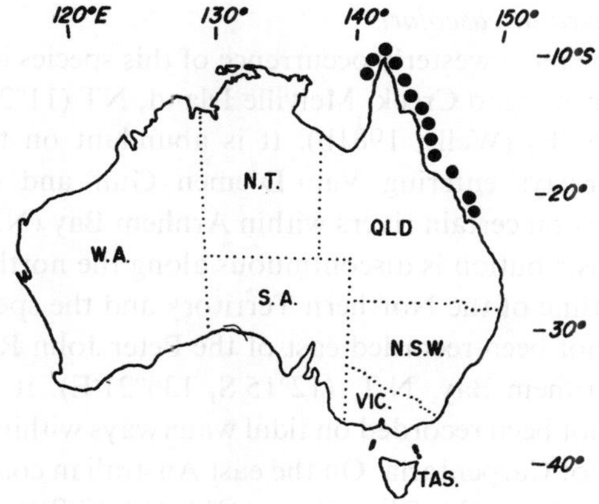

SONNERATIACEAE

Sonneratia alba

The southernmost limit of distribution of *Sonneratia alba* on the west Australian coast is at Cape Bossut (18°43′S, 121°38′E((Semeniuk et al., 1978). The species is common throughout the estuaries of tidal waterways in the Kimberleys W.A. (Wells, 1981a) but is largely absent from tidal waterways draining into Joseph Bonaparte Gulf.

The species is nearly continuous in occurrence from Port Paterson, N.T. (12°34′S, 130°31′E) around the northern coastline to Gove (12°13′S, 136°42′E) in the Northern Territory. The species has not been recorded on any tidal waterway entering the western or southern shores of the Gulf of Carpentaria (Wells, 1981b). *Sonneratia alba* occurs at least as far south as Weipa, Qld. (12°40′S, 141°50′E) on the eastern shores of the Gulf of Carpentaria (Love, pers. comm.). On the east Australian coast this species occurs as far south as Port Clinton, Qld. (22°30′S) (Duke and Bunt, pers. comm.). Work in progress at the Australian Institute of Marine Science suggests that a range of *Sonneratia* spp. occur along the east Australian coast (Duke and Bunt, pers. comm.).

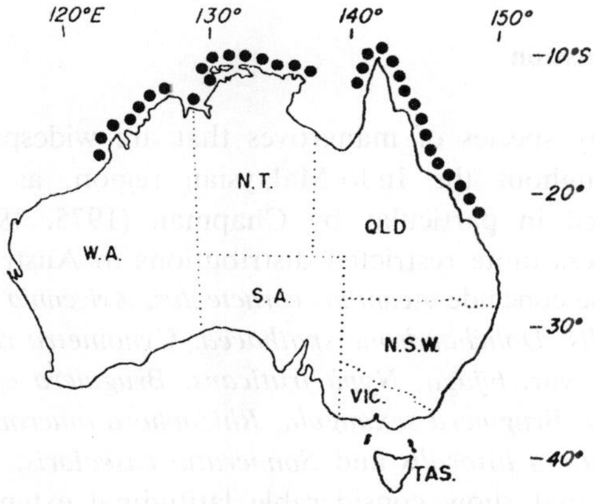

Sonneratia caseolaris

The most westerly occurrence of this species is on Andranangoo Creek, Melville Island, NT (11°21′S, 130°51′E) (Wells, 1981b). It is abundant on tidal waterways entering Van Diemen Gulf and also occurs on certain rivers within Arnhem Bay (N.T.). Its distribution is discontinuous along the northern coastline of the Northern Territory and the species has not been recorded east of the Peter John River in Arnhem Bay, N.T. (12°15′S, 136°21′E). It also has not been recorded on tidal waterways within the Gulf of Carpentaria. On the east Australian coast it occurs from the Olive River, Qld. (12°10′S) to the Murray River, Qld. (18°05′S) (Duke and Bunt, pers. comm.) although its distribution here is again discontinuous.

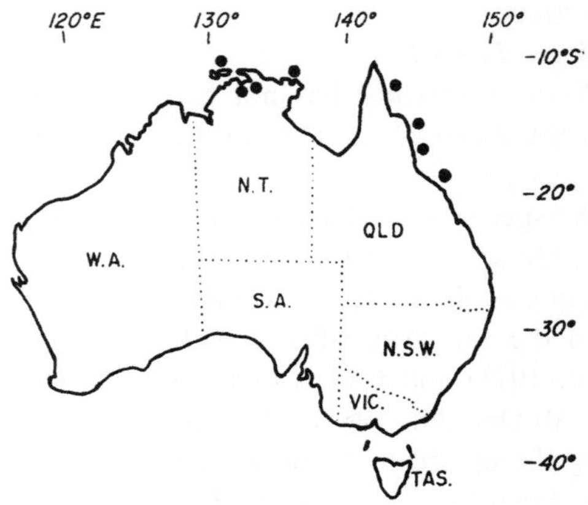

Discussion

Many species of mangroves that are widespread throughout the Indo-Malaysian region, as discussed in particular by Chapman (1975, 1976), possess quite restricted distributions in Australia. These conclude *Acanthus ebracteatus*, *Avicennia officinalis*, *Dolichandrone spathacea*, *Cynometra ramiflora* var. *bijuga*, *Nypa fruticans*, *Bruguiera cylindrica*, *Bruguiera sexangula*, *Rhizophora mucronata*, *Heritiera littoralis* and *Sonneratia caseolaris*. Species that show considerable latitudinal extension down the east as opposed to the west Australian coast include *Acanthus ilicifolius*, *Lumnitzera littorea*, *Lumnitzera racemosa*, *Excoecaria agallocha*, *Xylocarpus australasicus*, *Xylocarpus granatum*, *Aegiceras corniculatum*, *Bruguiera gymnorhiza*, *Ceriops decandra*, *Ceriops tagal* var. *australis*, *Ceriops tagal*, *Rhizophora apiculata*, *Rhizophora stylosa*, *Scyphiphora hydrophyllacea* and *Bruguiera parviflora*; while still other species *Aegialitis annulata*, *Bruguiera exaristata*, *Osbornia octodonta* and *Sonneratia alba* show no such uneven latitudinal extension between the eastern and western extremities of their distributions and occur at similar latitudes on both east and west coastlines. Only one species, *Camptostemon schultzii*, occurs at a higher latitude on the west as opposed to the east coast of the continent.

Considerable variation exists in species occurrences throughout their respective geographic ranges in Australia. Many species, *Acanthus ilicifolius*, *Aegialitis annulata*, *Aegiceras corniculatum*, *Avicennia marina*, *Bruguiera exaristata*, *Ceriops tagal* var. *australis*, *Excoecaria agallocha*, *Rhizophora stylosa* and *Xylocarpus australasicus*, are considered as having continuous distributions throughout their ranges. Other species, in particular *Bruguiera parviflora*, *Camptostemon schultzii* and *Ceriops decandra*, while reasonably continuous in occurrence around the north Australian coastline, are entirely absent from those regions within their geographic ranges that are subjected to high levels of seasonal aridity (i.e. Cambridge/Joseph Bonaparte Gulf; Gulf of Carpentaria). Additional species which are quite discontinuous in occurrence at sites across the Northern Territory – *Bruguiera sexangula*, *Ceriops tagal*, *Lumnitzera littorea*, *Nypa fruticans*, *Rhizophora apiculata*, *Rhizophora lamarckii*, *Scyphiphora hydrophyllacea* and *Xylocarpus granatum* – also are not recorded from sites in these gulfs but occur more frequently at localities along the north Queensland coastline.

Several species only recorded in Australia from north-east Queensland – *Bruguiera cylindrica*, *Cynometra ramiflora* var. *bijuga*, *Heritiera littoralis* and *Rhizophora mucronata* – shown reasonably continuous distributions throughout their Australian ranges, although *Dolichandrone spathacea* and *Lumnitzera rosea* present extremely restrictive distributions. However, this usual trend of uncommonly recorded mangrove species occurring in north-east Queensland does not include *Avicennia officinalis*, which is wholly limited in its Australian distribution to sites along the northern coastline of the Northern Territory.

Macnae (1966), Jones (1971), Gill (1975), Lear and Turner (1977), Saenger et al. (1977) and Semeniuk et al. (1978) all show that a gradual reduction in species number occurs with increasing latitude down both the east and west coasts of the continent from tropical to temperate regions. However, presence of species at sites within their geographic ranges depends on an interacting complex of factors. Many are discussed by Clark and Hannon (1969) in relation to zonation at a particular site whilst others, e.g. Thom et al. (1975), stress that the pool of species available are influenced by local environmental conditions which in turn are affected by geomorphic processes.

In this regard, in Australia, within the geographic range of a species, although edaphic and physiographic conditions of a site may be favourable for its occurrence, local climatic conditions may prevent establishment. Fosberg (1961), Zahran (1975) and Cintron et al. (1978) have observed this phenomenon elsewhere in the world.

Species distributions in northern Australia show certain trends. *Bruguiera sexangula*, *Ceriops decandra*, *Ceriops tagal*, *Nypa fruticans*, *Rhizophora apiculata*, *Rhizophora lamarckii*, *Sonneratia alba* and *Sonneratia caseolaris* all occur at sites across the northern coastline of the Northern Territory and down the east Queensland coastline, but remain unrecorded at any site within the larger portion of the Gulf of Carpentaria.

In addition, *Bruguiera gymnorhiza*, *Bruguiera parviflora*, *Camptostemon schultzii*, *Lumnitzera littorea*, *Scyphiphora hydrophyllacea* and possibly *Xylocarpus granatum* are represented only by straggly trees/shrubs at few sites within this Gulf.

Such a reduction in floristic diversity is not attributable to 'latitudinal sifting' of species as the Gulf of Carpentaria lies within the geographic ranges of all these species. Wells (1981b) has shown that the Arnhem Bay/Gove region at the north-western extremity of the Gulf with twenty-two species recorded, has the most diverse mangrove flora of any region in the Northern Territory. Likewise, the Port Musgrave region on the north-eastern shores of the Gulf of Carpentaria with twenty-seven recorded species (pers. obs.) is also particularly diverse in species when compared to other sites detailed by Bunt and Williams (1980) in north-eastern Queensland.

Such high levels of species numbers for mangroves at sites near the nothern limits of the Gulf of Carpentaria and the extremely low numbers of species recorded from sites further within this Gulf, frequently only eight to eleven species (Wells, 1981b), pose a number of problems.

The prevailing sea currents flow east along the northern coastline of the Northern Territory into and around the Gulf of Carpentaria during the 'wet' season (Australia Pilot, Vol. 5, 1972), when most fruiting of mangroves occurs. Such currents thus pass the Arnhem Bay/Gove region, deflect across the Gulf entrance to the Port Musgrave region in Queensland and then move southwards in a clockwise direction around the Gulf. Floating disseminules of the various mangrove species should then retain viability over such small sea journeys (when contrasted to the distributional spread of many of the species throughout islands in the Pacific Ocean), and should have little difficulty in reaching all sites within the Gulf of Carpentaria.

In this regard, throughout the southern portion of the Gulf, restricted sporadic occurrences of species like *Acanthus ilicifolius*, *Bruguiera parviflora*, *Camptostemon schultzii* and *Lumnitzera littorea* in extremely sheltered habitats and the absence of many other species that would have little difficulty in reaching these shores, appears to be largely attributable to the considerable climatic aridity experienced in this region during the 'dry' season (Climatic Averages, Qld., N.T., 1975).

In an analogous situation involving waterways entering into Cambridge and Joseph Bonaparte Gulfs on the north-west coast of Australia, only between seven to thirteen species are commonly recorded (Thom et al., 1975; Wells, 1981b) even though Semeniuk et al. (1978) and Wells (1981a) show that between thirteen to fifteen mangrove species are frequently recorded further south at sites throughout the Kimberley region of Western Australia.

Of the thirteen species occurring in these two gulfs, *Camptostemon schultzii, Rhizophora stylosa* and *Sonneratia alba* are infrequently encountered and then only as straggly trees or shrubs in contrast to their considerable abundance in adjacent tidal waterways of the Kimberleys (Wells, 1981a). An additional species, *Bruguiera parviflora*, is also uncommonly recorded at sites along this portion of the north-west coastline where it is at the southern end of its geographic range. However, both *Bruguiera parviflora* and *Bruguiera exaristata* are still less frequently encountered amongst the mangrove vegetation around Cambridge or Joseph Bonaparte Gulfs than at localities further south in the Kimberleys.

It might be expected that there would be fewer mangrove species in the Kimberley region, lying at higher latitudes than the Cambridge and Joseph Bonaparte Gulf regions. Such is not the case. The increase in number and abundance of species in the Kimberleys and the floristically and structurally depauperate mangrove scrubs fringing waterways entering the two latter mentioned gulfs may relate to subtle differences in seasonal aridity as shown in the Climatic Averages, N.T., W.A. (1975).

The south-east trade winds that blow rather consistently across the hot, dry interior of the continent during the 'dry' season (Gentilli, 1971) are considered by this writer to be most important in their effects on species numbers and structure of the mangrove vegetation along both the north-west coast of Australia and throughout the lower portions of the Gulf of Carpentaria. These drying winds blow straight into the mangrove scrubs fringing the low level plains abutting these gulfs and many species of mangroves with more humid climatic requirements are thus excluded from establishment in such seasonally arid regions. For the Kimberley region, which also receives considerable climatic aridity, the more diverse mangrove flora occurs in scattered pockets beneath cliffs of drowned river valleys. Such vegetation is often shielded from the sun's direct rays for a larger portion of the day (particularly during the winter, dry season months) due to the aspect of the overhead cliffs and is also protected from the desiccating effects of the south-east trade winds by the Mitchell Plateau.

Overall, distributions of the various mangrove species around the continent are considered to be the cumulative result of complex interactions of the biological, chemical and physical environments now and in the past. However, increased seasonal aridity on the west, compared with the east, Australian coast is considered as a major factor resulting in mangrove swamps on the west coast having only half the number of species present as have swamps at similar latitudes on the east Australian coastline.

Acknowledgements

I thank in particular Professor Harry Messel, Head of the School of Physics, University of Sydney, for assistance both in the field and in providing the considerable logistical and financial support (principally through the Science Foundation for Physics) that enabled all of the surveys along the northern coastline of Western Australia and the Northern Territory to be undertaken.

The field assistance of all members associated with the University of Sydney-Territory Parks & Wildlife Commission's Joint Crocodile Research project and in particular that of Fred Duncan, is gratefully acknowledged.

I also wish to thank Ralph Dowling and Norm Duke for providing information on species distributions along the east Australian coastline and Drs. Peter Myerscough, Bill Allaway and Professor Roger Carolin of the School of Biological Sciences, University of Sydney, for critical appraisal.

Especial thanks to Corinne Tuttlebee for typing this manuscript.

Literature cited

Anon 1972. Australia Pilot. North, north-west and west coasts of Australia from the west entrance of Endeavour Strait to Cape Leeuwin. Hydrographic Dept., Ministry of Defence, U.K. 5:211 pp.

Anon 1975. Climatic Averages, Queensland. Dept. of Science and Consumer Affairs, Bureau of Meteorology. Aust. Govt. Publ. Ser., Canberra 76 pp.

Anon 1975. Climatic Averages, South Australia and Northern Territory. Dept. of Science and Consumer Affairs, Bureau of Meteorology. Aust. Govt. Publ. Ser., Canberra 67 pp.

Anon 1975. Climatic Averages, Western Australia. Dept. of Science and Consumer Affairs, Bureau of Meteorology. Aust.

Govt. Publ. Ser., Canberra 82 pp.

Bunt, J.S. 1978. the mangroves of the eastern coast of Cape York Peninsula, north of Cooktown. In: Great Barrier Reef Marine Parks Authority workshop on northern sector. Townsville. Qld.

Bunt, J.S. and W.T. Williams. 1980. Studies in the analysis of data from Australian tidal forests ('Mangroves'). 1. Vegetational sequences and their graphic representation. Aust. Jour. Ecol. 5:385–90.

Butler, A.J., A.M. Depers, S.C. McKillup and D.P. Thomas. 1977. Distribution and sediments of mangrove forests in South Australia. Trans. Roy. Soc. S. Aust. 101:35–44.

Chai, P.K. 1973. The types of mangrove forest in Sarawak. Forest Dept. Sarawak, Malaysia. 34 pp.

Chapman, V.J. 1975. Mangrove Biogeography. In: Proceedings of the International Symposium on Biology and Management of Mangroves. Institute of Food and Agricultural Sciences, University of Florida, Gainesville, Florida 1:1–22.

Chapman, V.J. 1976. Mangrove Vegetation. Cramer. Lehr 425 pp.

Cintron, G., A.E. Lugo, D.J. Pool and G. Morris. 1978. Mangroves of arid environments in Puerto Rico and adjacent islands. Biotropica 10:110–21.

Clarke, L.D. and N.J. Hannon. 1969. The mangrove swamp and salt marsh communities of the Sydney district II. The holocoenotic complex with particular reference to physiography. Jour. Ecol. 57:231–4.

Dowling, R.M. 1979. The mangrove communities of Moreton Bay. Northern Moreton Bay Symposium. R. Soc. Qld. 54–62.

Dowling, R.M. and T.J. McDonald. 1981. Mangrove communities of Queensland. In: Structure and Function of Mangrove Ecosystems in Australia. A.N.U. Press, Canberra. A.C.T.

Duke, N.C. and J.S. Bunt. 1979. The genus *Rhizophora* (Rhizophoraceae) in northeastern Australia. Aust. Jour. Bot. 27: 657–78.

Floyd, A.G. 1977. Ecology of the tidal forests in the Kikori-Romilly Sound Area. Gulf of Papua. Ecol. Rept. No. 4. Divn. of Botany, office of For. Dept of Prim. Ind., Lae, Papua New Guinea 59 pp.

Floyd, A.G. 1980. N.S.W. Rainforest Trees. Forestry Commission of N.S.W. Sydney. Research note No. 41, 124 pp.

Fosberg, F.R. 1961. Vegetation-free zone on dry mangrove coasts. U.S. Geol. Surv., Prof. Pap., 424D:216–8.

Galloway, R. 1981 (In press). Distribution and patterns of Australian mangals. In: Structure and Function of Mangrove Ecosystems in Australia. A.N.U. Press. Canberra. A.C.T.

Gentilli, J. 1971. Climates of Australia. In: World Survey of Climatology 13:4–7 Elsevier Publ. Co. Amsterdam.

George, A.S. and K.F. Kenneally. 1975. The flora of the Prince Regent River Reserve, north-western Kimberley, Western Australia. In: A Biological Survey of the Prince Regent River Reserve, North-West Kimberley, Western Australia. Wildl. Res. Bull. West. Aust. 3:31–48.

Gill, A.M. 1975. Australia's mangrove enclaves: A coastal resource. In: Managing Aquatic Ecosystems. Proc. Ecol. Soc. Aust. 8:129–146.

Jones, W.T. 1971. The field identification and distribution of mangroves in Eastern Australia. Qld. Nat., 20:35–51.

Kenneally, K.F. 1981. Mangrove vegetation of Western Australia. In: Structure and Function of Mangrove Ecosystems in Australia. A.N.U. Press. Canberra. A.C.T.

Lear, R. and T. Turner. 1977. Mangroves of Australia. Univ. Qld. Press. St. Lucia, Qld. 84 pp.

Macnae, W. 1966. Mangroves in eastern and southern Australia. Aust. Jour Bot. 14:67–104.

Messel, H., A.A. Burbidge, A.G. Wells and W.J. Green. 1977. The status of the salt-water crocodile in some river systems of the north-west Kimberley, Western Australia. Dept. Fish. Wildl. West. Aust. Rept. 24:1–50.

Messel, H., A.G. Wells and W.J. Green. 1979a–g. Survey of tidal river systems in the Northern Territory of Australia and their crocodile populations. Series of seven monographs, Pergamon Press, Sydney.

Messel, H., G.C. Vorlicek, A.G. Wells and W.J. Green. 1980a–f. Surveys of tidal river systems in the Northern Territory of Australia and their crocodile populations. Series of six monographs. Pergamon Press, Sydney.

Messel, H., G.C. Vorlicek, A.G. Wells and W.J. Green. 1981a–c. Surveys of tidal river systems in the Northern Territory of Australia and their crocodile populations. Series of 3 monographs. Pergamon Press, Sydney.

Olsen, H.F., R.M. Dowling and D. Bateman. 1980. Biological resources survey. Round Hill Head to Tannum Sands, Queensland, Australia. Qld. Fish. Serv. Res. Bull. No. 2. 102 pp.

Percival, M. and J.S. Womersley. 1975. Floristics and ecology of the mangrove vegetation of Papua New Guinea. Bot. Bull. No. 8. Dept. Forestry, Divn. Bot. Lae, Papue New Guinea. 96 pp.

Saenger, P. and M.S. Hopkins. 1975. Observations on the mangroves of the southeastern Gulf of Carpentaria, Australia. In: Proceedings of the International Symposium on Biology and Management of Mangroves. Institute of Food and Agricultural Sciences, University of Florida, Gainesville, Florida 1:126–36.

Saenger, P. and J. Robson. 1977. Structural analysis of mangrove communities on the central Queensland coastline. Mar. Res. in Ind. 18:101–18.

Saenger, P., M.M. Specht, R.L. Specht and V.J. Chapman. 1977. Mangal and coastal salt-marsh communities in Australasia. In: Wet Coastal Ecosystems, Ecosystems of the World. Elsevier Scientific Publishing Company, New York, U.S.A. 1:293–345.

Semeniuk, V., K.F. Kenneally and P.G. Wilson. 1978. Mangroves of Western Australia. Handbook No. 11, Western Australian Naturalists Club. 89 pp.

Semeniuk, V. 1980. Mangrove zonation along an eroding coastline in King Sound, north-western Australia. Jour. Ecol. 68:789–812.

Specht, R.L. 1958. The climate, geology, soils and plant ecology of the northern portion of Arnhem Land. In: Records of the American-Australian Expedition to Arnhem Land, Botany and Plant Ecology. Melbourne University Press, Melbourne, Vic. 3:333–414.

Specht, R.L., R.B. Salt and S. Reynolds. 1977. Vegetation in the vicinity of Weipa, North Qld. Proc. Roy. Soc. Qld. 88:17–38.

Thom, B.G., L.D. Wright and J.M. Coleman. 1975. Mangrove ecology and deltaic-estuarine geomorphology, Cambridge Gulf-Ord River, Western Australia. Jour. Ecol. 63:203–232.

Tomlinson, P.B., J.S. Bunt, R.B. Primack and N.C. Duke. 1978. *Lumnitzera rosea* (Combretaceae) – its status and floral morphology. Jour. Arn. Arb. 59:342–351.

Wells, A.G. 1981a. A survey of riverside vegetation fringing tidal river systems of Kimberley, Western Australia. In: Biological Survey of Mitchell Plateau and Admiralty Gulf, Kimberley, Western Australia. Western Australian Museum, Publ. West. Aust. Mus. Publ. Part 3:95–121.

Wells, A.G. 1981b. Mangrove vegetation of Northern Australia. In: Structure and Function of Mangrove Ecosystems in Australia. A.N.U. Press Canberra. A.C.T.

Woolston, F.P. 1973. Ethnobotanical items from the Wellesley Islands, Gulf of Carpentaria. Occ. pap. Univ. Qld. Anthrop. Mus. 1:95–103.

Zahran, M.A. 1975. Biogeography of mangrove vegetation along the Red Sea coasts. In: Proceedings of the International Symposium on Biology and Management of Mangroves. Institute of Food and Agricultural Sciences, University of Florida, Gainesville, Florida 1:43–51.

CHAPTER 7

The New Zealand Mangrove Association

F.J. TAYLOR

Marine Research Laboratory, Leigh, New Zealand

New Zealand contains the southernmost mangals in the world. They are rather depauperate compared with those in the tropics and consist of only one species: *Avicennia marina* var. *resinifera*. Because the leaf tissues cannot withstand more than 12 hours of freezing conditions (Sakai and Wardle 1978) mangroves are found only in the northern part of the country. In general there is a decline in stature southwards. In the north they can be as high as 8 m, in the south, nearer 1 m.

Mangroves occur basically in three situations:

(1) Bay heads.
(2) River banks.
(3) In the shelter of sand or shell banks.

These are all protected situations and we have evidence from experimental plantings that when exposed to more than slight wave action *Avicennia* seedlings become broken, just leaving rooted stumps.

Küchler (1972) has divided the New Zealand mangrove communities into a series of associations defined in terms of the associated species. However, I do not consider this to be a very profitable approach as these associations do not appear to be correlated with environmental factors.

In many parts of New Zealand *Avicennia* is colonising new areas. In the Whangateau Harbour, a large tidal estuarine harbour which we have been studying in detail, the increase (from aerial photographs) has been from 28.7 ha in 1953 to 43.52 ha in 1973, a 52% increase in twenty years, though this has not been uniform over the whole area because some areas (2.55 ha) have been lost due to reclamations and natural erosion. Of the increase, 8.87 ha (60%) was due to the expansion of existing habitats and 6.63 ha (40%) to the colonization of new areas.

From an analysis of the structure of the mangrove areas in the Whangateau Harbour we have come to the conclusion that *Avicennia* has colonised the harbour in relatively recent years – probably within the last 100 years. We have recognised three types of habitat from transects (Cornforth 1975).

Type A: Here the canopy cover is more than 50% and represents mature bushes with an average height of 3–4 m with a canopy spread of some 8 m. Seedlings are rare and small. Pneumatophores are taller and denser than elsewhere (450 m^{-2}).

Type B: This has a canopy cover of 25–50% and the bushes are lower in height (2 m). Seedlings are more abundant, probably because of the more open situation, but the pneumatophore density is about the same as in Type A. This type is divided into two subtypes, depending on whether they are to seaward or landward of Type A. That to seaward (Type B1) is presumably more youthful and represents an intermediate stage in colonisation. That to landward (Type B2), although superficially similar in structure, is presumably growing under suboptimal conditions and is consequently more stunted. The absence of the remains of any large bushes or trees of Type A indicates that this is not a later stage in succession.

Tasks for vegetation science, Vol. 8 ed. by H.J. Teas
 ISBN 978-90-481-8526-9

Type C: has a canopy cover of 10–25% and again is divided into two types. Type C1 is newly colonising. Type C2 is more stunted than B2 and merges into upper salt marsh species such as *Juncus maritimus* var. *australiensis* and *Leptocarpus similis*. Chapman (1976) regards the mangrove succession in the Auckland area as starting with *Zostera*, but I find that this is the exception rather than the rule. *Zostera* has much more restricted ecological requirements than *Avicennia*.

In other parts of New Zealand the transects show a different pattern. In bay-heads and behind banks they usually start with C1, B1 and A, but these tend to be truncated compared with the Whangateau Harbour transects and pass quickly into a mangal composed of tall (3–4 m +) trees with trunks, rather than spreading bushes and with a 100% canopy cover. This I consider to be the mature type of mangal. Type B2 is absent or very reduced, but there is a C2 zone which is presumably formed from seedlings shed by the innermost trees.

Along river banks one usually finds that mangroves grow on a shelf above the permanent water channel. On the edge of the shelf are found the tallest trees, with occasional seedlings and pneumatophores growing down the banks of the channel. Further back the *Avicennia* is found as bushes, decreasing rapidly in size away from the channel.

The performance of the trees is modified in many places by grazing, mainly by cattle and goats, and by the firmness of the substratum. Grazing has marked effects and keeps the trees stunted. Stunted trees also grow on firm substrata. A soft, deep mud is necessary for maximum development.

As far as the animal life is concerned, there appear to be relatively few, if any, species confined to mangrove areas in New Zealand.

The Whangateau Type A and the fully mature mangals further north appear to have a relatively small interstitial animal population, but this has not yet been fully studied. The pneumatophores and trunks bear wefts of red and blue-green algae which may be classed as *Bostrychietum*. The blue-green algae (mainly *Calothrix* cf. *pulvinata*) are active nitrogen fixers (Hicks 1976). Amongst them are found isopods and amphipods. The main grazer in our study area in the Whangateau Harbour is the turbinid *Turbo smaragdus*, and the largest individuals are found in the mangrove areas. This snail is mainly found on rocky shores where it reaches its highest density in the lower regions of sheltered and semiexposed shores. In the mangrove swamps only large individuals are found. Individuals are up to 45 mm long and 90% are over 30 mm. They are probably of the order of 25 years old. No young ones are found, and the population is 0.5 km away from the nearest rocky shore population. This raises the problem of recruitment. The present evidence suggests that either it is a relict population or is the result of chance drift by rafting on a tree trunk. The population does not reproduce because this species needs the stimulus of rough water for spawning.

The rock oyster *Crassostrea glomerata* also occurs attached to the trunks and pneumatophores, and epizootically on the Turbo shells. The barnacle *Elminius modestus* is also characteristically found on pneumatophores, trunks and leaves.

Burrowing animals appear to be restricted in this area because of the dense pneumatophores and feeder roots of the *Avicennia* which appears to make burrowing difficult.

In more open places one finds the mud crab *Helice crassa*, an endemic member of the Grapsidae with a carapace 20 mm across. It has large chelae and in its habitat and attachment to the burrow is very similar to the fiddler crabs of tropical mangrove swamps, but the main difference is that the chelae do not appear to be used for signalling or sexual display, but as weapons of combat or threat to other males. Although they do not often move far from their burrows, they have been seen climbing *Avicennia* trees.

A common snail is the archaic pulmonate *Amphibola crenata*, which is at the 'lung-fish' stage of gastropod evolution. This is a detritus feeder which flourishes only on the finest grades of organic surface mud, passing this indiscriminately through the gut and triturating and straining it by a strong gizzard. After extracting the nutriment it passes the bulk of the meal in a continuous faecal string stretching for metres. A study of its energetics (Briggs 1972) has shown that it has the high annual energy flow of 450 k cals m^{-2} yr^{-1}, which is the second highest recorded value for molluscs. Its assimilation energy is low, because it has to ingest a large amount of detritus and extracts very little food value.

The other common detritus feeders and herbivores are *Diloma subrostrata*, *Zeacumantus lutulentus* and *Z. subcarinatus*. These are very common in mangrove areas where they appear to move about at random, yet always remain in the small very wet areas.

The major carnivore is *Cominella glandiformis* which is also a scavenger, consuming any dead flesh. Its major food organisms are crabs and *Amphibola*, but it will also eat polychaetes, and on the flats outside the mangrove areas the bivalve molluscs *Paphies australis* and *Chione stutchburyi*. It is a very efficient scavenger and can detect dead animals at 30 m in shallow channels. Normally it scavenges but when it cannot find dead food will seek out healthy food.

A large number of fish have been reported from mangrove areas. Recently we set a gill net across the channel draining an area of mangrove swamp at high tide at monthly intervals over a year. We found that of 948 fish caught 95% were yellow-eyed mullet (*Aldrichetta forsteri*). These were mostly adult fish. The juveniles remain in the mangrove channels at low tide. The mullet is an oportunistic feeder, feeding mainly on fine bottom sediments and detritus during the day, and at night on planktonic crustaceans, especially mysids and amphipods. These planktonic crustaceans dominate the zooplankton at night, but are scarce in the daytime (Davenport 1979). This fish is of little economic importance in New Zealand, because of its small bones, but forms about 50% of the Australian mullet fishery. However, we have collected a number of economically important fish during the survey, and have evidence that mangrove areas elswhere support populations of economically useful fish.

These studies all form part of an overall programme to produce an integrated account of the New Zealand mangrove association.

Literature cited

Briggs, I.W. 1972. The feeding, ecology and energetics of a population of *Amphibola crenata*. Unpublished M.Sc. thesis, Auckland University. 112 pp.

Chapman, V.J. 1976. Mangrove Vegetation. Vaduz: J. Cramer. 477 pp.

Cornforth, R.C.D. 1975. A biogeographical classification of coastal environments with particular reference to Pakiri-Leigh-Omaha. Unpublished M.A. thesis, Auckland University. 98 pp.

Davenport, M.W. 1979. Zooplankton and fish of a mangrove area of the Whangateau Harbour. Unpublished M.Sc. thesis, Auckland University. 67 pp.

Hicks, B.J. 1976. Nitrogen fixation associated with the New Zealand mangrove *Avicennia resinifera* (Forsb.). Unpublished M.Sc. thesis, Auckland University. 117 pp.

Küchler, A.M. 1972. The mangrove in New Zealand. New Zealand Geographer 28:13–129.

Sakai, A. and P. Wardle. 1978. Freezing resistance of New Zealand trees and shrubs. New Zealand Journal of Ecology. 1: 51–61.

CHAPTER 8

Mangroves in New Zealand

V.J. CHAPMAN†

Auckland University, New Zealand

Whilst mangrove vegetation is very extensive in certain tropical areas and in some countries is an important economic resource, the mangrove swamps in New Zealand, confined to North of Auckland and Auckland, have generally been regarded as so much wasteland. As a result considerable reclamations have taken place in the past and the swamps are still used in some localities for the dumping of town waste, e.g. Coromandel.

It is not suggested that all remaining mangrove swamps in New Zealand should be placed in a reserve status, and it is clear that in the future we should use them and manage them in the best possible manner, and this includes reservation. Since the first symposium on mangroves (1974) in Honolulu, a greater awareness of the importance of mangroves has developed and all governments have been requested not to permit further reclamations or alienation without very careful prior study. The significance of mangroves has been further emphasised by the establishment in 1978 of the SCOR working party 60 charged with surveying the world resources of mangroves and in drawing up plans for their management in different countries. In pursuit of this objective regional meetings have also been held in Columbia (Cali, 1978) Bangladesh and Thailand (1976), and Kuala Lumpur (1980) where local management plans have been discussed.

In New Zealand there is but a single mangrove species (*Avicennia marina* (Forsk.) Vierh. var. *resinifera* (Forst.) Bakh. (= *A. resinifera* Forst. f.), which so far has had no commercial value so far as timber is concerned. However, if timber becomes used as an energy resource I believe this species, with its capacity to regenerate and grow quickly, could be important, though its use in this manner would need to be weighed carefully against its value as a primary producer. In most harbours in New Zealand areas have been used as follows:

(1) Reclamation for farming. The enclosed mangroves are removed and the land sown for pasture after the salt has been leached out. The mangrove trees die if ponded.
(2) Substantial areas have been destroyed by use for sanitary land fill. However, the recent Auckland and Waitemata harbour studies have recommended that this use be discontinued.
(3) Mangrove and adjacent mud flats have been converted into oxidation ponds for the tertiary treatment of sewage effluent. In the Manukau harbour a significant area of mangrove was lost when the Mangere sewage works were established. Such effluent could be used for forest irrigation rather than lose more mangrove for more oxidation ponds. On the other hand if the ponds could be used for aquaculture then the entire situation would have to be re-evaluated.
(4) Oyster culture. Oyster farms are becoming increasingly prevalent in the Auckland and North Auckland area and they are associated with mangrove swamps. Mangrove detritus froms the basis of the food chain and without man-

† Deceased.

Tasks for vegetation science, Vol. 8 ed. by H.J. Teas

 ISBN 978-90-481-8526-9

groves oyster culture would not be so flourishing. If oyster culture is to be promoted then the mangrove swamps must be conserved.

(5) Roads. These are sometimes built across mangroves swamps. If, in the process, creeks are dammed then the mangroves die. Such roads should therefore be built on piles in order to leave the mangrove as little disturbed as possible.

(6) On-shore fisheries. Since mangrove detritus forms the basis of the food chain, culminating in the fish, preservation of adequate mangrove reserves is important if the fisheries are to survive. The creeks also provide spawning areas for some fish. At present there are inadequate data to enable one to determine the area of mangrove forest relevant to water surface to maintain a local fishery. In Australia, in Moreton Bay, a value of $ 800 USA per acre has been suggested for mangrove swamps for this purpose (Anonymous, Australian Conservation Foundation, 1972). Odum and Heald (1975) suggest four factors that indicate the highly productive nature of untouched mangrove swamps. These are:

(a) Heavy leaf fall (1.3–2.0 g/m^2/day have been recorded overseas).
(b) Export of leaf litter into channels.
(c) Transformation of fallen leaves into detritus particles covered with micro-algae, bacteria, fungi and protozoa.
(d) Use of this detritus by consumer organisms.

(7) Reclamation for cooling ponds. At least one proposal of this nature for a thermal power station has been made. This would not appear to be an appropriate usage.

Characteristically mangrove vegetation is developed on coastlines with certain physiographic features, e.g. estuaries, harbours, lagoons, behind spits and offshore islands and where the coast is protected from major wave action. Along a straight coastline the mangrove fringe is narrow (Chapman 1979). It attains a maximum in harbours and lagoons.

In the light of the foregoing it seems that the recommendations of the Australian Conservation Foundation (1972) should be adopted in relation to the New Zealand mangrove swamps. These are as follows:

(1) A very careful examination should be made by relevant state fisheries and/or fauna authorities and by plant ecologists before any area of mangroves is allowed to be altered in any way or allocated for any form of development. Its potential contribution to commercial and recreational fisheries as well as long-term environmental stability must be weighed against the proposed alternative land-use values for nearby centres of population. Applications to clear mangrove areas should be advertised publicly and state the intended purpose of the alteration.

(2) Mangrove swamps in bays and estuaries adjacent to developing areas should be reserved both for aesthetic reasons and for their value to fisheries. In Florida the importance of mangrove has resulted in a deliberate mangrove re-establishment programme, and in a recent marina proposal for Auckland Harbour the Nature Conservation Council recommended that the loss of a very small mangrove area be compensated for by a planting programme in the vicinity.

(3) More extensive mangrove areas, in which the habitat will be protected because of the fishing allowed in adjacent waters, should be reserved. Total protection should be given to areas of special interest by creating national parks or similarly restricted areas (see below).

(4) Research should be intensified by bodies including universities, DSIR, and state departments concerned with land and water resources. We need to know far more about our mangrove woodlands – about the animals and plants, about the food chain, about the saline environment, about the chemical processes, about the nutrient producing micro-organisms and the total productivity in varying types of physiographic situations.

Up to the present New Zealand has used its mangrove swamps with no attempt at conservation. At present there are only two reserved areas known to me: one in the Kaipara Harbour administered by

Auckland University and which was secured by myself many years ago when I foresaw the eventual disappearance of mangroves in the Waitemata and Manukau harbours. The second reserved area is some 500 ha, forming part of the Waitangi Trust reserve in the Bay of Islands, and in which a boardwalk has recently been constructed.

With the increasing emphasis generally being placed upon mangrove vegetation, and the need to conserve adequate proportions, the Auckland Lands Department requested me to make a survey of the swamps in their area and subsequently South Auckland Lands Department asked me to do the same for swamps in their area. Arising from this survey I have recommended the establishment of a mangrove marine park in Rangaunu Harbour in Northland where mangroves occupy some 45% of the harbour. Steps are being taken to make a formal application to have this area declared a marine reserve, and if granted, as I hope it will be, then New Zealand will have its third mangrove reserve.

Arising from the survey of the mangrove swamps many areas have been proposed as reserves in order to ensure sufficient primary production to meet the needs of inshore fisheries, oyster farms and general protection of the land. These proposals are currently being considered by the Lands Department. It is expected that all fringe mangrove, which is most important for shore protection, will be left untouched because the narrow belts do not justify reclamation for farming purposes.

The Kaipara was the first harbour to be surveyed (Chapman 1976). In the Hokianga harbour (Chapman 1978a), it has been proposed that 72% (21 km^2) of the remaining mangrove and salt marsh be conserved. The harbour originally possessed about 55 km^2 mangrove and salt marsh and now only 29 km^2 (55%) are left. It is considered that further reclamation could have serious effects upon the food chain. In the Herekino and Whangape harbours, where the mangrove areas are relatively small, no further reclamations should be permitted (Chapman 1978a). In the far north only 18% of Parengarenga harbour is occupied by mangrove and hence substantial areas have been recommended for preservation (Chapman 1978b). The setting aside of Rangaunu harbour as a mangrove maritime park has already been referred to. Major areas need preservation in the Taipa River and Mangonui harbour, whilst in Whangaroa harbour, as mangrove only occupies about 21.5% of the harbour, almost all the existing mangrove has been recommended for reservation (Chapman 1978c). The Bay of Islands is a major area so far as mangroves are concerned and substantial areas are proposed for conservation. There is already the Waitangi area. It is further suggested that some of the other areas should be added to the adjacent Russell State Forest. Some of the best mangal in New Zealand occurs in this region. More and more oyster farms are being permitted in this region and protection of the existing mangrove is again very important. In the case of Whangaruru Harbour the main mangrove there should be added to the Bay of Islands Maritime and Historic Park. The recent harbour studies for the Waitemata and Manukau harbours have both recommended no further interference with the mangroves. This is especially important in the case of the latter harbour where the amount of mangrove is relatively small. The last significant area on the West coast of New Zealand with mangrove is a small patch in Raglan harbour, which should be conserved.

On the East coast the shores in the Gulf of Thames from the Karito Canal to Thames itself all possess fine stands of mangroves most of which are recommended for conservation. Mangroves in Manaia and Coromandel harbours are recommended for conservation, especially as in the latter the swamps have been used for town waste dumping. In Colville Bay and Kennedy Bay mangrove areas have been proposed for reservation, as also in Whangapoura harbour where there is substantial colonising mangrove. In Whitianga harbour, where the mangrove occupies about 48%, mangrove preservation is important in order to maintain the good fishing. In Whangamata harbour significant areas are proposed for conservation and in Tauranga harbour, because the mangrove area is relatively small, most of it is recommended for conservation. The mangrove reaches its southern limit on the east coast in Ohiwa harbour and for this reason the two main areas here are both proposed for conservation.

If the recommendations referred to above are implemented by government action, then New Zealand can be listed as one of those countries that has paid attention to the international call to conserve mangroves.

Arising from this extensive survey of mangroves in New Zealand I believe that four ecosystems involving the mangrove can be recognised. These are as follows:

(1) Pure *Avicennia marina* var. *resinifera* community composed of tall or low trees (near southern boundary). Marine algae clothe the pneumatophores, and in places the eel-grass, *Zostera muelleri*, occupies the mud and in some places one may find the free-living fucoid, *Hormosira banksii* var. *libera* on the mud.
(2) *Juncus australiensis* – *Leptocarpus similis* savannah with scattered, large, isolated trees of *Avicennia*.
(3) *Salicornia australiensis* – low *Avicennia* community found on muddy sand flats, especially in areas subject to moderate frost action.
(4) Tall *Avicennia* associated with the sooty mould, *Capnodium fuliginodes*, which in its turn exists on the sugary excretion of the scale insect, *Ceroplastis sinensis*. This ecosystem is readily recognised by the black appearance of all the trees due to the mould. It was first recognised in the Hokianga harbour.

These four compare with Kuchler's (1972) recognition of some six communities associated with salt marsh plants:

(1) *Avicennia* + *Juncus* + *Leptocarpus*	= (2) above
(2) *Avicennia* + *Juncus maritimus* var. *australiensis*	
(3) *Avicennia* + *Leptocarpus similis*	
(4) *Avicennia* + *Salicornia australis*	= (3) above
(5) *Avicennia* + *Salicornia* + *Samolus repens*	
(6) *Avicennia* + *Scirpus cernuus*	

I do not believe the last species occurs in sufficient abundance to justify this community.

The New Zealand mangroves can also be classified according to their relationships to air temperature as follows:

(1) Tall *Avicennia* stands found in northern frost free areas and occupying channel banks and mud flats.
(2) Tall riverine *Avicennia* restricted to creek banks in areas of moderate winter frost (2–5 days with 2–3° C of frost).
(3) Low *Avicennia* restricted to rear mud flats in areas of moderate winter frost but found along the channels as well in southern areas near its biogeographical limit.

Finally one may note that throughout its distribution in New Zealand the *Avicennia* swamps exist alongside quite characteristic salt marsh communities e.g.:

(1) *Juncus maritimus* var. *australiensis* community. This is often monospecific. It may have *Leptocarpus similis* as an associate, and at higher levels patches of *Samolus repens* (on better drained areas), *Selliera radicans*, *Triglochin striatum*, *Scirpus cernuus* and *Muehlenbeckia complexa*.
(2) *Leptocarpus similis* community. This also can be monospecific or it may contain some or all of the associated *Juncus* species.
(3) *Salicornia australis* community. This is unusually monospecific but may have some *Samolus* associated with it in better drained areas.
(4) *Juncus* – *Leptocarpus* – *Typha* brackish marsh.
(5) General Salt Marsh community. This is a mixture of *Selliera radicans*, *Scirpus cernuus*, *Samolus repens*, *Triglochin striatum*, *Plantago coronopus* and *Cotula coronopifolia*.

Literature cited

Anonymous. 1976. Waitemata Harbour Plan. Report of ARA and Auckland Harbour Board.

Anonymous. 1972. Mangroves and Man. Viewpoint No. 7. Australian Conservation Foundation.

Anonymous. 1978. Manukau Harbour Plan. Part 1. Policy. Report of ARA and Auckland Harbour Board.

Chapman, V.J. 1976. Mangroves and Salt Marshes of the Kaipara Harbour. Report to Dept. Lands and Survey, Auckland, New Zealand.

Chapman, V.J. 1978a. Mangroves and Salt Marshes of the Herekino, Whangape and Hokianga Harbours. Report to Dept. Lands and Survey, Auckland, New Zealand.

Chapman, V.J. 1978b. Mangroves and Salt Marshes of the Parengarenga, Houhora, Rangaunu and Mangonui Harbours and the Taipa River. Report to Dept. Lands and Survey, Auckland.

Chapman, V.J. 1978c. Mangroves and Salt Marshes of the Whangaroa and Whangaruru harbours and the Bay of Islands. Report to Dept. Lands and Survey, Auckland.

Chapman, V.J. 1979. The place of Mangrove vegetation in the Coastal Zone. In: Proceedings 7. Coastal zone Workshop of Environment 77 Conference, Christchurch. Environmental Centre (Canterbury) Inc.

Kuchler, A.W. 1972. The Mangrove in New Zealand. N.Z. Geographer. 28:113–129.

Odum, W.E. and E.J. Heald. 1975. Mangrove forests and aquatic productivity. In: An Introduction to land water interactions. Springer-Verlag, pp. 129–136.

CHAPTER 9

An introduction to the nomenclature and taxonomy of the mangrove flora in Papua New Guinea and adjacent areas

J.S. WOMERSLEY*

82 Richmond Road, Westbourne Park, South Australia, 5041

The habitat characterized by the accumulation of mud and fine silt deposits and subject to regular, usually daily, inundation by tidal influence supports a variety of vegetation types of which that dominated by a tree flora with the adaptations of stilt roots and pneumatophores is the most widespread. This is the Mangrove Formation. Within the tropical and subtropical regions the mangrove formation is usually dominated by species of the family Rhizophoraceae. Species classified in other families occur and one or more of these may become dominant where certain edaphic limitations are in effect.

Most striking of these very special communities is that which is dominated by the stemless palm *Nypa fruticans*. In several vast areas along the southern coast of the island of New Guinea and in Sabah and Sarawak (north and north west Borneo) *Nypa* dominates, the sea of fronds broken only by the occasional tree crown.

The floristics of the mangrove formations in Papua New Guinea are very similar to those found elsewhere in equatorial (10° north and south of the equator) zone of southeast Asia. Eastward into the Pacific there is a paucity of suitable habitats and many species found in the mangroves in Papua New Guinea either do not occur or occupy different sites (e.g. fresh water swamp forest). In Papua New Guinea some 25 species of trees (classified in 7 families) contribute to the overall physiognomy of the mangrove formations. To these must be added the palm *Nypa fruticans*. These are enumerated in Table 1. Tidal inundation classes are shown in Table 2.

Other species of woody plants find a niche in the mangrove vegetation but only in edaphically distinct sites are these species of other than sporadic distribution. Among these species are: *Dolichandrone spathacea*, *Cynometra ramiflora*, *Cerbera floribunda*, *Cerbera manghas*, *Myristica hollrungii*, *Acanthus ilicifolius*, *Osbornia octodonta*, *Pemphis acidula*, *Aegialites annulata*, *Aegiceras corniculatum*, *Lumnitzera littorea*, *Lumnitzera racemosa* and *Scyphiphora hydrophyllacea*. Of special interest in the vicinity of Port Moresby is *Osbornia octodonta*, a member of the family Myrtaceae, which is restricted to a few localities along the south and east coast of Papua New Guinea and the east and west coasts of the Cape York peninsula of Australia. *Aegialites annulata* (Plumbaginaceae), while being more widely distributed than *Osbornia* (New Guinea to at least the Persian Gulf), does occur in a most interesting community at Idlers Bay where the individual plants have the structure of miniature trees complete with buttresses. *Pemphis acidula*, while being present in small numbers of individuals in most situations, can be seen abundantly on the seaward edge of coral rocks at Madang.

This complex of species can be very loosely grouped into three categories and the habitat sup-

*Formerly Assistant Director, Botany, Department of Forests, Papua New Guinea, 1946–1975.

Tasks for vegetation science, Vol. 8 ed. by H.J. Teas
 ISBN 978-90-481-8526-9

Table 1. Major components of the mangrove association in Papua New Guinea.

Nypaceae (sometimes included in Palmae)
Nypa fruticans – This species is present in distinct communities as a single species dominant.

Verbenaceae (Avicenniaceae)
Avicennia alba
A. eucalyptifolia – Localised and then not common.
A. marina
A. marina var. *officinalis*
A. officinalis

Rhizophoraceae
Rhizophora apiculata
R. lamarckii – (Localised but then abundant).
R. mucronata
R. stylosa
Bruguiera exaristata (rare)
B. gymnorhiza
B. hainesii – Very rare, possibly misidentified for Papua New Guinea.
B. cylindrica
B. parviflora
B. sexangula
Ceriops decandra
C. tagal

Sonneratiaceae
Sonneratia alba
S. caseolaris
S. ovata

Bombacaceae
Camptostemon schultzii – Apparently confined to the coastal region bordering Torres Strait.

Sterculiaceae
Heritiera littoralis

Euphorbiaceae
Excoecaria agallocha

Meliaceae
Xylocarpus granatum
X. moluccensis
Dolichandrone spathaceade (Bignoniaceae) – More frequently on the banks of fresh water streams subject to tidal backing up.
Cynometra ramiflora (Leguminosae, Caesalpiniodeae)
Cerbera floribunda Both species are found in mangrove but also in fresh water swamp forests (Apocynaceae).
Cerbera manghas
Myristica hollrungii (Myristicaceae)
Acanthus ilicifolius (Acanthaceae) – Very variable in leaf shape.
Osbornia octodonta (Myrtaceae) – An unusual plant of the mangrove strand to be seen near Port Moresby.
Pemphis acidula (Lythraceae) – Often associated with a coral or rocky substrate.
Aegialites annulata (Plumbaginaceae) – A curious plant occupying a variety of sites. Rare in Papua New Guinea.
Aegiceras corniculatum (Myrsinaceae) – More frequent along the banks of channels through the mangrove.
Lumnitzera littorea (Combretaceae) – Tree, flowers red.
Lumnitzera racemosa (Combretaceae) – Usually a shrub, flowers white. Common near Bootless Bay, Port Moresby.
Scyphiphora hydrophyllacea (Rubiaceae) – On the inland edge of the mangroves, usually in exposed positions.

Table 2. Inundation classes of tidal forest.

Class 1 – Inundated (at least partially) by all high tides
Avicennia spp.
Sonneratia alba J. Smith
Class 2 – Inundated by medium high tides
Avicennia spp.
Sonneratia alba
Rhizophora mucronata Lamk
Rhizophora apiculata Bl. & *R. stylosa*
Bruguiera parviflora (Roxb.) W. & A.
Class 3 – Inundated by normal high tides (but not necessarily by salt water)
Rhizophora mucronata
Rhizophora apiculata & *R. stylosa*
Ceriops tagal (Perr.) C. B. Rob.
Bruguiera spp. (excluding *B. parviflora*)
Xylocarpus granatum Koenig
Xylocarpus moluccensis (Lamk) Roem.
Class 4 – Inundated only by spring tides
Bruguiera spp.
Xylocarpus granatum
Xylocarpus moluccensis
Lumnitzera spp.
Class 5 – Occasionally inundated by exceptional or equinoctial tides
Sonneratia caseolaris
Lumnitzera spp.
Heritiera littoralis
Hibiscus tiliaceus

porting the mangrove association divided into three zones (Table 3). Descriptive accounts of the individual species can be found in Percival and Womersley (1975).

The taxonomy of the species of woody plants occurring in the mangrove association in Papua New Guinea is generally well known, particularly when compared with our state of knowledge of the taxonomy of other plant communities in this country. This is due largely to the apparently widespread occurrence of many of the species. There are problems, not least of which is the use of names originally typified by specimens collected outside the equatorial region of south east Asia. The recent paper by Duke and Bunt (1979) has drawn attention to the need to establish the correct nomenclature of the Australian species of *Rhizophora.* In these studies Duke and Bunt have used a variety of measurements to recognise four entities of Rhizophora along the north eastern Australian coast. Such studies should now be extended to other areas, particularly those from which the type collections of *Rhizophora mucronata* Lamk and *Rhizophora stylosa* Griff. were obtained. Percival and Womersley (1975) doubt the presence of *R. mucronata* Lamk. among the collections from New Guinea. Further studies similar to those of Duke and Bunt may confirm the presence of the plant being now called *R. mucronata* on the Australian coast. Duke and Bunt (1979, p. 677) have already pointed out the need to decide the real status of this entity.

The species of *Bruguiera* appear to be more distinct than those of *Rhizophora*, at least in Papua New Guinea, although the status of *Bruguiera exaristata* Ding Hou may change when study of a large number of collections of both putatively *Bruguiera exaristata* Ding Hou and *Bruguiera sexangula* (Lour.) Poir. can be made with correlation of phenological characters, measured as done by Duke and Bunt (1979) and the presence or absence of bristles in the sinus of the petals.

As everywhere in the range of the genus the classification of *Sonneratia* is less than satisfactory.

Literature cited

Duke, N.C. and J.S. Bunt. 1979. The genus *Rhizophora* (Rhizophoraceae) in northeastern Australia. Austr. Jour. Bot 27: 657–678.

Percival, M. and J.S. Womersley. 1975. Floristics and ecology of the mangrove vegetation of Papua New Guinea. Bot. Bull. No. 8. Department of Forestry, Div. Bot., Lae, Papua New Guinea. 96 pp.

Table 3. Distribution of species in the mangrove formation in Papua New Guinea.

ZONES	Group 1 Trees, or tall shrubs, abundant and contributing to the physiognomy of the association	Group 2 Shrubs, or small trees which may be locally abundant but contribute little to the overall physiognomy of the association	Group 3 Low shrubs or lianes frequently present in edaphically restrictive situations
OUTER EDGE	*Avicennia alba* *Nypa fruticans* *Rhizophora apiculata* *R. lamarckii* *R. mucronata* *R. stylosa* *Sonneratia alba* *S. ovata*		*Aegialites annulata*
MIDDLE	*Avicennia marina* *A. marina* var. *resinifera* *Bruguiera cylindrica* *B. parviflora* *Ceriops tagal*	*Aegiceras corniculatum* *Dolichandrone spathaceade* *Osbornia octodonta*	*Acrosticum aureum* *Derris* sp.
INNER ZONE	*Avicennia eucalyptifolia* *Bruguiera exaristata* *B. gymnorrhiza* *B. sexangula* *Ceriops decandra* *Sonneratia caseolaris* *Xylocarpus granatum*	*Bruguiera hainesii* *Camptostemon schultzii* *Cerbera floribunda* *C. manghas* *Cynometra ramiflora* *Excoecaria agallocha* *Heritiera littoralis* *Lumnitzera littorea* *L. racemosa* *Myristica hollrungii* *Scyphiphora hydrophyllacea* *Xylocarpus moluccensis*	*Acanthus ilicifolius* *Pemphis acidula*

CHAPTER 10

Mangrove fishes of New Guinea

BRUCE B. COLLETTE

National Marine Fisheries Service, Systematics Laboratory, Washington, D.C. 20560, U.S.A.

Abstract. Fishes were collected with rotenone from 14 mangrove swamps, 2 in northern Australia, 6 in Papua New Guinea and 6 in Irian Jaya. Over 200 species from 58 families were found. The 12 families found in the most mangrove swamps were Hemiramphidae (13 of 14), Gobiidae (13), Apogonidae (13), Tetraodontidae (10), Blenniidae (10), Lutjanidae (9), Ophichthidae (9), Eleotridae (8), Pomacentridae (8), Toxotidae (8), Ambassidae (8), and Siganidae (7). The number of families collected per mangrove swamp ranged from 9 to 25, the number of species from 12 to 57, and the number of specimens from 59 to 818. Among the most abundant species in some mangroves were two anchovies (*Thrissina baelama* and *Stolephorus commersonii*), a cardinalfish (*Apogon lateralis*), two ambassids (*Ambassis interruptus* and *Priopidichthys gymnocephalus*), an atherinid (*Atherinomorous endrachtensis*), a pomacentrid (*Neopomacentrus taeniurus*), and a goby (*Oligolepis moloanus*). Species found in the most mangroves were the archerfish *Toxotes jaculator*, a blenny *Omox biporos* and a cardinalfish *Apogon lateralis* (8 of 14); a cardinalfish *Fowleria* cf. *marmorata*, two gobies, *Acentrogobius* cf. *triangularis* and *Pandaka* cf. *pygmaea* (7); a pipefish *Hippichthys spicifer*, a third species of cardinalfish *Sphaeramia orbicularis*, three gobies, *Oligolepis moloanus*, *Parioglossus* sp., and *Yongeichthys* cf. *nebulosus*, and an eleotrid *Eleotris* cf. *fusca* (6). Comparisons are made between the New Guinea mangrove fish fauna and the mangrove fishes in 7 other studies from North Borneo, South Viet Nam, India, Madagascar, Florida, and Puerto Rico.

Fishes of mangrove swamps are generally poorly known, due perhaps to the difficulty of collecting in this habitat and also to the attractiveness of coral reefs in the same general areas. During the R/V *Alpha Helix* Moro Expedition to New Guinea and Indonesia in May–June, 1979, I devoted much collecting effort to this habitat. Fourteen stations were sampled with rotenone, 2 in the Northern Territory of Australia, 6 in Papua New Guinea, and 6 in Irian Jaya, the Indonesian western half of the island of New Guinea (see Appendix 1). Some mangrove swamps were at the back of bays, some along streams and small rivers; some on the main island of New Guinea or Australia, some on smaller offshore islands. The bottom was usually mud 0.3 m or more deep, but some mangroves were growing in sand and one mangrove swamp was mostly floored by coralline rock. (About 60 other collections of fishes were made in tide-pools, on coral reefs, along sandy shores, and offshore with rotenone, seines, and trawls.)

The collection was sorted to species and preliminary identifications made based largely on Munro's 'Fishes of New Guinea'. Material was then shipped to 28 specialists in various groups through the cooperation of the Smithsonian Oceanographic Sorting Center. Most of the material has been deposited in the U.S. National Museum of Natural

Tasks for vegetation science, Vol. 8 ed. by H.J. Teas
 ISBN 978-90-481-8526-9

History (USNM) in Washington, D.C. Representative material has been sent to the Australian Museum, Sydney and some specialists have retained some material for further study. Identifications have been confirmed for 56 families.

Results

About 204 species in 126 genera of 58 families were found at the 14 stations (see Appendix 2). From 9–25 families were found at the mangrove swamp stations (Fig. 1). The number of genera varied from 12–37, species 12–57, and specimens 59–818. The dozen families found at the most stations are:

(1) Hemiramphidae – 13 of 14
(2) Gobiidae – 13 of 14
(3) Apogonidae – 13 of 14
(4) Eleotridae – 10 of 14
(5) Tetraodontidae – 10 of 14
(6) Blenniidae – 10 of 14
(7) Lutjanidae – 9 of 14
(8) Ophichthidae – 9 of 14
(9) Pomacentridae – 8 of 14
(10) Toxotidae – 8 of 14
(11) Ambassidae – 8 of 14
(12) Siganidae – 7 of 14.

Some characteristic mangrove fishes were as follows:

Ophichthidae – *Muraenichthys* at 7 stations, 3 spp., only 1–11 specimens per station.
Hemiramphidae – *Zenarchopterus* at 13, 4 spp.; *Z. dunckeri* and *Z. gilli* at 5 stations each.
Syngnathidae – *Hippichthys spicifer* at 6, only 1–4 per station.
Carangidae – *Caranx sexfasciatus* at 5, together with *C. melampygus* at 3 stations.
Gerreidae – *Gerres* spp. at 5, 6 spp., 73 specimens of *G. abbreviatus* at BBC 1743.
Leiognathidae – *Leiognathus* spp. at 5, 54 specimens of *L. fasciata* at BBC 1743.
Lutjanidae – *Lutjanus* spp. at 9, 7 spp. represented.
Pomacentridae – *Neopomacentrus taeniurus* at 4, 249 specimens at BBC 1722.
Toxotidae – *Toxotes jaculator* at 8 stations, 1–14 specimens per station.
Apogonidae – *Apogon lateralis* at 8, 156 specimens at BBC 1714, 74 at BBC 1711.
– *Fowleria* cf. *marmorata* at 7, 1–13 specimens per station.
– *Sphaeramia orbicularis* at 6, 75 specimens at

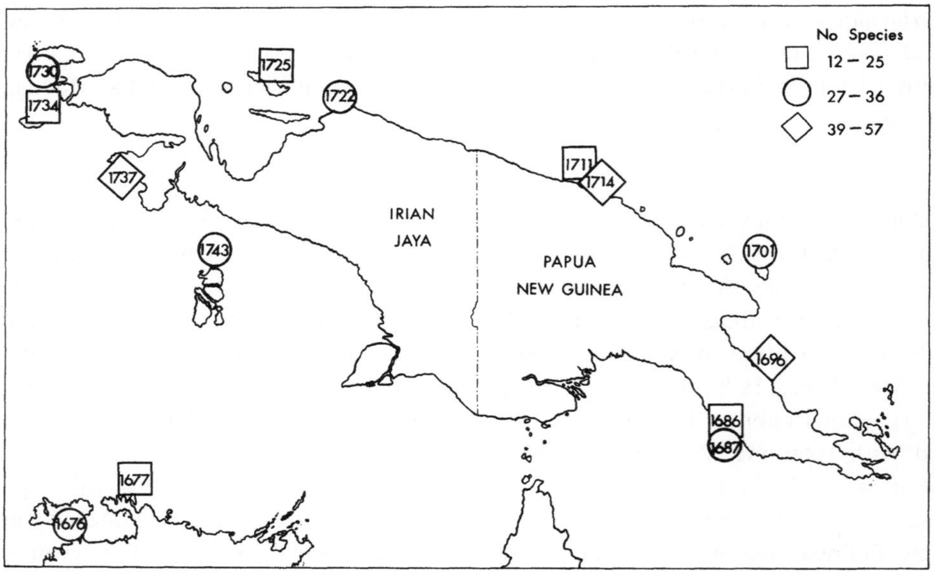

Fig. 1. Location of mangrove swamp fish collecting stations made about New Guinea. Numbers are station numbers, symbols indicate the number of species of fishes taken at each station (12–25, 27–36, and 39–57).

BBC 1714.

Ambassidae – present at 8 stations, 5 spp., occurring in large numbers at 3 stations, *A. interruptus* – 238 (BBC 1714). *Priopidichthys gymnocephalus* – 209 and 149 specimens (BBC 1677 and 1722).

Mugilidae – present at 6 stations, 5 spp.

Sphyraenidae – *Sphyraena barracuda*, juveniles at 5 stations, 1–9 specimens per station.

Siganidae – *Siganus* spp. at 7 stations, 4 spp.

Blenniidae – *Omox biporos* at 8 stations, 2–26 specimens per station.

– *Omobranchus* at 5 stations, 2 spp.

Eleotridae – *Butis* spp. at 6 stations, *B. butis* at 5.

Gobiidae – at all but 1 station, 27 genera and 49 species.

– *Acentrogobius* spp. at 9 stations, 9 spp., *A.* cf. *triangularis* at 7 stations, 1–73 specimens per station.

– *Pandaka* cf. *pygmaea* at 7 stations, 1–55 specimens per station.

– *Oligolepis moloanus* at 6 stations, 1–208 specimens per station.

– *Parioglossus* sp. at 6 stations, 1–22 specimens per station.

– *Yongeichthys* cf. *nebulosus* at 6 stations, 2–22 specimens per station.

Tetraodontidae – *Arothron* spp. at 6 stations, 2 spp.

– *Chelonodon* spp. also at 6 stations, 2 or 3 spp.

Discussion

Let us turn from the specifics of the mangrove swamp fishes of New Guinea to a consideration of some more general questions.

(1) How does the mangrove fish fauna of New Guinea compare with the mangrove fauna of other areas in terms of species and family composition?

(2) Is there a characteristic mangrove fish fauna or are the fishes found in mangroves part of the same fauna that inhabits other types of estuaries?

(3) What is the interest in the fish fauna of mangroves, scientifically and practically?

To gather comparative data to address these questions, 7 studies were selected that contain enough information on mangrove swamp fishes to make feasible a comparison with the New Guinea data. These studies, like mine, suffer from inadequate data on the habitat itself. Some of the studies list some of the mangrove species present, others do not. Little information is available on other important ecological characteristics – salinity, temperature, bottom type, etc. Two of the 7 comparative studies are from the western Pacific, 2 from the Indian Ocean, and 3 from the western Atlantic:

(1) Inger (1955) – North Borneo, 4 mangrove and nipa palm stations: 29 families, 33 genera, 40 species.

(2) de Sylva and Michel (1974) – South Viet Nam, about 12 stations in both wet and dry seasons in defoliated (by military use of herbicides) and undefoliated mangrove swamps: 27 families, 42 genera, 54 species, larvae, juveniles, and adults.

(3) Krishnamurthy and Prince Jeyaseelan (1981) – mangrove ecosystem of Pichavaram, southeast India; presence of juveniles, adults, larvae, and eggs: 35 families, 57 genera, 77 species.

(4) Kiener (1965) – Madagascar, showed range from freshwaters through brackish mangrove swamps and estuaries to coastal waters and the high seas: 46 families, 76 genera, 129 species.

(5) (a) Tabb and Manning (1962) – Florida Bay, several mangrove localities compared to freshwater and marine habitats: 29 families, 50 genera, 59 species.

(b) Odum and Heald (1972) – North River, tributary to Florida Bay: 27 families, 47 genera, 52 species.

(c) Total Florida Bay: 36 families, 65 genera, 76 species.

(6) Austin (1971) – Western Puerto Rico, 3 study areas on the south coast, 2 on the west, and one on the north coast: 28 families, 41 genera, 52 species.

Comparison of the mangrove fish fauna from the 7 areas shows that 14 families occurred in most of the areas (Table 1), 6 in all 7 areas: Gobiidae, Gerreidae, Clupeidae, Lutjanidae, Eleotridae, and Tetraodontidae. Four more families were found in most of the 5 Indo-Pacific areas but do not occur in the New World.

Another way of comparing the 7 areas is to list

Table 1. Families most often occurring in mangrove swamps in 7 regions (numbers in columns indicate number of genera and of species in region).

Families	New Guinea	N. Borneo	S. Viet Nam	SE India	Madagascar	Florida Bay	Puerto Rico
7/7 regions							
Gobiidae	27–49	3–3	7–8	2–3	7–15	3–3	3–3
Gerreidae	1–6	1–1	1–2	1–4	1–4	3–4	4–8
Clupeidae	3–3	3–3	2–2	4–4	2–2	3–3	3–3
Lutjanidae	1–7	1–1	1–1	1–4	1–3	1–2	1–3
Eleotridae	6–7	1–1	1–1	1–1	3–5	1–1	2–2
Tetraodontidae	2–5	1–2	1–1	3–3	4–6	1–2	1–1
5/5 regions							
Ambassidae	2–5	2–3	1–1	2–2	1–3	×	×
Leiognathidae	2–3	1–1	1–2	3–4	4–4	×	×
6/7 regions							
Carangidae	1–2	1–1	0	6–8	4–11	1–1	3–3
Mugilidae	4–5	2–3	0	3–6	3–5	1–2	1–2
Pomadasyidae	0	1–2	1–1	1–1	1–1	2–2	2–5
Engraulidae	2–3	3–4	2–2	0	1–1	1–2	2–2
Sphyraenidae	1–1	1–1	0	1–2	1–1	1–1	1–1
4/5 regions							
Platycephalidae	3–3	0	1–1	1–2	1–1	×	×
Sillaginidae	1–2	0	1–2	1–1	1–1	×	×
5/7 regions							
Sciaenidae	0	0	5–6	4–4	1–1	4–4	1–1
Syngnathidae	1–1	1–1	0	0	4–8	1–3	1–1
Hemiramphidae	1–4	2–2	0	0	3–4	1–1	1–1
No. families	58	22	27	35	46	29	28
No. genera	126	33	42	57	76	50	41
No. species	204	40	54	77	129	59	52

the families with the most fish species in mangroves in each area (Table 2). Three to 7 families contain at least one third of the species in each area (37.0–48.1%). The most speciose families contain 17–26% of the species. The Gobiidae ranks among the three most speciose families (5.8–24.3%) in all studies except Krishnamurthy and Prince Jeyasseelan (1981).

Turning from similarities to differences between the results from the New Guinea samples and other mangrove areas, several families were important in terms of number of species (and individuals in some cases) in New Guinea but not elsewhere.

(1) Apogonidae – 5 genera, 12 species in New Guinea mangrove swamps, 13/14 stations, samples of over 50 specimens in 4 cases. The only other report of Apogonidae – a single species – among the 7 comparative studies is the South Viet Nam Study.
(2) Pomadasyidae and Sciaenidae were absent at the 14 New Guinea stations but present at 6 and 5 of the other 6 areas respectively.
(3) Blenniidae were not reported from any of the other 6 areas but 5 genera and 8 species were present in New Guinea mangroves, 1 species, *Omox biporos* at 8/14 stations.

The latter two differences may reflect differences in sampling methods, my mostly 1-man rotenone collections were made while wading, and probably favored small species. Some other studies, e.g. de

Table 2. Most speciose families of fishes in mangrove swamps from 7 regions.

	W. Pacific			Indian O.		W. Atlantic	
	New Guinea	N. Borneo	S. Viet Nam	SE India	Madagascar	Florida Bay	Puerto Rico
No. families	58	22	27	35	46	36	28
No. species	204	40	54	77	129	76	52
Dominant families (%)	Gobiidae – 24.3 Apogonidae – 5.9 Ophichthidae – 4.0 Blenniidae – 4.0 Eleotridae – 3.5 Serranidae – 3.5 Lutjanidae – 3.5 Pomacentridae – 3.5	Engraulidae – 10.0 Gobiidae – 7.5 Clupeidae – 7.5 Mugilidae – 7.5 Ambassidae – 7.5	Gobiidae – 14.8 Sciaenidae – 11.1 Cynoglossidae – 11.1	Carangidae – 10.4 Mugilidae – 7.8 Clupeidae – 5.2 Lutjanidae – 5.2 Sciaenidae – 5.2 Leiognathidae – 5.2	Gobiidae – 11.6 Carangidae – 8.5 Syngnathidae – 6.2 Tetraodontidae – 4.7 Eleotridae – 3.9 Mugilidae – 3.9	Cyprinodontidae – 9.2 Gobiidae – 6.6 Gerreidae – 5.3 Sciaenidae – 5.3 Clupeidae – 3.9 Syngnathidae – 3.9 Poeciliidae – 3.9	Gerreidae – 15.4 Pomadasyidae – 9.6 Gobiidae – 5.8 Carangidae – 5.8 Lutjanidae – 5.8 Clupeidae – 5.8

Sylva and Michel in South Viet Nam, used trawls in deeper water and collected sciaenids and pomadasyids rather than blennies. They were unable to spend much time at any station due to the military situation and so usually made a 15-min trawl from a 12 m patrol boat and departed (D.P. de Sylva, pers. comm.). This explains the absence of very small species in their collections.

Many species of fishes move into mangroves at a particular stage in their life history – usually as juveniles. The success of fisheries in many tropical regions may be closely tied to the health of the mangrove swamps which serve as important nursery grounds for species such as snappers (Lutjanidae), jacks (Carangidae), and mullets (Mugilidae). Snedaker (1978) estimated that upwards of 90% of marine species in these regions are found in mangrove swamps during one or more periods of their life cycles. Much of the productivity of mangrove swamps is based on organic detritus of mangrove origin (Odum and Heald 1972, 1975) rather than on phytoplankton or sea grasses.

Conclusions

The study of mangrove swamp fishes deserves greater emphasis for two reasons:

(1) The fauna is poorly known and, on the basis of this study, is more diverse than previously thought.

(2) Many fishes of economic importance spend part of their life cycles in mangrove swamps. Throughout much of the tropics, expanding human populations are adversely affecting mangrove swamps through overcutting for firewood, land reclamation, pollution, use of herbicides, and changes in the supply of freshwater (Lindén and Jernelöv 1980). More knowledge is needed about the species composition of the fish fauna in mangrove swamps and their interactions and the biology of ecologically and economically important species found in mangrove swamps.

Acknowledgements

I am grateful for the opportunity of participating in the Moro Expedition of the R/V *Alpha Helix* sponsored by the National Science Foundation. The crew, the scientific party, and the chief scientist, Dr. A. Fleminger, supported my collecting efforts in many ways. Dr. Gordon Hendler of the Smithsonian Oceanographic Sorting Center was of special help in piloting me into and out of many of the mangrove swamps at the cost of numerous mosquito bites. When we visited Wewak, Mr. Jock Campbell, Marine Fisheries Advisor, found an ideal mangrove swamp for my sampling. My transportation expenses to Darwin and back from Ambon were paid by the L.P. Schultz Fund at the Smithsonian Institution. Drs. Frank Ferrari and Leslie W. Knapp of the Smithsonian Oceanographic Sorting Center provided collecting equipment and arranged for the return of specimens to the Sorting Center. Dr. Knapp facilitated the preliminary sorting and shipping of specimens to the specialists on the SOSC list.

The specialists who made or confirmed identifications of the mangrove swamp fishes made completion of this project possible. I am greatly indebted to them for their time and effort: Dr. G.R. Allen, Western Australian Museum, Perth (Lutjanidae); Dr. J.E. Böhlke, Academy of Natural Sciences, Philadelphia (Muraenidae); Dr. R.F. Cressey, USNM (Synodontidae); Mr. C.E. Dawson, Gulf Cost Research Laboratory, Ocean Springs, Miss., (Syngnathidae); Dr. D.P. de Sylva, University of Miami (Sphyraenidae); Dr. A. Emery, Royal Ontario Museum, Toronto (Pomacentridae); Dr. W.N. Eschmeyer, California Academy of Sciences, San Francisco (Scorpaenidae); Dr. M.F. Gomon, National Museum of Victoria, Melbourne (Labridae); Mrs. J. Gomon, USNM (Plotosidae); Dr. P.C. Heemstra, J.L.B. Smith Institute of Ichthyology, Grahamstown; (Serranidae); Mr. J.B. Hutchins, Western Australian Museum (Batrachoididae, Monacanthidae); Dr. W. Ivantsoff, Macquarie University, Sydney (Atherinidae, Pseudomugilidae); Dr. L.W. Knapp, Sorting Center, USNM (Platycephalidae); Dr. E.A. Lachner, USNM (Apogonidae, Mullidae); Dr. J.E. McCosker, California Academy of Sciences, San Francisco (Moringuidae, Ophichthidae); Mr. R.J. McKay, Queensland Museum, Brisbane (Sillaginidae); Mr. J.F. McKinney, Grice Marine Biological Laboratory, College of Charleston (Gobiidae, Eleotridae, Kraemeriidae, Periopthalmidae); Dr. A.G.K. Menon, University of Dar Es Salaam, Tanzania (Cynoglossidae, Soleidae); Dr. J. Nielsen, Zoological Museum, University of Copenhagen (Bothidae); Dr. T.W. Pietsch, University of Washington (Antennariidae); Dr. T. Sato, University of Tokyo (Pentapodidae, Scolopsidae); Dr. W. Smith-Vaniz, Academy of Natural Sciences, Philadelphia (Carangidae, Opistognathidae); Dr. V.G. Springer, USNM (Blenniidae); Dr. J.M. Thomson, University of Queensland, Brisbane (Mugilidae); Dr. J. Tyler, National Science Foundation (Balistidae, Tetraodontidae); Dr. R. Vari, USNM (Ambassidae, Teraponidae); Dr. P.J.P. Whitehead, British Museum (Natural History), London (Clupeidae, Engraulidae); and Dr. D.J. Woodland, University of New England, Armidale, New South Wales (Gerreidae, Leiognathidae, Siganidae).

The paper was presented for me at the International Symposium on the Biology and Management of Mangroves by Dr. Lamarr Trott, National Marine Fisheries Service, Washington. Comments on drafts of the manuscript have been provided by Drs. D.M. Cohen, D.P. de Sylva, L. Trott, and R. Vari.

Literature cited

Austin, H.M. 1971. A survey of the ichthyofauna of the mangroves of western Puerto Rico during December, 1967–August, 1968. Carib. J. Sci. 11(1–2):27–39.

de Sylva, D.P. and H.B. Michel. 1974. Effects of mangrove defoliation on the estuarine ecology and fisheries of South Vietnam. Natl. Acad. Sci., The effects of herbicides in South Vietnam, Part B, Working Papers, 126 pp.

Inger, R.F. 1955. Ecological notes on the fish fauna of a coastal drainage of North Borneo. Fieldiana Zool. 37:47–90.

Kiener, A. 1965. Contributions a l'étude écologique et biologique des eaux saumâtres malgaches. Les poissons euryhalins et leur rôle dans le développement des peches. Vie Milieu 16C(1):1013–1149.

Krishnamurthy, K. and M.J. Prince Jeyaseelan. (1981). The early life history of fishes from Pichavaram Mangrove Ecosystem of India. Intl. Council Exploration Sea. Intl. Symp., ICES Rapp. Proc. Verb. 178:416–423.

Lindén, O. and A. Jernelöv. 1980. The mangrove swamp – an ecosystem in danger. Ambio 9(2):81–88.

Munro, I.S.R. 1967. The fishes of New Guinea. Dept. Agr., Stock, Fish., Port Moresby, 651 pp.

Odum, W.E. and E.J. Heald. 1972. Trophic analyses of an estuarine mangrove community. Bull. Mar. Sci. 22(3):671–738.

Odum, W.E. and E.J. Heald. 1975. Mangrove forests and aquatic productivity. In: A.D. Hasler, ed., Coupling of land and water systems, pp 129–136, Ecological Studies, vol. 10, Springer-Verlag, New York.

Snedaker, S.C. 1978. Mangroves: their value and perpetuation. Nature & Resources 14(3):6–13.

Tabb, D. and R.B. Manning. 1962. A checklist of the flora and fauna of northern Florida Bay and adjacent brackish waters of the Florida mainland collected during the period July, 1957 through September 1960. Bull. Mar. Sci. Gulf Carib. 11(4): 552–649.

Appendix 1 – Station list

BBC 1676. Australia, Northern Territory, E. Vernon Is., creek in mangrove swamp emptying into backwater; 31 May 1979. 18 families, 27 genera, 29 species, 360 specimens.

BBC 1677. Australia, Northern Territory, Croker Is., Cape Croker, inlet behind barrier reef that is covered at high tide;

1 June 1979. 16 fam., 19 gen., 22 spp., 350 spec.

BBC 1686. Papua New Guinea, Port Moresby, Fairfax Harbour, across from Dept. Primary Industry 'Pearl Camp', small mangrove stand on exposed sandy point: 12 June 1979. 16 fam., 18 gen., 20 spp., 181 spec.

BBC 1687. Papua New Guinea, Port Moresby, Fairfax Harbour, mud-bottomed mangrove swamp behind Dept. Primary Industry 'Pearl Camp'; 12 June 1979. 16 fam., 18 gen., 34 spp., 390 spec.

BBC 1696. Papua New Guinea, Cape Ward Hunt, sand overlain with mud; 17 June 1979. 24 fam., 27 gen., 42 spp., 382 spec.

BBC 1701. Papua New Guinea, Umboi I, Higgins Point, stream draining mangrove swamp; 18 June 1979. 22 fam., 33 gen., 35 spp., 371 spec.

BBC 1711. Papua New Guinea, Muschu I., Kairiru swamp, mud buttom, sea grasses offshore; 3° 23.1′ S, 143° 33.2′ E; 21 June 1979. 14 fam., 20 gen., 25 spp., 396 spec.

BBC 1714. Papua New Guinea, Wewak, Cape Wom, Nambogoru Cr., sand-bottomed channels; 3° 32.5′ S, 143° 36′ E; 23 June 1979. 25 fam., 37 gen., 57 spp., 818 spec.

BBC 1722. Irian Jaya, mainland E of Palau Yapen; small patch of mangrove along shore of inlet; 1° 30.8′ S, 137° 20.7′ E; 29 June 1979. 17 fam., 22 gen., 27 spp., 536 spec.

BBC 1725. Irian Jaya, Schouten Is., Biak I., Korido Bay; mangrove over coralline rock; 0° 50.1′ S, 135° 30.3′ E; 30 June 1979. 9 fam., 12 gen., 12 spp., 59 spec.

BBC 1730. Irian Jaya, Batanta I., Marchesa Bay; 0° 48.2′ S, 130° 52.8′ E; 2 July 1979. 13 fam., 23 gen., 29 spp., 298 spec.

BBC 1734. Irian Jaya, Misool I., 3 July 1979. 13 fam., 19 gen., 19 spp., 278 spec.

BBC 1737. Irian Jaya, mainland in lee of Semai I.; mangrove swamp along small river; 3° 5.0′ S, 132° 29.8′ E; 4 July 1979. 19 fam., 36 gen., 39 spp., 430 spec.

BBC 1743. Irian Jaya, Aru Arch., Boear I. off Kola I.; 5° 27′ S, 134° 27′ E; 6 July 1979. 18 fam., 30 gen., 37 spp., 615 spec.

Appendix 2. List of species

Acanthuridae

Acanthurus xanthopterus Valenciennes, 1835
1722 – (6, 26–184).

Ambassidae

Ambassis buruensis Bleeker, 1856
1714 – (37, 35–45).

Ambassis commersoni Cuvier, 1828
1686 – (13, 7–14); 1687 – (12, 18–46); 1701 – (61); 1737 – (2, 11–15).

Ambassis interruptus Bleeker, 1856
1714 – (238, 45–60).

Ambassis nalua (Hamilton-Buchanan, 1822)
1677 – (2, 38–39).

Priopidichthys gymnocephalus (Lacepède, 1802)
1676B – (20, 18–36); 1677 – (209, 17–29); 1722 – (149, 32–44).

Antennariidae

Antennarius dorehensis Bleeker, 1859
1701 – (1, 17).

Apogonidae

Apogon amboinensis Bleeker, 1858
1701 – (9, 12–29).

Apogon ceramensis Bleeker, 1852 = *Apogon lateralis* Valenciennes, 1832
1686 – (5, 12–20); 1687 – (5, 21–30); 1696 – (5, 31–57); 1711 – (74, 20–30); 1714 – (156, 45–105); 1730 – (2, 27–38); 1734 (1, 30); 1737 – (6, 11–33).

Apogon darnleyensis (Alleyne and Macleay, 1877)
1676B – (6, 27–72).

Apogon hyalosoma Bleeker, 1852
1676B – (8, 90–125); 1696 – (5, 18–88); 1714 – (63, 20–110); 1737 – (7, 27–37).

Apogon cf. *leptacanthus* Bleeker, 1856
1677 – (1, 46).

Apogon rueppelli Günther, 1859
1676B – (68, 22–71); 1677 – (3, 57–61).

Apogon sangiensis Bleeker, 1857
1687 – (15, 26–43).

Archamia buruensis (Bleeker, 1856)
1696 – (9, 59–63).

Fowleria cf. *aurita* (Valenciennes, 1831)
1676B – (1, 41).

Fowleria cf. *marmorata* (Alleyne & Macleay, 1877)
1687 – (2, 18–22); 1711 – (7, 9–37); 1714 – (3, 35–60); 1725 – (13, 13–51); 1734 – (1, 24); 1737 – (2, 31–33); 1743 – (10, 10–34).

Pseudamia cf. *polystigma* (Bleeker, 1859–60)
1711 – (1, 20); 1730 – (1, 31).

Sphaeramia orbicularis (Cuvier, 1828)
1686 – (13, 10–64); 1687 – seen; 1711 – (5, 14–61); 1714 – (75, 16–80); 1725 – (4, 18–42); 1730 – (2, 8–20).

Atherinidae

Atherinomorus endrachtensis (Quoy and Gaimard, 1824)
1686 – (57, 14–64); 1687 – (106, 7–68); 1701 – (54, 17–29).

Hypoatherina ovalaua (Herre, 1935)
1677 – (31, 16–37).

Balistidae

Abalistes stellaris (Bloch and Schneider, 1801)
1725 – (2, 24–25).

Batrachoididae

Halophryne diemensis (LeSueur, 1824)
1676B – (4, 51–161).

Belonidae

Strongylura strongylura (van Hasselt, 1823)
1676B – (12, 170–245 mm BL).
Strongylura urvillii (Valenciennes, 1846)
1701 – (2, 200–240 mm BL; 11, 87–134).

Blenniidae

Istiblennius sp.
1725 – (1, 82)
Meiacanthus anema (Bleeker, 1852)
1701 – (1, 29); 1737 – (9, 8–31)
Omobranchus elongatus (Peters, 1855)
1686 – (13); 1711 – (3); 1730 – (6).
Omobranchus ferox (Herre, 1927)
1676B – (1, 47); 1696 – (1).
Omox lupus Springer, 1981
1696 – (2, 28–30); 1737 – (1, 26).
Omox biporos Springer, 1972
1677 – (2, 27–29); 1686 – (11); 1696 – (3); 1701 – (4, 18–30); 1711 – (9, 24–35); 1730 – (26, 10–35); 1734 – (8, 15–39); 1737 – (3, 14–22).
Petroscirtes thepassi Bleeker, 1853
1701 – (3, 23–40).
Salarias spaldingi Macleay, 1878
1676B – (4, 28–62).

Bothidae

Bothus pantherinus (Rüppell, 1828)
1714 – (1, 61).
Pseudorhombus jenynsii (Bleeker, 1855)
1743 – (5, 43–53).

Callionymidae

Callionymus spp.
1686 – (2, 15–21); 1687 – (3, 16–26); 1700 – (8, 16–28); 1743 – (3, 13–32).

Carangidae

Caranx melampygus Cuvier, 1833
1701 – (2, 85–86); 1714 – (10, 67–82); 1722 – (6, 67–230).
Caranx sexfasciatus Quoy and Gaimard, 1824
1696 – (1, 115); 1701 – (4, 62–117); 1711 – (1, 100); 1714 – (11, 66–114); 1722 – (32, 62–102); 1727 –

Clupeidae

Anodontostoma cf. *selangkat* (Bleeker, 1852)
1722 – (21, 44–49).
Sardinella melanura (Cuvier, 1829)
1696 – (2, 90–96); 1701 (13, 82–89); 1714 – (1, 93); 1722 – (13, 50–70).
Spratelloides delicatalus (Bennett, 1831)
1686 – (26, 16–37); 1687 – (1, 30).

Cynoglossidae

Cynoglossus puncticeps (Richardson, 1846)
1696 – (3, 16–33).

Eleotridae

Bostrichthys sinensis (Lacepède, 1802)
1730 – (3, 34–58).
Butis cf. *amboinensis* (Bleeker, 1853)
1701 – (3, 17–27).
Butis butis (Hamilton-Buchanan, 1822)
1696 – (9, 15–71); 1714 – (6, 29–68); 1722 – (11, 20–57); 1730 – (6, 15–72); 1737 – (17, 13–99).
Butis sp.
1737 – (1, 9).
Eleotris cf. *fusca* (Bloch and Schneider, 1801)
1696 – (7, 15–48); 1701 – (1, 48); 1711 – (23, 12–40); 1714 – (2, 35–39); 1730 – (5, 12–32); 1737 – (9, 11–41).
Ophiocara porocephala (Valenciennes, 1837)
1701 – (1, 84); 1714 – (2, 73–92).
Oxyeleotris cf. *lineolatus* Steindachner, 1867
1714 – (1, 53).
Prionobutis koilomatodon Bleeker, 1849
1722 – (5, 14–36).

Engraulidae

Stolephorus commersonii Lacepède, 1803
1696 – (175, 24–75); 1737 – (1, 60).
Stolephorus indicus (van Hasselt, 1823)
1687 – (2, 39–44); 1701 – (12, 22–62).
Thrissina baelama (Forsskål, 1775)
1711 – (163, 21–24); 1734 – (191, 39–75).

Gerreidae

Gerres abbreviatus Bleeker, 1850
1677 – (10, 36–117); 1743 – (73, 17–70).
Gerres argyreus (Bloch and Schneider, 1801)
1743 – (2, 27–58).
Gerres filamentosus Cuvier, 1829
1696 – (4, 32–101); 1743 – (1, 98).
Gerres kapas Bleeker, 1851
1677 – (1, 54).
Gerres macracanthus Bleeker, 1854
1714 – (1, 83).
Gerres macrosoma Bleeker, 1854
1676B – (1, 24); 1677 – (47, 26–110); 1743 – (3, 50–54).

Gobiidae

Acentrogobius ornatus (Rüppell, 1828)
1730 – (1, 32).
Acentrogobius puntang (Bleeker, 1852)
1696 – (1, 24); 1711 – (2, 31–58); 1714 – (29, 37–108); 1730 – (2, 34–67).
Acentrogobius cf. *reichei* (Bleeker, 1853)

1677 – (11, 16–23); 1687 – (13, 9–43); 1714 – (6, 13–20); 1743 – (16, 11–23).

Acentrogobius cf. *triangularis* Weber, 1909
1676B – (36, 11–40); 1687 – (1, 8); 1696 – (1, 14); 1714 – (1, 12); 1730 – (7, 11–33); 1737 – (9,17–25); 1743 – (73, 7–36).

Acentrogobius sp. No. 1
1714 – (1, 73); 1730 – (32, 9–52).

Acentrogobius sp. No. 2
1687 – (3, 16–23).

?*Acentrogobius* sp. No. 3
1696 – (1, 17).

?*Acentrogobius* sp. No. 4
1714 – (5, 15–16).

?*Acentrogobius* sp. No. 5
1687 – (5, 17–36).

Amblygobius cf. *bynoensis* (Richardson, 1844–1848)
1676B – (62, 25–71); 1743 – (1, 51).

Amblygobius linki Herre, 1927
1701 – (1, 30); 1711 – (1, 21).

Amblygobius cf. *perpusillus* (Seale, 1909)
1687 – (20, 21–55).

Apocryptodon taylori Herre, 1927
1730 – (1, 18).

Bathygobius cf. *fuscus* (Rüppell, 1828)
1701 – (1, 23); 1722 – (2, 18–20); 1737 – (1, 33).

Bathygobius sp.
1743 – (2, 20–27).

Brachygobius sp.
1701 – (7, 14–23); 1714 – (7, 14–21); 1737 – (2, 15).

Callogobius cf. *hasselti* (Bleeker, 1851)
1687 – (1, 22); 1711 – (1, 18); 1730 – (19, 14–37); 1737 – (2, 14–17); 1743 – (2, 13–14).

Callogobius n.sp.?
1737 – (2, 16–17).

Cristatogobius lophius Herre, 1927
1711 – (7, 14–21); 1737 – (1, 21).

?*Cryptocentroides* sp.
1687 – (2, 34–39).

Cryptocentrus cf. *callopterus* H.M. Smith, 1945
1687 – (1, 64).

Cryptocentrus cf. *russus* (Cantor, 1849)
1743 – (11, 13–40).

Cryptocentrus sp. No. 1
1687 – (65, 12–53).

Cryptocentrus sp. No. 2
1725 – (1, 35).

Glossogobius biocellatus (Valenciennes, 1837)
1677 – (1, 33); 1696 – (15, 12–73); 1701 – (2, 23–26); 1714 – (6, 17–44); 1737 – (1, 48).

Glossogobius celebius (Valenciennes, 1837)
1696 – (1, 46).

Gnatholepis knighti Jordan and Evermann, 1903
1676B – (15, 14–25); 1701 – (1, 20).

Gobiopsis macrostoma Steindachner, 1861
1722 – (2, 33–36).

Mahidolia mystacina (Valenciennes, 1837)
1687 – (3, 14–31); 1714 – (3, 26–33); 1734 – (1, 49).

Mars caeruleomaculatus Herre, 1933
1676B – (1, 32).

Oligolepis moloanus (Herre, 1927)
1676B – (1, 41); 1687 – (4, 9–16); 1711 – (9, 15–31); 1730 – (87, 11–37); 1734 – (2, 10–39); 1737 – (208, 14–36).

?*Oplopomus* sp.
1696 – (2, 27–28).

Oxyurichthys ophthalmonema Bleeker, 1856–1857
1711 – (28, 7–44); 1734 – (5, 17–51); 1737 – (23, 17–44).

Pandaka cf. *pygmaea* Herre, 1927
1687 – (1, 7); 1701 – (7, 10–12); 1714 – (2, 10); 1725 – (3, 8–9); 1730 – (9, 9–11); 1737 – (55, 7–14); 1743 – (1, 11).

Parapocryptes serperaster (Richardson, 1845)
1737 – (2, 16–43).

Parioglossus sp.
1696 – (1, 15); 1711 – (1, 17); 1725 – (22, 9–17); 1730 – (20, 13–21); 1734 – (4, 15–27); 1737 – (9, 9–24).

Pleurosicya sp.
1701 – (1, 16); 1734 – (2, 21–24).

?*Rhinogobius suluensis* Herre, 1927
1711 – (7, 12–24); 1730 – (1, 29); 1737 – (17, 24–34).

?*Rhinogobius* sp.
1722 – (3, 15–32).

Stigmatogobius sp. No. 1
1730 – (2, 14–15).

Stigmatogobius sp. No. 2
1696 – (2, 12).

?*Stigmatogobius* sp. No. 3
1696 – (5, 17–40).

?*Stigmatogobius* sp. No. 4
1730 – (14, 9–18).

?*Stigmatogobius* sp. No. 5
1737 – (1, 17).

Valenciennea longipinnis (Bennett, 1839)
1743 – (12, 36–65).

Yongeichthys cf. *nebulosus* (Forsskål, 1775)
1676B – (2, 24–45); 1687 – (22, 14–38); 1711 – (12, 16–26); 1714 – (3, 25–36); 1737 – (4, 31–54); 1743 – (8, 24–35).

Yongeichthys sp.
1696 – (8, 14–22); 1701 – (4, 15–29).

Zonogobius semidoliatus (Valenciennes, 1837)
1722 – (2, 12–16).

?new genus and species
1687 – (2, 18–21).

Hemiramphidae

Zenarchopterus buffonis (Valenciennes, 1846)
1701 – (9, 124–154).

Zenarchopterus dispar (Valenciennes, 1846)
1696 – (16, 13–113).

Zenarchopterus dunckeri Mohr, 1926
1686 – (20, 11–52); 1687 – (52, 25–87); 1701 – (54, 27–97); 1711 – (17, 17–78); 1714 – (27, 40–90).

Zenarchopterus gilli Smith, 1945
1676B – (20, 39–86); 1677 – (50, 26–69); 1722 – (3, 43–71); 1734 – (3, 53–73); 1743 – (3, 71–73).

Zenarchopterus sp.
1725 – (4, 28–57); 1730 – (5, 17–74).

Kraemeridae

Gobiotrichinotus sp.
1701 – (1, 30).

Kuhliidae

Kuhlia cf. *marginata* (Cuvier, 1829)
1686 – (1, 22).

Kurtidae

Kurtus gulliveri Castelnau, 1878
1686 – (3, 13–17).

Labridae

Halichoeres binotopsis (Bleeker, 1849)
1725 – (1, 34).
Halichoeres dussumieri (Valenciennes, 1839)
1722 – (3, 37–68).
Halichoeres sp.
1722 – (4, 20–44).

Latilidae

Lates calcarifer (Bloch, 1790)
1676B – (2, 266–273).

Leiognathidae

Gazza minuta (Bloch, 1797)
1687 – (5, 20–33); 1696 – (1, 75); 1734 – (6, 33–49); 1737 – (1, 54).
Leiognathus equulus (Forsskål, 1775)
1696 – (11, 45–66).
Leiognathus fasciatus (Lacepède, 1803)
1687 – (2, 63–64); 1711 – (1, 72); 1734 – (1, 49); 1743 – (54, 44–83).

Lutjanidae

Lutjanus argentimaculatus (Forsskål, 1775)
1676B – (2, 204–212); 1714 – (6, 93–139).
Lutjanus ehrenbergi (Peters, 1869)
1711 – (2, 27–31); 1714 – (14, 25–82); 1725 – (2, 23–47).
Lutjanus fulviflamma (Forsskål, 1775)
1722 – (1, 39).
Lutjanus fulvus (Schneider, 1801)
1696 – (1, 33); 1714 – (7, 25–101); 1722 – (1, 47).
Lutjanus fuscescens (Valenciennes, 1830)
1696 – (2, 60–87); 1714 – (1, 88).
Lutjanus russelli (Bleeker, 1849)
1676B – (21, 38–98); 1677 – (8, 40–88); 1743 – (27, 35–76).
Lutjanus nishikawae Smith and Pope, 1906
1696 – (1, 45); 1714 – (1, 94); 1722 – (1, 127); 1737 – (2, 53–70).

Monacanthidae

Monacanthus chinensis (Osbeck, 1765)
1743 – (1, 42).

Monodactylidae

Monodactylus argenteus (Linnaeus, 1758)
1696 – (1, 31); 1737 – (2, 19–21).

Moringuidae

Moringua sp.
1714 (1, 160); 1730 – (1, 350); 1734 – (1, 220).

Mugilidae

Cestraeus plicatilis Valenciennes, 1836
1686 – (1, 23).
Liza subviridis (Valenciennes, 1836)
1677 – (1, 83); 1714 – (1, 128); 1743 – (1, 92).
Liza vaigiensis (Quoy and Gaimard, 1824)
1701 – (3, 14–18); 1722 – (9, 10–22).
Valamugil buchanani (Bleeker, 1853)
1677 – (2, 13–124); 1714 (4, 71–109); 1743 – (52, 13–?).
Valamugil seheli (Forsskål, 1775)
1696 – (1, 37).

Mullidae

Mulloidichthys samoensis (Günther, 1874)
1714 – (1, 130).
Parupeneus indicus (Shaw, 1803)
1687 – (2, 92–95).
Parupeneus sp.
1722 – (5, 34–36); 1737 – (1, 35).
Upeneus tragula Richardson, 1846
1687 – (2, 110).

Muraenidae

Echidna rhodochilus Bleeker, 1863
1696 – (1, 19); 1700–1701 (1, 53); 1714 (17, 105–350).

Notograptidae

Notograptus guttatus Günther, 1867
1676B – (6, 90–130).

Ophichthidae

Bascanichthys sp.
1696 (8, 305–440); 1714 (5, 230–425).
Cirrhimuraena (*Jenkinsiella*) sp.
1714 – (4, 94–210).
Lamnostoma orientalis (McClelland, 1845)
1701 – (77, 76–210); 1714 – (9, 185–330).
Muraenichthys (*Muraenichthys*) *gymnopterus* (Bleeker, 1853)
1730 – (1, 230); 1734 – (1, 190); 1737 – (1, 69); 1743 – (1, ?).

Muraenichthys cf. *laticaudata* (Ogilby, 1897)
1714 - (2, 107–135).
Muraenichthys (*Scolecenchelys*) *macropterus* Bleeker, 1857
1676B - (11, 106–170); 1677 - (1, 156); 1743 - (7, 66–195).
Ophichthus (*Centrurophis*) *cephalozona* (Bleeker, 1864)
1730 - (1, 305).
Ophichthus sp.
1714 - (1, 105).

Opistognathidae

Opistognathus darwiniensis (Macleay, 1870)
1676B - (14, 64–133).

Pentopodidae

Pentapodus cf. *caninus* (Cuvier, 1830)
1686 - (2, 93–146); 1687 - (22, 34–95).

Periophthalmidae

Periophthalmus sp.
1714 - (1, 38); 1730 - (31, ca. 20–70); 1737 (4, ca. 20–50).

Platycephalidae

Cociella crocodila (Tilesius, 1812)
1743 - (2, 94–103).
Cymbacephalus nematophthalmus (Günther, 1860)
1743 - (11, 76–151).
Platycephalus indicus (Linnaeus, 1758)
1743 - (2, 116–120).

Plectorhynchidae

Pseudopristipoma nigra (Cuvier, 1830)
1696 - (1, 54).

Plotosidae

Paraplotosus albilabris (Valenciennes, 1840)
1722 - (3, 46–69).

Pomacentridae

Abudefduf septemfasciatus (Cuvier, 1830)
1686 - (4, 15–17).
Dischistodus fasciatus (Cuvier, 1830)
1676B - (18, 24–80).
Neopomacentrus taeniurus (Bleeker, 1856)
1696 - (1, 29); 1701 - (8, 20–52); 1714 - (4, 17–44); 1722 - (92, 23–52 + 157, 11–13 = 249).
Pomacentrus bankanensis Bleeker, 1853 complex
1714 - (3, 15–42).
Pomacentrus emarginatus Cuvier, 1830
1714 - (6, 65–74).
Pomacentrus tripunctatus Cuvier, 1830
1676B - (1, 18); 1711 - (1, 11).
Pomacentrus sp.
1737 - (1, 14).

Pseudomugilidae

Pseudomugil sp. nov. No. 1 Ivantsoff
1676B - (3, 10–15); 1696 - (1, 32).
Pseudomugil sp. nov. No. 2 Ivanstoff
1730 - (8, 17–30); 1734 - (47, 12–27).

Scatophagidae

Scatophagus argus (Linnaeus, 1760)
1676B - (2, 128–133); 1714 - (3, 70–80).
Solenotoca multifasciata (Richardson, 1844–8)
1677 - (1, 93).

Scolopsidae

Scolopsis cf. *dubiosus* Weber, 1913 or *personatus* (Cuvier, 1830)
1687 - (33, 33–47.5).

Scorpaenidae

Scorpaenodes guamensis (Quoy and Gaimard, 1824)
1714 - (1, 60).
Synanceia horrida (Linnaeus, 1758)
1714 - (1, 190).
Tetraroge barbata (Cuvier, 1829)
1696 - (6, 17–57); 1714 - (2, 27–58); 1737 - (2, 27–47).

Serranidae

Centrogenys vaigiensis (Quoy and Gaimard, 1824)
1676B - (6, 22–67); 1743 - (2, 50–51).
Epinephelus cf. *caeruleopunctatus* (Bloch, 1790)
1722 - (3, 60–285).
Epinephelus corallicola (Valenciennes, 1828)
1676B - (5, 136–226).
Epinephelus fario (Thunberg, 1793)
1714 - (4, 94–180).
Epinephelus slacksmithi (Whitley, 1959)
1714 - (1, 40).
Epinephelus tauvina (Forsskål, 1775)
1676B - (13, 56–156).
Epinephelus sp.
1686 - (3, 10–12); 1722 - (1, 230); 1743 - (2, 48–70).

Siganidae

Siganus canaliculatus (Park, 1797)
1714 - (1, 28); 1743 - (3, 51–74).
Siganus guttatus (Bloch, 1787)
1737 - (1, 44); 1743 - (3, 72–77).
Siganus lineatus (Valenciennes, 1835)
1676B - (1, 41); 1677 - (2, 51–86); 1687 - (1, 26); 1714 - (2, 52–82).
Siganus sp. nov. Woodland
1711 - (4, 26–27); 1714 - (5, 77–82).

Sillaginidae

Sillago maculata (Quoy and Gaimard, 1877)
1743 – (5, 54–94).
Sillago sihama (Forsskål, 1775)
1701 – (2, 37–44); 1722 – (1, 42).

Soleidae

Pardachirus poropterus (Bleeker, 1851)
1696 – (40, 9–28); 1700 + 1701 (5, 19–22).
Phyllichthys sclerolepis (Macleay, 1878)
1676B – (1, 40).

Sparidae

Acanthopagrus berda (Forsskål, 1775)
1686 – (1, 164).

Sphyraenidae

Sphyraena barracuda (Walbaum, 1792)
1677 – (7, 43–182); 1687 (1, 112); 1711 – (2, 61–68); 1737 (1, 131); 1743 (9, 91–144).

Syngnathidae

Hippichthys spicifer (Rüppell, 1838)
1686 – (2, 40–102); 1696 – (1, 84); 1701 – (2, 91–95); 1714 – (4, 110–123); 1725 – (1, 80); 1730 – (1, 105).

Synodontidae

Saurida gracilis (Quoy and Gaimard, 1824)
1687 – (6, 58–143).

Taenioididae

?Brachyamblyopus sp.
1734 – (1, 29).
Taenioides gracilis (Valenciennes, 1837)
1677 – (1, 139); 1737 – (1, 73).

Teraponidae

Amniataba caudavittatus (Richardson, 1844)
1677 – (6, 54–109).
Pelates quadrilineatus (Bloch, 1790)
1743 – (4, 38–67).
Terapon jarbua (Forsskål, 1775)
1696 – (2, 86–97); 1730 – (2, 13–16); 1734 – (1, 12); 1743 – (12, 33–115).

Tetraodontidae

Arothron immaculatus (Bloch and Schneider, 1801)
1677 – (3, 31–44); 1686 – (1, 104); 1687 (5, 41–84); 1722 – (2, 41–63); 1743 – (4, 40–55).
Arothron reticularis (Bloch and Schneider, 1801)
1696 – (2, 100–110); 1743 – (1, 64).
Chelonodon cf. *laticeps* (Smith, 1948)
1701 (1); 1714 (1, 80).
Chelonodon patoca (Hamilton – Buchanan, 1822)
1696 – (9, 21–117); 1734 – (1, 20).
Chelonodon sp. ? *patoca*
1722 – (7, 13–41); 1737 – (2, 14–18).

Toxotidae

Toxotes jaculator (Pallas, 1766)
1696 – (14, 23–127); 1701 – (2, 28–79); 1711 – (1, 152); 1714 – (8, 60–70); 1722 – (1, 128) + sev. seen; 1730 – (2, 19–68); 1734 – (1, 90) + sev. ads. seen; 1737 – (11, 18–47).

CHAPTER 11

The faunal communities of Australian mangroves

P.A. HUTCHINGS and H.F. RECHER

The Australian Museum, Sydney, N.S.W., Australia

Abstract. Mangroves occur in Australia from Corner Inlet, Victoria (38° 45′ S) to Cape York, Queensland (19° S). South of Merimbula (36° 50′ S) on the east coast and south of Carnarvon (25° S) on the west coast, only one species *Avicennia marina* occurs sporadically. North of these latitudes the number of mangrove species increases, and in Cape York over 30 species occur. A synopsis of the diverse fauna is given and an attempt is made to describe the fauna according to the habitat occupied. The necessity for detailed information on the life cycles and feeding strategies of the dominant faunal organisms is stressed, before estimates of secondary productivity can be made. These are necessary for assessments of the importance of Australian mangrove communities and the inherent dangers of extrapolating data from one geographical region to another.

Introduction

Mangroves occur in Australia from the upper shores of Spencer and St Vincent gulfs, South Australia, as isolated pockets in Victoria, with the most southerly known occurrence of mangroves in Australia at Corner Inlet (38°45′S) and as extensive areas along the coasts of New South Wales, Queensland, and Northern Territory. In Western Australia extensive mangroves occur north of Carnarvon (25° S); southwards, distribution is sporadic. *Avicennia marina* is the most widespread mangrove in Australia and is the only species occurring south of Merimbula (36° 50′ S) on the east coast and south of Carnarvon (25°) on the west coast. North of these latitudes the diversity of the flora gradually increases and in far north Queensland over 30 species occur (Saenger *et al.* 1977).

Precise figures on the area occupied by mangroves in Australia are not available, but mangrove communities are well developed in tropical regions and often form extensive stands with connecting corridors. By comparison, stands are often discrete in subtropical and temperate regions. The plant communities have been relatively well studied (Chapman 1975; Macnae 1966; Saenger *et al.* 1977). In contrast, the faunal components of the mangrove communities have been relatively neglected. This neglect is most pronounced in tropical areas, especially in north west Western Australia and Northern Territory. For an understanding of the mangrove ecosystem we need information on the distribution and abundance of the fauna, life cycles and feeding strategies of the dominant organisms. This will enable food webs to be constructed, and estimates of secondary productivity to be made. Only then can the widely quoted statement, that mangroves are important in Australian estuarine ecosystems, be substantiated. Currently, this statement is largely based on limited overseas data (Odum and Heald 1975) obtained from relatively simple mangrove systems. Such data may hold for

Tasks for vegetation science, Vol. 8 ed. by H.J. Teas
 ISBN 978-90-481-8526-9

subtropical mangrove regions in Australia, but the situation may be far more complex in our tropical mangroves. For example, Bunt and his co-workers are suggesting that terrestrial sources of nutrients play a major role in the mangrove system of Hinchinbrook, North Queensland (Boto 1982).

Within this paper, we have attempted to summarise the existing data on mangrove fauna which illustrates its great diversity. A more detailed review of Australian mangrove fauna together with marine invertebrate species lists and distribution patterns appears in (Hutchings and Recher, 1982).

Mangroves represent a transitional zone between terrestrial and marine ecosystems and this is reflected in their fauna. In addition, in upstream mangrove areas some freshwater species may occur. For many terrestrial species mangroves provide an extension to their normal forest habitat, a corridor linking two habitats, or a refuge in an otherwise unsuitable environment. For example, where mangroves abut onto arid zones in north west Australia and South Australia, such an extension is possible for species living in the mangrove canopy, which resembles other forest canopies, so that no special adaptations are required for animals to live in it. Animals feeding on the floor of the mangrove forest, however, may require adaptations and will be affected by the daily tidal inundation. Similarly for marine animals living on intertidal muddy shores, mangroves provide an extension of habitat plus additional habitats such as surfaces for encrusting organisms, and browsers, dead wood, etc. Mangroves also provide protection from desiccation and predation, resulting in a greater diversity of marine fauna in comparison to bare muddy shores. This will be most marked in tropical areas where desiccation problems are more pronounced.

Although for convenience we discuss the fauna in terms of arboreal and ground-living animals, this is somewhat artificial as many animals move between the two habitats. In cases where the data on a particular group are well documented, just the reference has been cited, as one of the purposes of this synoptic paper is to discuss the inter-relationships and roles of the various groups in the mangrove ecosystem.

Arboreal faunal communities

The arboreal fauna of mangroves is diverse and abundant. However, for the majority, mangroves are secondary habitat with nearby terrestrial forests being primary habitat. This may explain the relatively few endemic species occurring.

Birds are the most conspicuous component of this community. Saenger *et al.* (1977), for example, record over 200 species from Australian mangrove habitats, but fewer than half occur regularly (Schodde *et al.*, 1982) Of these, 14 species are virtually restricted to mangroves, 12 species use mangroves as a primary habitat in part of their range, and 60 species use mangroves regularly throughout the year or in particular seasons (Schodde *et al.*, 1982). This is a rich mangrove bird fauna compared to other parts of the world (Macnae 1968). Endemic bird species are absent from temperate mangroves and first appear around the latitude of Sydney (34° S) (e.g. the mangrove heron *Butorides striatus*), and the number increases with decreasing latitude. For example, on the Cape York Peninsula (19° S) 19 species of birds are considered mangrove specialists and seven of these are endemic to mangrove habitats (Schodde *et al.*, 1982). None of these species show major morphological adaptations to the mangrove environment. However, there is a tendency for bills to be longer than in non-mangrove congeners (Schodde *et al.*, 1982).

Birds which regularly use mangroves include aquatic and wading birds (herons, ibis, spoonbills) which forage among the trees and often use them as high tide roosts. Other species nest in mangroves. The Lesser Noddy *Anous tenuirostris* is restricted to nesting in mangroves of the Albrolhos Islands, Western Australia. Other species use mangroves seasonally, such as honey eaters which commonly visit mangroves for nectar when blossom is abundant. Other species utilise mangroves and associated wetlands (salt marsh and seagrass beds) for feeding and roosting during migration. Such areas in temperate and sub-tropical areas assume critical importance to these species, as here wetlands occur as isolated discrete stands. In contrast, in tropical regions, mangroves and associated wetlands often occur as wide bands along estuaries. Mangroves are also used by some inland bird species as refuges during drought conditions. Many of these birds are feeding on the marine invertebrates, but others are insect or nectar feeders.

A large number of species of mammals have been recorded from Australian mangroves. Many of these are casual sightings but a few such as the rodents (*Rattus colletti*, *Hydomys chrysogaster*, *Mus musculus*, *Melomys* spp., *Mesembriomys* spp. and *Conilurus* spp.), bandicoots (*Perameles*, *Isoodon*), the northern brush-tailed possum (*Trichosurus arnhemensis*), regularly utilise mangroves as secondary habitat (Frith 1973). Their use may be seasonal, such as the flying foxes *Pteropus poliocephalus* and *P. alecto* which come into mangroves to feed on nectar or camp. Some tree roosting species of bats (*Tadarida planiceps* and *P. loriae*) sometimes occur. Only one mammal utilises mangroves as primary habitat, the rat *Xeromys myoides* in Queensland and Northern Territory (Magnusson *et al.* 1976). *Xeromys* feeds on crabs and is at least partially arboreal.

Mammals are generally more prevalent in tropical mangroves than in temperate ones. This may just be a reflection of the greater area occupied by mangroves in the tropics, and/or the lack of other suitable habitats in some coastal regions.

Reptiles too are more common in tropical mangroves than in temperate ones. Like the mammals, many use mangroves only as secondary habitat (Cogger 1979). The pythons (*Liasis fuscus* and *L. olivaceus*) are attracted to the mangroves by large camps of flying foxes on which they feed. A few species such as the mangrove monitor *Varanus indicus*, the file snake *Acrochordus granulatus*, the bockadam *Cerberus rhynchops*, the white-bellied mangrove snake *Fordonia leucobalia*, and the mangrove snake *Myron richardsoni* occur regularly in tropical mangroves. These animals move freely over the forest floor at low tide, some feeding on crabs or the mammals.

Amphibians are rarely seen in Australian mangroves.

The majority of the insects are arboreal although some such as the biting midges and mosquitoes lay their eggs in semi-permanent pools of water. The insect fauna is incredibly diverse but has been little studied. Most of these studies are purely taxonomic. The ecological studies are mainly restricted to potential pests such as termites (Hutchings and Recher, 1982) or groups which may be involved in the transmission of disease (mosquitoes, for review, see Lee *et al.* 1980: and biting midges, Ceratopogonidae, for review see Debenham 1979).

Mangrove insects show varying degrees of dependence on mangroves, from casual visitors such as grasshoppers, to some butterflies which commonly feed on mangroves (Common and Waterhouse 1972). Other insects have larval stages dependent on restricted species of mangroves, and some insects like the Diptera *Melanagromyza avicenniae* feed exclusively on *Avicennia marina* (Hutchings and Recher 1974). Presumably much of the insect fauna is seasonal and may be subjected to periodic wipe-outs during flooding in the wet season.

The ant fauna is abundant but almost totally neglected. Most are arboreal but a few species in tropical areas live intertidally. The ant *Pheidole myrmecodiae* is restricted to the ant house plants *Myrmecodia antoinii* and *Hydnophytum formicarum* in north east Queensland.

The role of insects in Australian mangrove communities is poorly understood, but some are important in the breakdown of dead wood which tends to accummulate in mangroves. Perhaps insect leaf damage facilitates the dropping and subsequent breakdown of leaf litter (Onuf *et al.* 1977). Insects are important sources of food for the reptiles and birds and in pollinating some mangrove species. Moths pollinate species whose flowers open only at night.

Another very visible component of this community is the spider. However, so far no species restricted to mangroves have been recorded. The dominant groups are the orb web weavers (*Tetragnatha* and *Eriophora* spp.), the wolf spiders (*Geolycosa* sp.), and the allied hunting spiders of the family Pisauridae (*Dolomedes* sp.). The orb weavers feed by catching flying prey and the other two groups are opportunistic hunters on the exposed mud at low tide.

The gastropod *Littorina scabra* is commonly seen

grazing on the surfaces of leaves throughout Australian mangroves. Several species of crabs (*Sesarma* spp. and *Macrophthalmus* spp.) regularly move between the forest floor and the lower levels of the canopy.

Ground-living faunal communities

Far more data are available on these communities which can therefore be discussed in terms of habitat. However, this discussion is restricted to macrofauna as virtually no information is available on the microfauna of Australian mangroves (Dacraemes and Coomans 1978).

Although this fauna has been assigned to particular habitats, these are not rigid categories, as much of the fauna regularly moves between habitats. Faunal lists of the molluscs, crustaceans, and polychaetes occurring in Australian mangroves are given by Hutchings and Recher (1982).

a. Infauna

The infauna is restricted to the upper layers of the mud; below this the dense root mat of the mangroves forms an almost impenetrable barrier. The majority of this fauna lives in burrows or is tubicolous. Distinct zonation of this fauna occurs. The soft-bodied groups such as polychaetes sipunculans, and nemerteans, and the filter feeding bivalves tend to be restricted to the seaward margins of the mangroves and along the banks of tidal creeks. In tropical mangroves, fish belonging to the gobiid subfamily Oxcidercinae also occur in these frontal areas, and are well represented by five species in two genera, *Periophthalmus* and *Periophthalmodon*. Nursall (1981) has recently discussed the zonation of these mudskippers in the Townsville, Queensland region. Less visible than these mudskippers, are members of the genus *Boleophthalmus* and *Scartelaos* (S.F. Oxcidercinae), which burrow in the mud in Northern Territory and Queensland. Other gobies (*Mugilogobius*, *Taenioides*, and *Arenigobius*) are common in Australian mangroves but also occur in other parts of the estuary.

By comparison, infaunal crustaceans occur throughout the mangroves, but individual species show marked zonation patterns. George and Jones (1982) have documented the zonation of *Uca*'s in north west Australia according to sediment grain size. Many of the shrimps and amphipods are restricted to lower levels of the mangrove shore but crabs occur throughout the mangrove zone with species restricted to particular zones (Yates 1978).

In temperate Australian mangroves, seagrass beds often abut onto the seaward margins of the mangroves and some species occur in both zones. However, in tropical regions, because of the large tidal amplitude and the amount of sediment carried down the rivers during the wet season, extensive bare mud flats abut onto the mangroves, rather than seagrass beds. The relationship between the faunas of these two habitats is unknown.

Although marine animals dominate this infauna, some ant species as already mentioned live intertidally in tropical mangroves. They build their nests on the mud, and an undescribed capitellid polychaete is commonly found in the walls. The ants probably plug the entrance to the nest during high tide, trapping air.

b. Substrate epifauna

Many of the casual mammal visitors forage over the surface of the mud at low tide, as do some of the snakes and goannas. Similarly, some of the arboreal insects and spiders may feed on the forest floor, or on leaf litter. In tropical mangroves, the salt water crocodile *Crocodylus porosus* is common, and juveniles feed on crabs (especially sesarmids) prawns, and fish. The most diverse group is the gastropod molluscs (Hutchings and Recher, 1982) which are zoned according to tidal inundation, which may be related to feeding strategies. Algal grazers, we suspect, are restricted to the wetter frontal margins, whereas deposit or carnivorous species may have wider distribution patterns, but little work has been done in this area.

c. Wood-boring fauna

This fauna is not restricted to mangroves, but as

dead wood tends to accumulate in them, they typically have a rich and diverse fauna. The fauna can be considered in terms of borers (teredinid molluscs and gnathiid isopods) which attack and break down the wood, and nestling species (polychaetes, nemerteans, gastropods, small crabs, and amphipods) which utilise the tunnels created by the true boring organisms. This latter group will be most diverse in the wetter frontal margins of the mangroves. Recently, there has been strong evidence that before boring can occur, the wood must be colonised by fungi (Cragg and Swift 1980). As yet, in Australia, no evidence exists that live wood is attacked by marine borers as in Florida (Simberloff *et al.* 1978). Dead wood in the drier parts of the mangroves is often heavily infested by terrestrial animal borers such as beetles and weevils.

d. Encrusting fauna

This fauna, which is restricted to the frontal margins of the mangroves is dominated by the oyster *Saccostrea commercialis* and barnacles. Between these shells occurs a humid, protective environment for many species of polychaetes, crustaceans, and gastropods. Recently, Saenger *et al.* (1979) have documented a series of successional stages in the development of this community.

e. Pelagic communities

This community is a transient one, as it comes into the mangroves on a rising tide and leaves, albeit somewhat changed, on a falling tide. It includes many species of juvenile and adult fish and prawns. However, the actual species involved and the relative importance of mangroves to these species (i.e. the percentage of time spent in the mangrove during their life history or the relative amounts of food consumed in this environment) has not been well documented. In addition, estuarine plankton will be carried into the mangroves providing food for the filter feeding organisms (barnacles, bivalves, and serpulid polychaetes). This potential food supply will be seasonal and modified by local weather patterns.

Discussion

From this brief overview, it is apparent that Australian mangroves have a very diverse fauna, represented by many groups of animals. Although we lack detailed geographical distribution data for many animals, there is a trend clearly evident in the molluscs and crustaceans, but probably shown by all groups, of increasing diversity with decreasing latitude, at least along the east coast, which parallels the increasing diversity of the plant communities. An exception to this simple relationship exists on the north west coast of Australia, where more endemic bird species occur in a less floristically diverse community than in mangrove communities occurring at similar latitudes on the east coast of Australia. Schodde *et al.* (1982) attempt to explain this discrepancy by the different origins of the bird fauna in the two regions.

Although there is a general increase in the total number of species along the east coast, this is the result of some species reaching their limits of distribution and being replaced, often by several species, so that the fauna of South Australian mangroves is very different from that of Queensland mangroves, although a few species, such as *Littorina scabra*, *Balanus amphitrite* and *Mictyris longicarpus* occur in all areas.

As well as a general increase in species with decreasing latitude, there is a similar increase in the number of mangrove endemic marine species from zero in temperate mangroves to many in tropical areas. In temperate areas, the marine fauna is basically an estuarine community. Whether similar patterns exist among the terrestrial fauna is unknown, but this seems likely at least for the insect fauna.

Whether this increase in diversity is also manifest in an increase in the total number of individuals present is less clear. Superimposed on any trends of increasing diversity from temperate to tropical mangroves, are reductions in diversity along an estuary with reductions in salinity.

The zonation of fauna within the mangrove community is poorly documented. For the arboreal community, probably few if any of the casual visitors show any zonation patterns, and many may not be an integral part of the mangrove ecosystem. However, we suspect that much of the insect fauna shows distinct zonation determined by the zonation of the mangroves themselves, the height and form of the trees, proximity to and type of adjacent terrestrial habitats and sources of permanent pools of fresh water in nearby terrestrial systems. The zonation of prey organisms may be important for carnivorous species. Species feeding on the forest floor may be restricted to certain tidal zones of the mangrove, to allow sufficient time for feeding. Small islands of mangroves may have reduced arboreal communities compared to those occurring on mainland sites (Simberloff and Wilson 1970; Wilson and Simberloff 1969a, 1969b).

The fauna living on the floor of the mangroves will be affected by tidal amplitude and by tidal cycle, which may be modified by diurnal periodicity. Hutchings and Recher (1974) observed maximum densities of some crab species at night on a rising tide. Other species change their distribution according to their reproductive cycle. Yates (1978) found that *Sesarma erythrodactyla* and *Paragrapsus laevis* move to lower zones in the mangroves to spawn.

Sediment distribution is critical to the distribution of infaunal animals, both within a particular stand of mangroves, and latitudinally. In general, with decreasing latitude, sediment size decreases at least on the mainland. However, offshore stands of mangroves often associated with coral reef areas tend to have coarser sediments. Much of the fine sediment is brought down by the rivers in flood during the wet season and falls out of suspension as it flows through the mangroves. Great changes in sediment distribution may occur after cyclones which periodically affect all subtropical and tropical regions in Australia.

Other factors include the salinity regime, and the salinity of the interstitial water. These regimes may vary markedly seasonally and between years. Snelling (1959), working in the Brisbane River, Queensland, observed changes in crab distribution following floods. Similar changes must occur among other groups. Many of the sedentary species must be wiped out during a flood; others may be washed further downstream and perhaps survive. A few species can take avoiding action, such as bivalves or sipunculans which burrow deeper to avoid salinity changes (Green 1975b). In areas which have very limited freshwater input such as South Australia and north west Australia, very high salinities and temperatures may occur spasmodically in the interstitial water during low tide in the summer, causing localised wipe-out of populations.

Information on the seasonal or between-year variations in faunal populations is scarce and almost restricted to temperate mangroves (Mc Cormick 1978; Yates 1978). We suspect that, for some reasons noted above, faunal populations may fluctuate widely over a year and between years. Such fluctuations may be greatest in tropical regions. Arboreal communities may fluctuate less, but McCormick (1978) found some correlation between their diversity and the regularity of flooding.

Although describing the fauna of mangroves is important, of far more importance is the study of the role of this diverse mangrove fauna. Life history and feeding strategy data are scant especially in tropical areas. Yates (1978) has some data on the life cycles of the common crabs in the Sydney region, and Muggeridge (1979) has similar data for the gastropod *Littorina scabra*. Some information is available for the sipunculan *Phascolosoma arcuatum* (Green 1975a), and recently, Hill (1979, and pers. comm.) has been studying the importance of mangroves to the commercially important species of crab *Scylla serrata*. For the arboreal communities we only have detailed information for some of the mosquitoes and biting midges. Feeding strategy studies have been restricted to juvenile crocodiles (Taylor 1979), mangrove creek fishes (Beumer 1978; Blaber 1980), and Goulter and Allaway's (1979) study on the rate of shredding of fallen leaves by amphipods. Until we have obtained this data for all the dominant animals in the mangroves we cannot construct food chains or begin to assess the secondary productivity of these communities.

As Australian mangroves occur over such a wide latitudinal gradient, with such different faunas and floras, extrapolating figures obtained from one area to another may be hazardous. Similar caution should be applied to utilizing overseas data in Australia. For overseas, too, faunal studies are limited and comparative data over a similar latitude to Australia are unavailable.

Finally we should like to consider the relationships between the plants and animals in mangrove communities. Some concept of the importance of the relationship between plants and animals is apparent in the literature with emphasis on the production of detritus from mangrove sources (Goulter and Allaway 1979; Odum and Heald 1975; Malley 1978). The importance of mangroves to animals may vary between species and during the life cycle of the animal. This is a reflection of the transitional nature of the environment, a mergence of two types of habitats, a forest and an intertidal area, and the dynamic state of animal populations, both seasonally and during a tidal cycle. Although many of the animals are using the detritus, we suspect that for others the protection from desiccation and predation afforded by the mangroves may be of similar or greater value. Also the mangroves tend to trap and stabilise the sediment which will have consequences not only for the mangrove animals but perhaps for other estuarine communities.

A question rarely asked is, are the mangroves in any way dependent on the fauna? Perhaps insect damage to leaves encourages leaf fall and facilitates breakdown into detritus, which may release trace elements necessary for plant growth. Pollination and perhaps dispersal of mangrove seeds is facilitated by animals, although from casual observations in polluted areas, mangroves often appear 'relatively healthy' while the fauna is depauperate, perhaps suggesting that mangroves are not dependent upon the fauna, at least in the short term.

In Australia, not only from a biological but from an economic point of view (Hutchings and Recher 1977) we are in a unique position to study different types of mangrove communities and to assess the value of them, to estuarine ecosystems. Only when this is done can these resources be properly managed.

Literature cited

Beumer, J.P. 1978. Feeding ecology of fishes from a mangrove creek in north Queensland, Australia. Jour. Fish. Biol., 12: 475–490.

Blaber, S.J.M. 1980. Fish of the Trinity Inlet system of north Queensland with notes on the ecology of fish faunas of Tropical Indo-Pacific Estuaries. Aust. Jour. mar. Freshwat. Res. 31:137–146.

Boto, K.G. 1982. Nutrient and organic fluxes mangroves. In: Structure Function and Management of Mangrove Ecosystems in Australia. ANU Press. 239–257.

Chapman, V.J. 1975. Mangrove Vegetation. J. Cramer, Leutershausen, Germany. 447 pp.

Cogger, H.G. 1979. Reptiles and Amphibians of Australia. Sydney, Wellington, London. 2nd Edition, 584 pp.

Common, I.W.B. and D.F. Waterhouse. 1972. Butterflies of Australia. Angus and Robertson, 498 pp.

Cragg, S.M. and M.J. Swift. 1980. The contribution of fungi and marine borers to wood decay of some mangrove communities of Papua New Guinea. Second Int. Sym. Biol. Management of Mangroves and Tropical Shallow Water Communities. (Abstract only) p. 22.

Debenham, M.L. 1979. An annotated checklist and bibliography of Australasian region Ceratopogonidae, (Diptera, Nematocera). Entomology Monograph, No. 1. Aust. Government Publishing Service Canberra, 671 pp.

Decraemer, W. and A. Coomans. 1978. Scientific report on the Belgian expedition to the Great Barrier Reef in 1967. Nematodes XII. Ecological notes on the nematode fauna in and around mangroves on Lizard Island. Aust. Jour. mar. Freshwat. Res. 29:497–508.

Frith, H.J. 1973. Wildlife Conservation. Angus and Robertson. Sydney, 414 pp.

George, R.W. and D.S. Jones. 1982. West Australian and Northern Territory Fiddler crabs (subfamily Ocypodinae). Rec. West Aust. Mus. Suppl. No. 14. 99 pp.

Goulter, P.F.E. and W.G. Allaway. 1979. Litter fall and decomposition in a mangrove stand *Avicennia marina* (Forsk.) Vierh., in Middle Harbour, Sydney. Aust. Jour. mar. Freshwat. Res. 30:541–546.

Green, A.W. 1975a. The annual reproductive cycle of *Phascolosoma lurco* (Sipuncula) In: Proc. Int. Symp. Biol. Sipuncula and Echiura, Kotcr 1970, pp. 161–169.

Green, A.W. 1975b. *Phascolosoma lurco*: A semi-terrestrial sipunculan. In: Proc. Int. Symp. Biol. Sipuncula and Echiura, Kotor. 1970, pp. 267–281.

Hill, B.J. 1979. Aspects of the feeding strategy of the predatory crab *Scylla serrata*. Mar. Biol. 55:209–214.

Hutchings, P.A. and H.F. Recher. 1974. The fauna of Careel Bay with comments on the ecology of mangroves and sea-grass communities. Aust. Zool. 18:99–128.

Hutchings, P.A. and H.F. Recher. 1977. The management of mangroves in an urban situation. Mar. Res. Indonesia. 18: 1–13.

Hutchings, P.A. and H.F. Recher. 1982. The fauna of Australian Mangroves. Proc. Linn. Soc. N.S.W.

Lee, D.J., M.M. Hicks, M. Griffiths, R.C. Russell and E.N. Marks. 1980. The Culicidae of the Australasian region. Commonwealth Inst. of Health Monograph Series, Entomology Monography No. 2. Vol. 1, Aust. Government Publishing Service Canberra, 248 pp.

Macnae, W. 1966. Mangroves in Eastern and Southern Australia. Aust. Jour. Bot. 14:67–104.

Macnae, W. 1968. A general account of the fauna and flora of mangrove swamps and forests in the Indo-West Pacific region. Advances in Marine Biology 6:73–270.

Magnusson, W.E., G.J.W. Webb and J.A. Taylor. 1976. Two new locality records, a new habitat and a nest description for *Xeromya myoides* Thomas (Rodentia: Muridae). Aust. Wild. Res. 3:153–157.

Malley, D.F. 1978. Degradation of mangrove leaf litter by the tropical sesarmid crab *Chiromanthes onychophorum*. Mar. Biol. 49:377–386.

McCormick, W.A. 1978. The ecology of benthic macrofauna in New South Wales mangrove swamps. MSc thesis, University of New South Wales, 333 pp. Unpublished.

Muggeridge, P. 1979. The reproductive biology of the mangrove littorinids *Bembicium auratum* (Quoy Gaimard) and *Littorina scabra* (Linne) (Gastropoda, Prosobranchiata) with observations on the reproductive cycle of rocky shore littorinids of N.S.W. PhD thesis, Sydney University, 253 pp. Unpublished.

Nursall, J.R. 1981. Habitat and habitat changes as they affect mudskippers, *Periophthalmus* spp. and *Periophthalmodon*. Bull. Mar. Sci. 31:730–736.

Onuf, C.P., J.M. Teal and I. Valiela. 1977. Interactions of nutrients, plant growth and herbivory in a mangrove ecosystem. Ecology 58:514–526.

Saenger, P., M.M. Specht, R.L. Specht and V.J. Chapman. 1977. Mangrove and coastal salt-marsh communities in Australasia. In: Ecosystems of the World, Vol. 1. Wet Coastal Formations. Elsevier Publ. Co., Amsterdam, pp. 293–345.

Saenger, P., W. Stephenson and J. Moverly. 1979. The subtidal fouling organisms of the Calliope River and Auckland Creek, Central Queensland. Mem. Qd. Mus. 19:399–412.

Schodde, R., I.J. Mason and H.B. Gill. 1982. The avifauna of the Australian mangroves – a brief review of composition, structure and origin. In: Structure, Function and Management of Mangrove Ecosystems in Australia. ANU Press. 141–150.

Simberloff, D.S., B.J. Brown and S. Lowrie. 1978. Isopod and insect root borers may benefit Florida. Science 201:630–632.

Simberloff, D.S. and E.O. Wilson. 1970. Experimental zoogeography of Islands. A two-year record of colonization. Ecology 51:934–937.

Snelling, B. 1959. The distribution of intertidal crabs in the Brisbane River. Aust. Jour. Mar. Freshwat. Res. 10:67–83.

Taylor, J.A. 1979. The foods and feeding habit of subadult *Crocodylus porosus* Schneider in Northern Australia. Aust. Wildl. Res. 6:347–359.

Wilson, E.O. and D.S. Simberloff. 1969a. Experimental zoogeography of Islands. Defaunation and monitoring techniques. Ecology 50:267–278.

Wilson, E.O. and D.S. Simberloff. 1969b. Experimental zoogeography of Islands. The colonisation of empty islands. Ecology 50:278–296.

Yates, R. 1978. Aspects of the ecology and reproductive biology of crabs in a mangrove swamp at Patonga Creek, NSW. MSc thesis, University of Sydney, 135 pp. Unpublished.

CHAPTER 12

Distribution of Phycomycetes in mangrove swamps with brackish waters and waters of high salinity

ANNEMARIE ULKEN

Institut für Meeresforschung Bremerhaven, West Germany

Abstract. Mangrove swamps at Cananéia (Brazil), Heeia, Pearl Harbour (Hawaii), Cockroach Bay (Florida), Veracruz (Mexico), Sinai (Israel) and Bermuda were investigated with respect to Phycomycetes that degrade cellulose, chitin and keratin. The fungi isolated and maintained in pure culture were: Strains of *Phlyctochytrium mangrovis* Ulken, *Catenaria anguillulae* Sorokin, and some species of *Thraustochytrium*, *Schizochytrium* and *Ulkenia*. In the swamps of Cananéia, Heeia, and Pearl Harbour with salinities between 5 and 30‰ a great variety of Phycomycetes, uniflagellate as well as biflagellate, were found. At Veracruz many uniflagellate chytrids, which are known to occur in fresh waters, but which are also known to be halotolerant, could be observed. In the mangrove swamps of Sinai and Bermuda, where the salinity usually is high, mainly representatives of the biflagellate genera *Thraustochytrium*, *Schizochytrium* and *Ulkenia* were found. The degradative activity of these fungi at different combinations of light and temperature was investigated.

Introduction

Odum and Heald (1975) pointed out that '... mangrove detritus serves as the energy base for an extensive food web. The first link is provided by fungi and bacteria which convert the relatively undigestible lignin and cellulose of vascular plant tissue into a protein source that can be digested by the organisms in the second link of the food web.' A great part of this activity is certainly due to the degradative activity of many higher fungi as was shown by Kohlmeyer and Kohlmeyer (1979). An investigation of the lower fungi (aquatic Phycomycetes) in mangrove muds has been carried out since 1969. Some of the fungi isolated from mangrove muds were found to degrade keratin of hair and chitinous materials besides cellulose. Only a very few Phycomycetes degraded cellulose. These studies were made on mangrove muds and parts of mangrove trees collected from North and South America, Mexico, Bermuda and Sinai.

Samples were collected either by cruising with a boat through channels and creeks (Cananéia Brazil, Pearl Harbour Hawaii) or by walking along the shore at low tide. They were obtained by scraping mud from aerial roots or by taking the mud of the rhizosphere of mangrove trees. Each sample contained between 10 and 50 ml of wet mud. Samples of parts of aerial roots, leaves or seedlings were also collected. All samples were incubated in water corresponding to the salinity of the environment from which they were collected and provided with pine pollen or other materials such as cattle hair, chitin of crab skeleton etc. as substrates. Samples from Bermuda were flooded with water of different salinities in order to obtain all fungi present. Growing fungi were then isolated and treated with antibacterial antibiotics until pure cultures were obtained. The following experiments were done with pure cultures of strains of *Phlyctochytrium*

Tasks for vegetation science, Vol. 8 ed. by H.J. Teas

 ISBN 978-90-481-8526-9

mangrovis Ulken, *Thraustochytrium* sp., and *Ulkenia visurgensis* (Ulken) Gaertner which have been maintained in culture since 1969, 1974 and 1979 respectively (Bermuda and Sinai). Experiments on the combined effect of light and temperature were done using 100 ml Erlenmeyer flasks with 40 ml of nutrient broth, which was inoculated with a platinum loop from a well growing culture and kept in incubators at 15, 20, 25 or 30° C. Light and dark were each of 12 h duration. In addition, Petri dishes of 5 cm diameter were filled with brackish water, inoculated with a pure culture of the Phycomycete being studied, and sterilized crab skeleton, hair, purified chitin or keratin dusted on the surface of the water in the Petri dish; light dark periods were 12 hours. After one week the culture was observed, and micrographs were prepared to show the growth of the fungi.

Results

The Phycomycetes found in the Laguna de Mandinga, Veracruz, are shown in Table 1. The area has been described by Vasquez-Yanes (1971). He found the salinity changing during a year from oligo- to polyhaline. *Catenaria anguillulae* Sorokin (Fig. 1), a member of the Blastocladiales, was isolated from here as were different species of chytrids as well as species of the biflagellate genera *Thraustochytrium*, *Schizochytrium* and *Ulkenia*; the last named appears to be euryhaline.

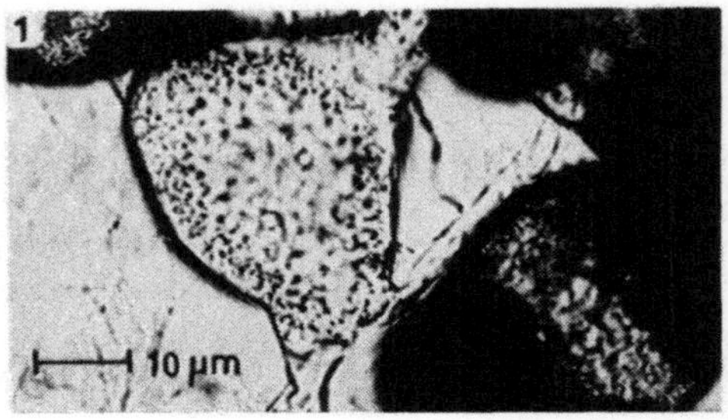

Fig. 1. Catenaria anguillulae from the Laguna de Mandinga, Veracruz, Mexico. The rhizoids are penetrating the pine pollen grains.

In Table 2 the Phycomycetous flora from the saline area of the Wadi Kid, Shura el Manqata, Sinai is shown. Only species of the biflagellate genera could be found. The salinity was 41‰ at the day of collection. The environment of the mangal of the Sinai has been described by Por *et al.* (1977) as 'Temperature and salinity conditions were found to be extreme, obviously reducing the diversity of the ecosystem and presumably setting the geographical limit of its distribution'. This was stated for the macroflora and -fauna. Thus the results on lower fungi were not surprising. Even in samples which were flooded with water of low salinity no uniflagellate chytrids could be found.

In Table 3 results on fungi from the Bermuda mangrove swamps are presented.

At station 1 and 2, a chytrid, a *Phlyctochytrium* sp. was found in samples which were flooded with tap water or with water of low salinity (8–12‰)

Table 1. Phycomycetes found in mangrove mud from the Laguna de Mandinga, Veracruz, Mexico.

Sample nr.	Fungi found
1	*Catenaria anguillulae* Sorokin, *Phlyctocytrium mangrovis* Ulken, *Anisolpidium* sp. ? *Ulkenia visurgensis* (Ulken) Gaertner, *T.* sp. *Schizochytrium* sp.
2	*T.* sp. *T. pachydermum* Scholz
3	*Olpidium* sp. *T. pachydermum*, *T.* sp. *S.* sp.
4	*T.* sp.
5	*Rhizophydium* sp., *Olpidium* sp., *T.* sp. *T. pachydermum*
6	*Chytridium* sp. *T.* sp. *R.* sp. *S.* sp.
7	*T. pachydermum*
8	*Chytridium proliferum* Karling, *T. multirudimentale* Goldstein, *T.* sp. *S.* sp.
9	*T.* sp. *T. pachydermum*
10	*T. multirudimentale*, *Chytriomyces multioperculatus* Sparrow and Dogma, *P. mangrovis*, unidentified chytrid

Table 2. Phycomycetes found in mangrove mud from the Wadi Kid, Shura el Manqata.

Sample nr.		Fungi found
1	Mud scraped from a pneumatophore	*Ulkenia visurgensis* (Ulken) Gaertner, ? *Schizochytrium* sp.
2	Mud from leaves	*U. visurgensis*, ? *Thraustochytrium pachydermum* Scholz
3	Mud from pneumatophores	*T. striatum* Schneider, *U. visurgensis*
4	Particles from living and dead pneumatophores	*T. pachydermum*, *U. visurgensis*, *T.* sp., *S.* sp.
5	Sediment from a depth of 50 cm	*S.* sp., *U. visurgensis*
6	Sediment from a depth of 100 cm	*T. pachydermum*, *U. visurgensis*

Table 3. Fungi found in mangrove mud from Bermuda[a]

Station nr.	Tap water	Brackish water (8–12‰)	Brackish water (25‰)	Sea water (32‰)	Sea water (35‰)
1	Chytrid	Chytrid	Thraust.[b] sp.	—	? Spores
2	Chytrid	Chytrid	Thraust. sp.	? Schizo.[b]	—
3	? Thraust	—		Thraust. sp.	—

[a] Samples were sent by air mail.
[b] Thraust. = Thraustochytrium, Schizo. = Schizochytrium.
Station 1 = Blue Grotto, Station 2 = Walsingham Pond, Station 3 = Lover's Lake.

(Figs. 2–5). Another fungus isolated was a thraustochytrid which appears to be *Thraustochytrium aureum* Goldstein (Fig. 6). On agar the fungus had a yellowish colour and often a persistent membrane of an empty sporangium could be seen (Fig. 7); however, no sporulation could be observed. In this species the spores are supposed to become fully motile before leaving the sporangium. Because we had treated the cultures with antibiotics in order to free them from bacteria it is possible that a mutation had occurred the affected sporulation.

Figures 11–14 show the results of a combined effect of temperature and light on the growth of different isolates of *Phlyctochytrium mangrovis*. At the first mangrove symposium in Honolulu (Ulken, 1975) I reported some results on the influence of light on *P. mangrovis*. The present investigations, conducted in a climatically controlled room at 16° C, were in conflict with some earlier investigations at room temperature (20–21° C). Sometimes light had a stimulating and sometimes an inhibitory effect on growth of the fungi, depending on the temperature. This variation caused me to repeat the experiments under carefully controlled conditions.

The isolate from Pearl Harbour grew better in the dark at 15° C than in the light; however, light stimulated growth between 20 and 30° C. The isolate from Veracruz grew better in the dark at 15° C and at 30° C than in the light, whereas at 20 and 25° C light had a stimulating effect. The isolate from Heeia did not grow at 15° C; at 30° C growth was poorer in the light than in the dark, while at 20 and 25° C light had a stimulating effect. The isolate from Cananéia showed only slight growth at 15° C in either the light or dark. The growth at 30° C was unlike that of the other isolates, being nearly the same in the dark as in the light. At 20 and 25° C light had a small stimulatory effect. Growth of *Ulkenia visurgensis* (Fig. 15) was stimulated by light at 20 and 25° C, while at 15° C growth was better in the dark and at 30° C no growth occurred. All results are the means of three replicates; the light intensity used was between 2000 and 3000 lux.

Lower fungi seem to be only weak degraders of cellulose. Although *Phlyctochytrium mangrovis* showed a luxurious growth on cellulose agar, the presence of cellulolytic enzymes remains doubtful as no clear zones appeared on the agar in the

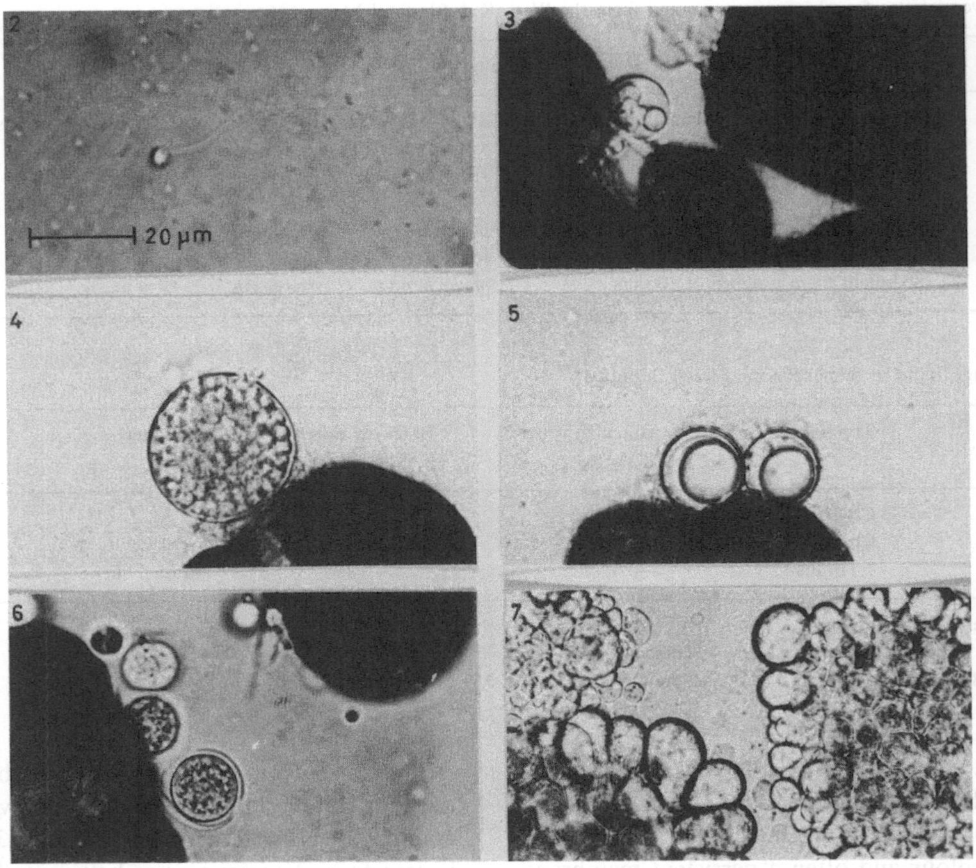

Fig. 2–5. Phlytochytrium sp. Fig. 2: zoospore. Fig. 3: young thallus with apophysis within the pollen grain. Fig. 4: mature sporangium with 2 pores. Fig. 5: 2 resting spores.
Figs. 6 and 7. Thraustochytrium? *aureum* Goldstein. Fig. 6: sporangia on pollen grains. Fig. 7: growth on agar.

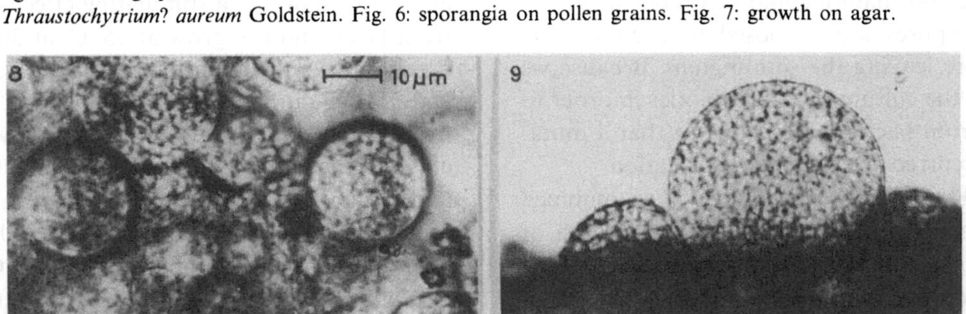

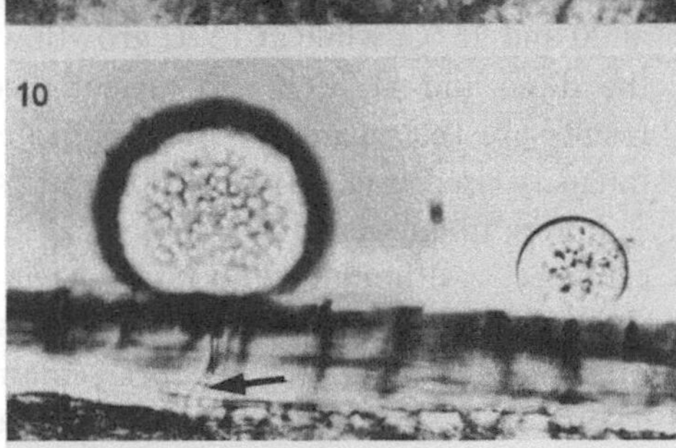

Figs. 8–10. Growth of *P. mangrovis* Ulken, isolate from Pearl Harbour. Fig. 8: growth on crab skeleton at 25° C. Fig. 9: growth on the broken end of cattle hair at 25° C. Fig. 10: growth on the side of the hair at 30° C; arrow indicates penetrating rhizoid.

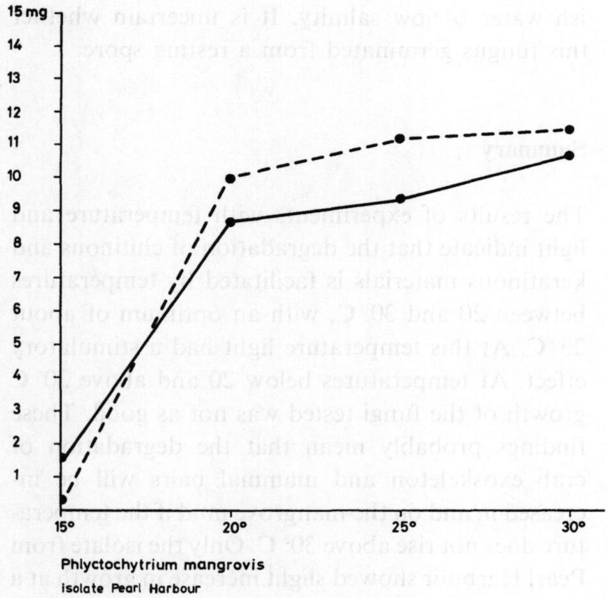

Fig. 11. Isolate Pearl Harbour.

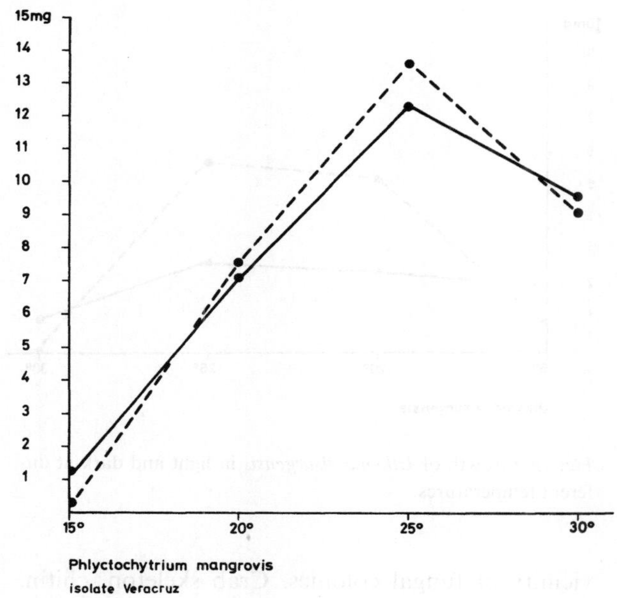

Fig. 12. Isolate Veracruz.

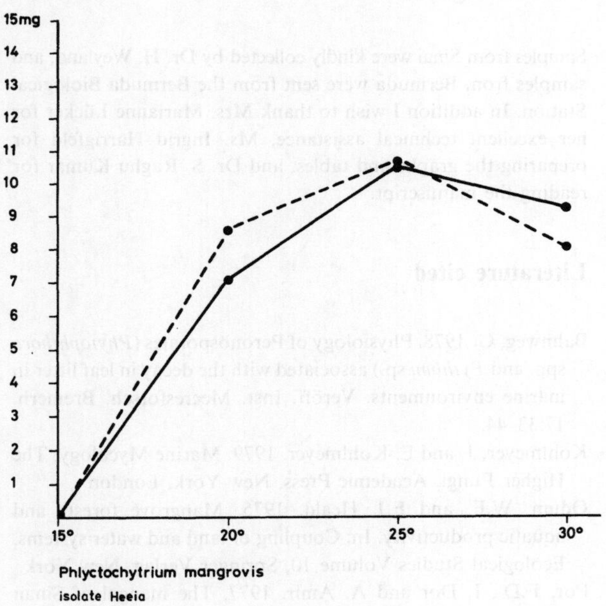

Fig. 13. Isolate Heeia.

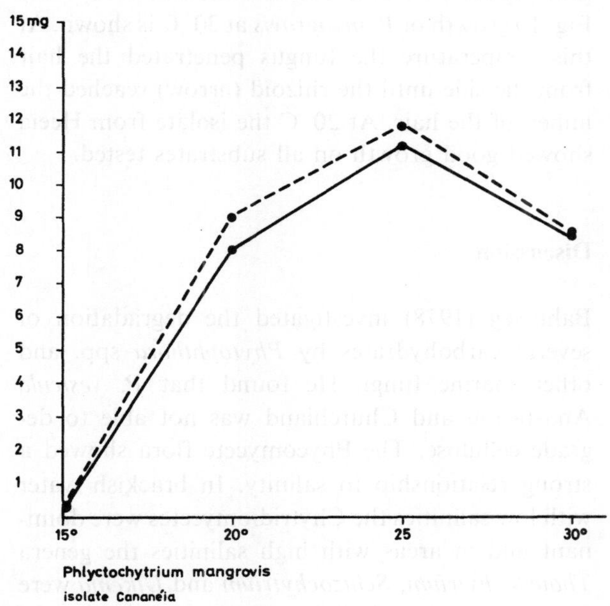

Fig. 14. Isolate Cananéia.

Figs. 11–14. Growth of different isolates of *P. mangrovis* in light (dotted line) and dark at different temperatures.

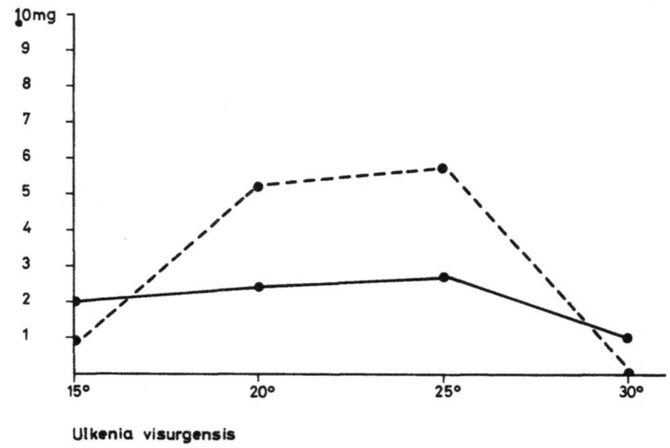

Fig. 15. Growth of *Ulkenia visurgensis* in light and dark at different temperatures.

vicinity of fungal colonies. Crab skeleton, chitin, hair and keratin of hair served well as nutrients. In Fig. 8 growth of *P. mangrovis* on crab skeleton at 25° C is shown; in Fig. 9 *P. mangrovis* is shown growing on the broken end of cattle hair at 25° C. In Fig. 10 growth of *P. mangrovis* at 30° C is shown. At this temperature the fungus penetrated the hair from the side until the rhizoid (arrow) reached the lumen of the hair. At 20° C the isolate from Heeia showed good growth on all substrates tested.

Discussion

Bahnweg (1978) investigated the degradation of several carbohydrates by *Phytophthora* spp. and other marine fungi. He found that *P. vesicula* Anastasiou and Churchland was not able to degrade cellulose. The Phycomycete flora showed a strong relationship to salinity. In brackish water with low salinities the Chytridiomycetes were dominant and in areas with high salinities the genera *Thaustochytrium*, *Schizochytrium* and *Ulkenia* were more frequent, which could be shown for the samples from Sinai. Similar results were expected for the samples from Bermuda because fresh water there comes only from rainfall; however, there was a *Phlyctochytrium* sp. found in the Bermuda samples that were flooded with tap water or brackish water of low salinity. It is uncertain whether this fungus germinated from a resting spore.

Summary

The results of experiments with temperature and light indicate that the degradation of chitinous and keratinous materials is facilitated by temperatures between 20 and 30° C, with an optimum of about 25° C. At this temperature light had a stimulatory effect. At temperatures below 20 and above 30° C growth of the fungi tested was not as good. These findings probably mean that the degradation of crab exoskeleton and mammal hairs will be increased in and on the mangrove mud if the temperature does not rise above 30° C. Only the isolate from Pearl Harbour showed slight increase in growth at a temperature of 30° C.

Acknowledgements

Samples from Sinai were kindly collected by Dr. H. Weyland, and samples from Bermuda were sent from the Bermuda Biological Station. In addition I wish to thank Mrs. Marianne Lücker for her excellent technical assistance, Ms. Ingrid Harrigfeld for preparing the graphs and tables, and Dr. S. Raghu Kumar for reading the manuscript.

Literature cited

Bahnweg, G. 1978. Physiology of Peronosporales (*Phytophthora* spp. and *Pythium* sp.) associated with the decay in leaf litter in marine environments. Veröff. Inst. Meeresforsch. Bremerh. 17:33–44.

Kohlmeyer, J. and E. Kohlmeyer. 1979. Marine Mycology: The Higher Fungi. Academic Press, New York, London.

Odum, W.E. and E.J. Heald. 1975. Mangrove forests and aquatic productivity. In: Coupling of land and water systems. Ecological Studies Volume 10, Springer Verlag, New York.

Por, F.D., I. Dor and A. Amir. 1977. The mangal of Sinai: Limits of an ecosystem. Helgoländer wiss. Meeresunters. 30:295–314.

Ulken, A. 1975. Further studies on *Phlyctochytrium mangrovis* Ulken. In: Proc. Int. Symposium on Biology and Management of Mangroves. Univ. Florida, Gainesville, pp. 680–687.

Vasquez-Yanes, C. 1971. La vegetación de la Laguna de Mandinga, Veracruz. A. Inst. Biol. Univ. Nal. Autón. Mexico, Ser. Botánica 42:49–94.

CHAPTER 13

Inheritance of albinism in the red mangrove, *Rhizophora mangle* L.

SHIRLEY H. HANDLER* and HOWARD J. TEAS**

* *Biology Department, East Texas Baptist College, Marshall, Texas 75670, U.S.A.*
** *Biology Department, University of Miami, Coral Gables, Florida 33124, U.S.A.*

Abstract. Occasional trees of *Rhizophora mangle* L. have been found in Puerto Rico and South Florida that bear approximately one fourth albino propagules, apparently a consequence of self-pollination and segregation for a recessive gene. Counts of propagule types on a series of trees that produce albino propagules showed that the albino class is sometimes smaller than one fourth of the total, suggesting that there may be a variable degree of outcrossing.

Introduction

Albinism

The red mangrove, *Rhizophora mangle*, is the dominant mangrove species in south Florida (Teas, 1977). This species, so named because of its red inner bark, is viviparous, that is, the seed germinates within the fruit while it is still attached to the parent tree. The pencil-like seedlings, referred to as propagules, are normally green. The authors have found albino propagules in the shoreline wrack and have observed occasional red mangrove trees bearing albino and normal green propagules in a ratio of approximately 1:3 in south Florida and Puerto Rico. A report on the occurrence of such albino propagules has appeared (Teas and Handler, 1979).

Albinism, the absence of chlorophyll in a plant part that is normally green, has been well known for many years. In maize the occurrence of albino seedlings may be due to a simple Mendelian recessive gene in which the selfing or intercrossing of two heterozygotes produces green and white offspring in a ratio of 3:1. A number of non-allelic loci that affect chlorophyll development are known in maize (Sinnott *et al.*, 1958). In addition to the simple recessive albino character, albinism due to epistasis has been reported. In one case heterozygotes produce green and white offspring in a ratio of 9:7 and in another, green and white seedlings produced in a ratio of 15:1 (Srb and Owen, 1952).

Genetic albinism has been reported in a variety of plants. Butany and Singh (1965) have observed both simple Mendelian recessive genes and epistatic genes for chlorophyll deficiency in cotton, *Gossypium hirsutum*. In the slash pine tree, *Pinus elliottii* var. *elliotii*, Squillace and Kraus (1963) reported that controlled self-pollination of heterozygotes produced a ratio of 3 normal to 1 albino seedlings, whereas wind pollinated progenies of heterozygotes produced variable smaller proportions of albino offspring. Johnson (1948) reported albinism in the Austrian pine, *Pinus nigra* var. *austriaca*. He obtained 3 normal to 1 albino seedlings from one tree which he concluded was heterozygous and self-pollinated. Anderson *et al.* (1974) obtained ratios of 3 normal to 1 albino seedlings in the Norway spruce, *Picea abies*, with controlled pollination. Franklin (1969) reported chlorophyll

Tasks for vegetation science, Vol. 8 ed. by H.J. Teas
 ISBN 978-90-481-8526-9

deficiencies in the loblolly pine, *Pinus taeda*. Inherited chlorophyll deficiency was found by Gabriel (1962) in the sugar maple, *Acer saccharum*. Snyder (1951) reported that albino and normal green seedlings were produced regularly by trees of the same species and in some cases normal and albino seedlings were produced in an approximate 3:1 ratio. Squillace and Kraus (1963) noted that chlorophyll deficiencies had been reported in cacao, *Theobroma cacao*; Chinese chestnut, *Castanea mollissima*; Siberian pea-tree, *Caragana arborescens*; quaking aspen, *Populus tremuloides*; red pine, *Pinus resinosa*; Serbian spruce, *Picea omoriaka*; blue spruce, *Picea pungens*; and western white pine, *Pinus monticola*. Venkatesh and Emmanuel (1976) reported finding a series of chlorophyll mutations in the Kapok tree, *Bombax ceiba*; Venkatesh *et al.* (1978) reported the occurrence of an albino chlorophyll mutant in *Gmelina arborea*; and Filho (1969) reported finding albino seedlings in the baobab, *Adansonia digitata*.

Rhizophora mangle *reproduction*

The red mangrove flower has been described by Gill and Tomlinson (1969). The pale yellow flowers are borne in axillary clusters, usually three-flowered with a single terminal and lateral pair of flowers. The terminal flower may develop later than the lateral pair or be absent. The flower has four persistent and valvate sepals, four petals, eight stamens with septate pollen saces that dehisce within the flower bud, and an inferior ovary with four ovules, only one of which usually develops.

Gill and Tomlinson (1969) noted that little was known about pollination biology in the red mangrove; subsequently Tomlinson *et al.* (1979) concluded, on circumstantial evidence, that *Rhizophora* species are wind pollinated. Guppy (1906) had earlier concluded that *R. mangle* and *R. mucronata* are self-fertilized because no open flowers were found with undehisced stamens. [Guppy was actually dealing with *R. samoensis* and *R. stylosa* (Tomlinson, 1978).] Guppy suggested, however, that cross-fertilization could occur because he observed flies, ants, and small coleoptera visiting the opened flowers at a time when pollen was retained by the hairs of the inner faces of the petals. Teas and Handler (1979) reported evidence that insects are not required for pollination.

The red mangrove fruit is ca. 2.5–3.5 cm long. The seed germinates within the fruit while it is on the parent tree. The seedling or propagule elongates to a length of ca. 8–24 cm during the four to seven months after fertilization while it is 'maturing' (Gill and Tomlinson, 1971). When fully developed, the propagule drops from the parent tree and may be dispersed by tidal waters.

Normal propagules are green with a brownish base; albino propagules are yellow or white and brown toward the base (see black and white photograph, Fig. 1). Albino propagules exposed to the sun may develop a red or pink color. In nature or in the laboratory albino seedlings live for a few weeks and die, apparently, when their reserves are exhausted.

The albino propagules that occur on trees in south Florida and Puerto Rico are of two general types: white albinos (that become pink to red in strong light) and yellow or xantha. Both types of albinos are consistent for a given tree from year to year. Comparison of albino propagules from widely

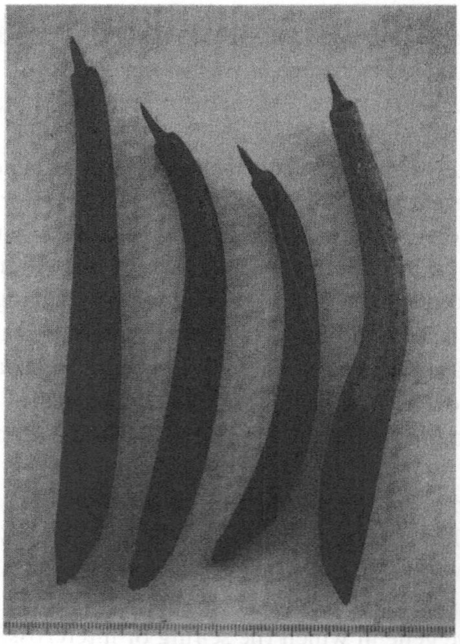

Fig. 1. Normal and an Albino propagule from a *Rhizophora mangle* tree at Miami, Florida, U.S.A. (scale in mm).

spaced locations suggest that there may be several subtypes.

Sidhu (1968) reported that several mangrove species of the family Rhizophoraceae have small chromosomes and perfect meiotic divisions and also that the diploid chromosome for three genera and six species of mangroves in the family was 36. Chromosome numbers in the two Asian species of the genus *Rhizophora* from India (*R. conjugata* = *R. apiculata*, and *R. mucronata*) that he studied were identical, $N = 18$ (Sidhu, 1968). No information is available on the chromosomal number of *R. mangle*.

Materials and methods

Propagule counts were made by enumerating the green and albino propagules limb by limb from the ground or by climbing the tree, or in some cases by harvesting the propagules.

Results and discussion

Over a period of five years eleven hundred ten green propagules and three hundred thirty-one albino propagules on 21 trees were counted at three localities in the Miami area of south Florida and in Puerto Rico. In most cases where trees were noted as bearing albino propagules, complete counts could not be made because the trees were too tall or propagule fall had begun. The data in Table 1 are separated into 'complete counts' and 'partial counts'. 'Complete counts' are counts of all or a substantial fraction of the propagules on a tree before appreciable propagule fall. 'Partial counts and small trees' are counts made after propagule fall had begun, counts on tall trees where counting precision was reduced or counts on small trees with few propagules. As can be seen in Tables 1 and 2, counts of propagules on five trees where all or a substantial fraction of the total propagules could be counted, did not differ significantly from a 3:1 ratio. However, the albinos in the pooled partial counts were significantly fewer than 3:1.

Table 1. Green and Albino Propagules on *Rhizophora mangle* Trees.

Location	Numbers of propagules		
	Green	Albino	Total
Complete counts of large trees[a]			
Chapman Field Park, Miami, FL	101	37	138
Matheson Hammock, Miami, FL	228	88	316
	87	27	114
Saga Bay, Miami, FL	150	41	191
Punta Verraco, Puerto Rico	189	58	247
Total Complete Counts	755	251	1006
Partial counts and small trees[a]			
Matheson Hammock Park, Miami, FL	5	3	8
	50	10	60
	11	3	14
	17	3	20
	6	3	9
	14	5	19
	34	10	44
	81	11	92
	4	2	6
	15	2	17
	3	1	4
	16	3	19
	8	1	9
Chapman Field Fark, Miami, FL	21	10	31
Saga Bay, Miami, FL	17	6	23
	53	7	60
Total partial counts	355	80	435
Total all counts	1110	331	1441

[a] See text.

The data thus suggest a 3:1 ratio from self-fertilization with a deficiency of albino propagules. This recorded deficient number of albinos might be an artifact from earlier fall of albino compared to green propagules. Data are not available to evaluate this possibility.

Another possible explanation for the deficient numbers of albino propagules is that, although self-pollination ordinarily occurs, there is some outcrossing. The difference between trees with 3:1 ratios and

Table 2. Observed numbers of green and albino propagules and tests for goodness-of-fit to an expected 3:1 ratio.

Location	Number of Propagules				
	Green	Albino	X^2	P	Conclusion[a]
Complete counts					
Chapman Field	101	37	0.196	.50–.70	NS
Matheson Hammock	228	88	1.367	.20–.30	NS
	87	27	0.105	.70–.80	NS
Saga Bay	150	41	1.282	.20–.30	NS
Punta Verraco	189	58	0.304	.50–.70	NS
Total complete counts	668	224	0.006	.80–.95	NS
Partial counts					
Total	355	80	10.134	>.01	SIG

[a] NS = not significant as deviation from 3:1 ratio.
SIG = significant as deviation from 3:1 ratio.

ones with a deficient number of albinos may thus correspond to the ratios of albino seedlings that Squillace and Kraus (1969) obtained in *Pinus elliottii* where progenies of heterozygous trees showed 3:1 ratios of normal to albino following controlled self-pollinations, but variable smaller fractions of albinos after wind pollination, i.e. natural self-fertilization and outcrossing. On a forest-wide basis most pollen carried by the wind or insects in a *Rhizophora* mangrove forest would bear the normal rather than albino allele. Thus the reduction in the fraction of albino propagules on a heterozygous mother tree should be proportional to the amount of outcrossing, and the number of albino propagules would be reduced by approximately the percentage of outcrossing. However, the relatively small reduction in albino propagules from the theoretical one fourth that was recorded in the present study suggests that one of the following is the explanation: that fertilization usually occurs before the flowers open, that pollen is not often carried from one tree to the next by insects or wind, or that both of these explanations are effective. Guppy's (1906) finding that anthers dehisce before flowers open in two Fijian *Rhizophora* species may be the basis for the 3:1 ratios that we have obtained in *R. mangle*. The finding of 3:1 ratios of normal and albino propagules itself established the self-compatibility of the species.

Acknowledgments

The authors wish to thank Sally Black and Dean Richardson for field assistance, and The Oak Ridge Associated Universities for support of studies in Puerto Rico.

Literature cited

Anderson, E., R. Jansson and D. Lindgrin. 1974. Some results from second generation crossings involving inbreeding in Norway spruce (*Picea abies*). Silvae Genetica 23:34–43.

Butany, W.T. and M. Singh. 1965. The inheritance of chlorophyll deficiency in cotton. Indian Journal of Genetics and Breeding 25:85–90.

Filho, L.E. dem. 1969. Albinism em *Adansonia digitata* L. (Bombacaceae). Boletim do Museu Nacional No. 38 (Nov.), pp. 1–3.

Franklin, E.C. 1969. Mutant forms found by self-pollination in Loblolly pine. Journal of Heredity 60:315–320.

Gabriel, W.J. 1962. Inbreeding experiments in sugar maple (*Acer saccharum* Marsh.) – early results. Ninth Northeastern Forest Tree Improvement Conference Proceedings.

Gill, A.M. and P.B. Tomlinson. 1969. Studies on the growth of red mangrove (*Rhizophora Mangle L.*) 1. Habit and general morphology. Biotropica 1:1–9.

Gill, A.M. and P.B. Tomlinson. 1971. Studies on the growth of the red mangrove. (*Rhizophora mangle L.*) 3. Phenology of the shoot. Biotropica 3:109–124.

Guppy, H. 1906. Observations of a naturalist in the Pacific between 1896 and 1899. Vol. II. Plant dispersal. MacMillan, New York.

Johnson, A.G. 1948. Albinism in the Austrian Pine. Journal of Heredity 39:9–10.

Sidhu, S.S. 1968. Further studies on the cytology of mangrove species of India. Caryologia 21:353–357.

Sinnott, E.W., L.C. Dunn and T. Dobzhansky. 1958. Principles of Genetics. McGraw-Hill, New York.

Snyder, L.H. 1951. The Principles of Heredity, D.C. Heath, Boston.

Squillace, A.E. and J.F. Kraus. 1963. The degree of natural selfing in slash pine as estimated from albino frequencies. Silvae Genetica 12:46–50.

Srb, A.M. and R.D. Owen. 1952. General Genetics. W.H. Freeman, San Francisco.

Teas, H.J. 1977. Ecology and restoration of mangrove shorelines in Florida. Environmental Conservation 4:51–58.

Teas, H.J. and S.H. Handler. 1979. Notes on the Pollination Biology of *Rhizophora mangle L.* Proceedings of International Symposium on Marine Biogeography and Evolution in the Southern Hemisphere 2:357–361.

Tomlinson, P.B. 1978. Rhizophora in Australasia, some clarification of taxonomy and distribution. Journal of the Arnold Arboretum 59:156–169.

Tomlinson, P.B., R.B. Primack and J.S. Bunt. 1979. Floral morphology in relation to pollination biology in mangrove Rhizophoraceae. Biotropica 11:256–277.

Venkatesh, C.S. and C.J.S.K. Emmanuel. Spontaneous chlorophyll mutations in *Bombax* L. Silvae Genetica 25:137–138.

Venkatesh, C.S., R.S. Arya and R.C. Thapliyal. 1978. An albina type natural chlorophyll mutant in *Gmelina arborea* Roxb. Silvae Genetica 27:40–41.

CHAPTER 14

Significance of cryptovivipary in *Aegiceras corniculatum* (L.) Blanco

LEELA J. BHOSALE and L.S. SHINDE

Department of Botany, Shivaji University, Kolhapur, 416004, India

Abstract. Cryptovivipary was studied in the mangrove *Aegiceras corniculatum* (L.) Blanco from western India. Na, K, Ca, Cl, Mg and P were measured in seedling parts at three developmental stages. Chloride ion uptake and distribution was followed by use of the radioisotope ^{36}Cl, taken up through the plant stem. Photosynthate distribution was followed from leaves to fruits and seedling parts using the radioisotope ^{14}C, applied to an adjacent leaf as a buffered bicarbonate solution.

It was found that the cryptoviviparous seedlings of *A. corniculatum* were nutritionally dependent on the parent and that this dependence was greatest during early seedling development. At all stages the translocated carbon was most concentrated in the embryo. Evidence from ^{36}Cl experiments indicated that there are tissue barriers on the ion transport at two sites: between the stalk/calyx and the fruit, and between the seed coat and embryo. Chloride analyses of seedlings in nature supported the ^{36}Cl uptake studies. Translocation of Cl to the seedlings was maximal toward maturity when they are developing salt tolerance against the time when they are shed and must grow on saline soil.

Introduction

Mangroves have acquired modified characteristics as a response to saline environment. The effect of salinity as a specific and dominant factor of the saline environment determines, to a great extent, the ability of halophytes to reproduce. The germination of seeds is usually retarded by high concentration of salt solutions both in glykophytes and halophytes (Adriani 1958; Henckel 1963; Chapman 1976). Toxic effects of certain ions have been investigated by number of workers. Stewart (1898) and Harris and Pittman (1918, 1919) reported that the chloride salts are more toxic for seed germination as compared to sulphate. Mangroves appear to have undergone modifications with respect to phenology and mode of germination as a means of overcoming the toxic effect of chlorides on seed germination. The adaptations for germination of mangrove seeds proceed along different lines in different species (Henckel, 1963). Variously they exhibit vivipary, cryptovivipary or no vivipary. The physiology of vivipary has been investigated amply by Pannier (1962), Pannier and Rodriguez (1967), Lotschert and Liemann (1967) and Joshi *et al.* (1972), but cryptovivipary has been little studied.

The present investigation deals with physiology of cryptovivipary in *Aegiceras corniculatum* (L.) Blanco. In the case of vivipary it was reported by Haberlandt in 1895 that the seedling is dependent on the mother plant for the nutrients. What happens in the case of cryptovivipary was not clear. However, Joshi *et al.* (1972) have shown that for both viviparous and cryptoviviparous seedlings salt tolerance increases with the age of the seedlings. The present study reports the results of ^{36}Cl-

Tasks for vegetation science, Vol. 8 ed. by H.J. Teas
 ISBN 978-90-481-8526-9

translocation from the mother plant to the seedlings and the translocation of ^{14}C-photosynthates from the leaves to the seedlings.

Material and methods

The fruits of *Aegiceras corniculatum* were collected from Ratnagiri (17° 0′ N and 73° 3′ E) and brought to the laboratory in air-tight polyethylene bags. The fruits contain the germinated seed, here referred to as the seedling, which was used for analysis. An attempt has been made to find out the nature of the barrier to Na and Cl translocation; for this purpose different parts of the fruit were analysed, including the seedling. The stages were established based on the length (age) of the seedling inside the fruit. Stage I, less than 1 cm; stage II, from 1–3 cm; and stage III, greater than 3 cm length. The subtending leaves were also analysed. For translocation studies entire twigs were used; a separate twig was utilized for each stage of seedling growth.

Inorganic elements were determined from oven dried material. Na, K and Ca were estimated flame photometrically. The method described by Richards (1954) was used for magnesium determination. Chloride and phosphorus estimations were carried out by following the methods of Volhard (1956) and Sekine (1965) respectively.

Translocation studies

The experiments were designed according to the method used by Pannier and Pannier (1975). Healthy twigs with seedlings were selected and the basal ends cut under water. The experiment was arranged as shown in Fig. 1 for ^{36}Cl studies and as shown in Fig. 2 for translocation of photosynthates. The twig was allowed to take up ^{36}Cl for 24 hours under natural conditions of day and night. For the study of translocation of photosynthates the plant was exposed to natural sunlight for 6 hours. The $NaH^{14}CO_3$ (sp. act. 50 mCi/mmole) was added to the Tris-HCl buffer, pH 8.0 and the twig base was dipped into tap water.

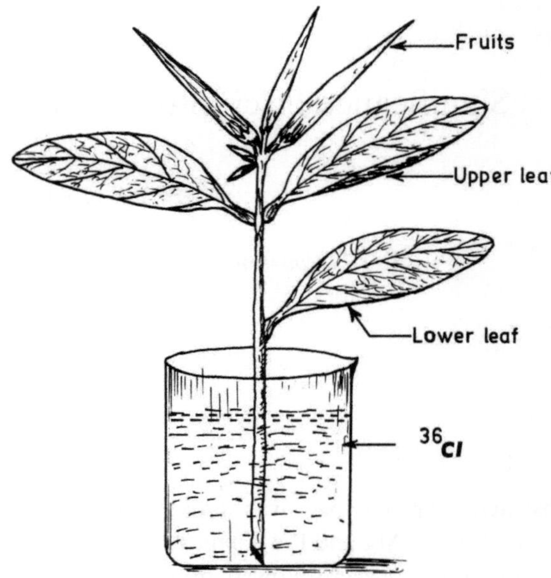

Fig. 1. Design of the experiment for uptake of 36 Cl.

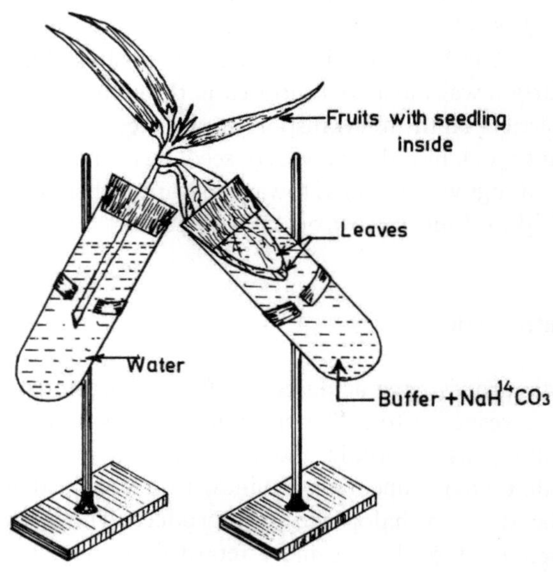

Fig. 2. Design of the experiment for translocation of photosynthates.

Results and discussion

The results of analysis of *Aegiceras corniculatum* are presented in Tables 1 and 2. Table 1 shows the inorganic constituents in different parts of the seedling as well as the subtending leaves. Table 2

Table 1. Translocation of mineral elements in cryptoviviparous seedlings of *Aegiceras corniculatum* (values expressed as g/100 g dry tissue).

Stages	Na	K	Ca	Cl	P	Na/K
Leaf	2.30	0.53	0.35	2.39	0.094	4.34
Stage I						
Stalk and Calyx	3.40	0.64	0.31	2.22	0.062	5.31
Fruit wall	3.00	1.44	0.18	2.92	0.075	2.08
Seed coat	2.50	1.62	0.09	1.21	0.105	1.54
Embryo (seedling)	1.55	0.88	0.04	0.86	0.125	1.78
Stage II						
Stalk and Calyx	3.00	1.20	0.35	2.46	0.076	2.50
Fruit wall	2.45	1.42	0.12	2.91	0.063	1.72
Seed coat	2.75	1.78	0.11	1.21	0.175	1.56
Funiculus	2.55	1.22	0.14	1.80	0.240	2.09
Embryo (seedling)	1.75	1.12	0.06	0.86	0.100	1.56
Stage III						
Stalk and Calyx	3.50	0.65	0.27	0.92	0.063	5.38
Fruit wall	2.85	0.84	0.14	2.91	0.075	3.39
Seed coat	2.95	0.85	0.08	1.23	0.175	2.76
Funiculus	1.98	0.80	0.10	1.21	0.236	2.97
Embryo (seedling)	1.60	0.81	0.06	1.20	0.112	1.98

Table 2. Translocation of ^{14}C-photosynthates and ^{36}Cl to seedlings parts at different growth stages.

Stage	parts	Percent of total activity left with		Percent activity translocated		Relative percentage of activity translocated	
		^{14}C-photo-synthates	^{36}Cl	^{14}C-photo-synthates	^{36}Cl	^{14}C-photo-synthates	^{36}Cl
I	Leaf	72.77	—	—	Lower 9.26 Upper 25.31	—	Lower 21.86 Upper 58.32
	Stem	—	57.62	1.55	—	16.80	—
	Calyx and stalk	—	—			9.56	13.76
	Fruit wall	—	—	25.68	7.81	8.14	2.46
	Seed coat	—	—			7.32	1.84
	Embryo	—	—			58.18	1.76
II	Leaf	69.24	—	—	Lower 4.18 Upper 7.48	—	Lower 24.21 Upper 43.32
	Stem	—	83.72	4.99	—	16.30	—
	Calyx and stalk	—	—			8.62	17.09
	Fruit wall	—	—	25.77	5.62	2.57	6.77
	Seed coat	—	—			8.71	6.18
	Embryo	—	—			63.80	2.40
III	Leaf	81.74	—	—	Lower 8.89 Upper 22.36	—	Lower 18.35 Upper 46.14
	Stem	—	51.54	7.79	—	42.64	—
	Calyx and stalk	—	—			5.93	17.97
	Fruit wall	—	—	10.47	17.21	10.32	4.98
	Seed coat	—	—			6.45	7.54
	Embryo	—	—			34.66	5.12

— = not applicable

presents the translocation of ^{14}C-photosynthates and uptake and translocation of ^{36}Cl, by the stem to the leaves and the seedlings. It is evident from Table 1 that Na and Cl concentrations in the embryo remained low compared to the leaves, although they slowly increase with age. Sodium accumulation was found in the stalk and calyx, the values being higher than in the leaves. From stalk and calyx to fruit wall there was a drop in Na content, suggesting the possibility of a barrier between the stalk and calyx and the fruit wall.

There was a second drop in the values of sodium from seed coat to embryo. This must be the other position for a barrier of Na. A similar trend was observed with chlorides. The chloride percentage in the initial stage of development was 0.86 which increased to 1.2. This shows that the barrier (tissue) behaves in a special manner so that as the seedling ages, greater amounts of Na and Cl are allowed to pass to the seedling therby increasing the internal salt concentration, as well as the tolerance to soil salinity, in the mature seedling. It is important to note that the Na/K ratio in the leaves, which was 4.34, dropped to 1.78 in the embryo during the first stage of development.

Calcium remained more or less constant in the embryos (seedlings) and the level was very low, possibly due to a reduced mobility of this cation. Magnesium (values not reported) increases slightly, but to a lesser extent than calcium. It is interesting to note that phosphorus was higher in the embryos than in the leaves. During early growth more phosphorus was translocated to the embryos; however, with increasing age more P was left with the seed coat and at a later stage was retained by the funiculus.

It has been shown (Walter, 1961) that viviparous seedlings of *Rhizophora mangle* absorb considerable amounts of water through the glandular tissue. There is difference in osmotic potential of the cell sap of the mother plant and that of the seedlings: translocation of water is against the osmotic gradient, and therefore the moisture percentage of the seedlings remains higher. The dry matter production, which is low during initial stages of growth, increases with seedling maturity. Similar observations were reported by Pannier and Pannier (1975) for *R. mangle* propagules at progressive developmental stages.

Translocation of mineral elements

The present work reports increases in sodium and chloride ions with the age of the seedlings. This is in support of an earlier result of Joshi *et al.* (1972). Rains and Epstein (1967) have reported that Na/K ratio of sea water which is 40 becomes reduced to 7 in the leaves.

In the present paper the Na/K ratio was 4.5 in the leaves of *A. corniculatum*. As this ratio increases with the developmental stage of the seedling, induction of salt tolerance is evident. Sutcliffe (1962) has stated that halophytes have the capacity to accumulate potassium in preference to sodium. Greenway (1967) claimed that NaCl reduces K-uptake. So far as cryptoviviparous seedlings of *A. corniculatum* are concerned, there is no deficiency of K, which supports Sutcliffe's observation. Epstein (1965) has shown a preferential absorption mechanism of K in *Avicennia marina*. The problem of mangroves is not the uptake of sodium but it is the need to take up K from the environment in which there are high Na/K ratios. Joshi *et al.* (1972) have shown that *Ceriops tagal* (*C. candolleana*) has the most efficient K-uptake mechanism. According to Onal (1964) K content in coastal halophytes decreases with age. Similar observations were made by Lotschert and Liemann (1967) who reported that Na, K and Ca are partly increased and partly decreased during various stages of growth. Pannier and Rodriguez (1967) found that in *R. mangle* seedlings glandular tissue is responsible for more translocation of K than Na.

Attention is drawn to the works of Toole *et al.* (1956), Crocker and Barton (1953), Koller (1955) and Waisel (1958) who reported that germination of seeds is a critical period in the plant life cycle and inhibition of seed germination by high salt concentration may limit species ranges. According to Stewart (1898) and Harris and Pittman (1918, 1919) chloride salts are more toxic for seed germination than is sulphate. A similar observation was made by Chapman (1976). Therefore one of the objects of a plant species resorting to the viviparous or cryptoviviparous type of germination must be to keep the developing seedling away from the high Cl content of the substratum.

Walter and Steiner (1936) postulated the presence of a glandular tissue between the fruit wall and seed coat in *Rhizophora* seedlings which checks the entry of harmful ions to the seedlings from the mother plant. Pannier (1962) showed that such a specialized tissue in *R. mangle* had high acid phosphatase activity, characteristic of glandular tissue.

The present investigation indicates that a tissue similar to those reported by Walter and Steiner (1936) and Pannier (1962) exists at two sites in cryptoviviparous seedlings of *A. corniculatum*. The first site appears to be between the stalk and calyx and the fruit wall and the second one between the seed coat and the embryo. The chloride analysis of cryptoviviparous seedlings in nature support the ^{36}Cl-uptake and translocation experiments.

^{36}Cl translocation

^{36}Cl uptake from the stem is translocated to the leaves and to the seedlings. Large amount of ^{36}Cl is retained in the stem, a major quantity is translocated to the leaves and a very minor amount to the seedlings. Mothes (1961) and Pannier (1962) suggested that higher chloride values in the vegetative parts than in the hypocotyl represented a decrease in Cl content from the pericarp to the seedlings. The present investigation also indicates a drop in Cl contents from the fruit wall to the embryo (seedlings). The highest amount was found in the stalk and calyx. The amount of uptake was increased during development, the greatest translocation being observed at the mature stage. This observation clearly shows that the permeability of the barrier tissue varies with the age of the growing seedling. It is interesting to note that during stages II and III the Cl translocation to different parts of the seedlings, except the embryo, is more or less the same (30.0 and 30.5 respectively); however, the amount of Cl translocated to the embryo is double in case of stage III (Table 2, last column, sum of the radioactivity shown against calyx and stalk, fruit wall and the seed coat). This is suggestive of a tissue barrier that is sensitive to the age of the seedling, one allows more chloride to pass as the embryo ages.

Thus the experiment performed with ^{36}Cl confirms the conclusions drawn from analyses of the inorganic elements.

Translocation of photosynthetic products

Because it has been shown that viviparous seedlings exhibit nutritional dependency, it was decided that translocation of photosynthates in the cryptoviviparous seedlings of *A. corniculatum* should be investigated.

The nutritional inter-relationship between the components of two living systems as represented by the young developing *A. corniculatum* seedling and the mother plant can be derived from the data in Table 2. $^{14}CO_2$ provided to the subtending leaves was distributed in different parts. This clearly indicates that there exists a transport of labelled assimilates from the mother plant to the seedling which gives experimental proof of the existence of nutritional dependency in cryptoviviparous seedlings. That this relationship is more intimate during early stages of development of the cryptoviviparous seedling is illustrated in Fig. 3.

Brown (1968) showed in cotton (*Gossypium*) that from the upper leaf the assimilates descend through the stem into the nearby bolls (fruits). Pannier and Pannier (1973) have investigated translocation of photosynthetic products in viviparous seedlings of *R. mangle*. They reported that most of the asssimilates are translocated to the seedlings which are in the immediate vicinity of the treated leaf, irrespective of the age of the seedling. *A. corniculatum* is different from *R. mangle* in that in *A. corniculatum* the seedlings in the axil of one leaf are all of the same age, thus for different age groups different twigs must be selected. In the case of *R. mangle* each seedling is supported by a separate leaf.

The results presented in Table 2 clearly show that translocation takes place from the leaf to the seedlings and that it is age dependent. The Table shows that if the relative percentage of radioactivity translocated is calculated (column 5 of the Table), very little activity moves to the stem from the leaves (16.8 and 16.3 percent), the major part being

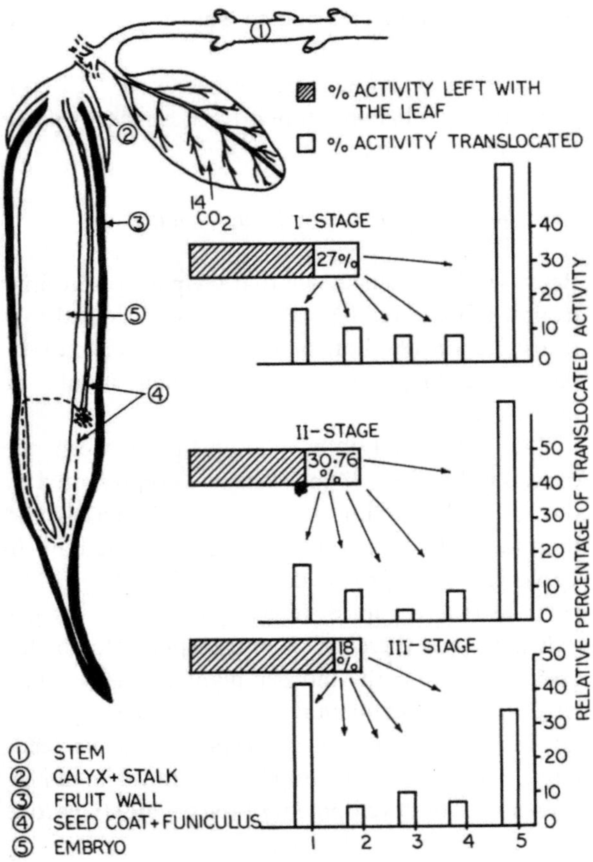

Fig. 3. Translocation of photosynthates from the leaf to different parts; total translocated activity has been split into relative percentage.

translocated to the seedlings in the axils of the leaves, 83.2 and 83.7 percent during stages I and II respectively. This is also shown in Fig. 3. This value decreases to 57.4 at maturity when the stem shows 42.6% accumulation of photosynthates. When different parts of the fruit are considered it is evident that stage II is the most active stage of growth when the supply of food material from the mother plant is maximum (63.8%). At maturity this drops to 34.7%).

It should be noted that the translocation of chloride is maximum (5.12%) during the mature stage while that of assimilates is minimal. During stage III the seedling inside the fruit becomes mature and is ready to be shed. It is obvious that at this stage the seedling must be independent of the mother plant and should have acquired the property of salt tolerance so that the salinity of the soil can be tolerated.

Pannier and Pannier (1973) have noted the accumulation of reserve food during the developmental stages of *R. mangle*. Bhosale (1974) and Joshi *et al.* (1972) have reported a similar observation where the percentage of starch is maximum at the mature stage of the seedling. Thus, enough food, possibly, accumulated starch, checks the translocation of photosynthates during mature stage. Wareing and Patrick (1975) have shown that net assimilation must always be in balance with net consumption and the relationship between the source and sink of assimilate is dynamic.

Humphries (1967) reported that a leaf photosynthetic potential may be limited by the degree to which photosynthate transport and utilization occur. Thorne and Ronald (1974) reported that the assimilate demand has a marked influence on source leaf photosynthesis and carbohydrate formation and export. These two reports probably explain the lower rate of translocation to the seedlings during the mature stage that we have found in *A. corniculatum*.

Acknowledgements

The authors are grateful to the late Prof. G.V. Joshi for his encouragement during this study, which was supported by DST project No. 7(15)/76-SERC, Government of India.

Literature cited

Adriani, M.J. 1958. Halophyten. In: Encyclopaedia of Plant Physiology Vol. IV:709–736, Springer Verlag, Berlin.

Bhosale, L.J. 1974. Physiology of salt tolerance of plants. Ph.D. Thesis, Shivaji University, Kolhapur, India, 293 pp.

Brown, K.J. 1968. Translocation of carbohydrates in cotton: movement to the fruiting bodies. Ann. Bot. 32:703–707.

Chapman, V.J. 1976. Coastal vegetation. 2nd Ed., Pergamon Press. Oxford, N.Y., 292 pp.

Crocker, W. and L.V. Barton. 1953. Physiology of seeds. Chronica Botanica, Waltham, Massachusetts.

Epstein, E. 1965. Mineral metabolism. In: Plant Biochemistry, pp. 348–366. Academic Press, N.Y.

Greenway, H. 1965. Plant response to saline substrates. VII. Growth and ion uptake through plant development in two varieties of *Hordeum vulgare*. Aust. Jour. Biol. Sci. 18:763–779.

Haberlandt, G. 1893. Ueber die Ernaehrung der Keimlinge und die Bedeutung des Endosperms bei viviparen Mangrovepflanzen. Ann. Jard. Bot. Buitenzorg 12:91–116.

Harris, F.S. and D.W. Pittman. 1918. Soil factors affecting the toxicity of alkali. Jour. Agr. Res. 15:287–319.

Harris, F.S. and D.W. Pittman. 1919. Relative resistance of various crops to alkali. U.S. Utah Agr. Exp. Sta. Bull. 168.

Henckel, P.A. 1963. On the ecology of mangrove vegetation. Mitt. Flor. Soz. Arb. Gemeinsch. N.F. 10:Stolzenau/Weser: 201–205.

Humphries, E.C. 1967. The dependence of photosynthesis on carbohydrate sink: current concepts. Proc. First Symp. Tropical Root Crops, Univ. West Indies, St. Augustine, Trinidad, 2:34–45.

Joshi, G.V., M. Pimplaskar and L.J. Bhosale. 1972. Physiological studies in germination of mangroves. Bot. Mar. 15:91–95.

Koller, D. 1955. Regulation of germination in seeds. Bull. Res. Counc., Israel 5D:85–105.

Lotschert, W. and F. Liemann. 1967. Accumulation of salt in the embryo of *Rhizophora mangle* L. developing on the mother plant. Planta. 77:142–156.

Mothes, K. 1961. Aktiver Transport als regulatives Prinzip fur gerichete Stoffverteilung in hoheren Pflanzen. 12 Koll Ges: fur Physiol Chemie 5:189–207.

Onal, M. 1962. Zusammensetzung des Zellsaftes einiger Salzmarshen und Dunenpflanzen in der Umgebung von Neapel. Beit Phytol. (Walter-festschrift). 30:89–100.

Pannier, F.P. 1962. Estudio fisiologico sobre la viviparia de *Rhizophora mangle* L. Acta Cient Venez. 13:184–197.

Pannier, R.F. and F. Pannier. 1973. Determination de substancias de tipo gibberellina en tejidas de *Rhizophora mangle* L. en diferentes etapas de desarrollo. Acta Cient. Venez. 24 (Suppl. 1):33–34.

Pannier, F. and R.F. Pannier. 1975. Physiology of vivipary in *Rhizophora mangle* L. Proc. Int. Sym. Biol. Mgt. Mangroves. Vol. II:632–639.

Pannier, F. and M.P. Rodriguez. 1967. The β-complex inhibitor and its relation to vivipary in *Rhizophora mangle*. L. Int. Res. Ges. Hydrobiol. 52:783–792.

Rains, D.W. and E. Epstein. 1967. Preferential absorption of potassium by leaf tissue of the mangrove *Avicennia marina*. An aspect of halophytic competence in coping with salt. Aust. Jour. Biol. Sci. 20:847–857.

Richards, L.A. 1954. Diagnosis and improvement of saline and alkali soils. USDA Book No. 60(56):130–131.

Sekine, T. 1965. Photoelectric colorimetry in biochemistry, Part II. Nanko-d-Publishing Co. Tokyo, 242 pp.

Stewart, J. 1898. Effect of alkali on seed germination. 9th Ann. Rept. Utah. Agr. Exp. Sta.:26–35.

Sutcliffe, J.F. 1962. Mineral salts absorption in plants. *Pergamon* Press. Oxford, London, 194 pp.

Toole, E.H., S.B. Borthwick and V.K. Toole. 1956. Physiology of seed germination. Ann. Rev. Pl. Physiol. 7:299–324.

Volhard, A. 1956. Chlorides. Methods of plant analysis. Springer-Verlag, Berlin, Vol. I, 487 pp.

Waisel, Y. 1958. Germination behaviour of some halophytes. Bull. Res. Counc. Israel, 6D:187–189.

Walter, H. 1961. The adaptation of plants to saline soils. Arid Zone Research, Vol. XIV, Salinity problem in the arid zone, Published by UNESCO.

Wareing, P.F. and J. Patrick. 1975. Source-sink relation and the partition of assimilates in the plants. In: Photosynthesis and Productivity in Different Environments, pp. 481–499, Cambridge Univ. Press, London.

CHAPTER 15

Succession in zoned mangrove communities: where is the climax?

I.M. JOHNSTONE

NZ Electricity, P.O. Box 445, Hamilton, New Zealand

Abstract. Data are presented which suggest the mangal 'climax' is not necessarily at the rear of the mangrove swamp. Biomass and soil nutrients are greatest in the *Bruguiera* and/or *Ceriops* zone and leaf area ratio and environmental fluctuations (temperature) are least in the *Bruguiera* zone. Supporting observations are also discussed. It is suggested that if the mangrove climax is separated from the land by a series of sub-climax zones, mangrove and terrestrial succession are independent and it is not justified to view the mangal solely in terms of a pioneer assemblage leading to a terrestrial climax; but, rather, the mangal should be considered as an entire community with its own climax.

Introduction

The generally accepted model of succession for mangrove communities parallels that of the classical freshwater hydrosere, that is:

open lake → aquatic macrophyte system → swamp → terrestrial system

sea → (seagrass) → mangroves → strandline or freshwater swamp → terrestrial system.

In the classical freshwater hydrosere, inorganic and organic sedimentation paves the way for the terrestrial community. For example, many lakes ultimately fill with sediment and, thus, the aquatic communities in these lakes can be considered transitory.

Numerous authors, summarized by Chapman (1976), have described mangrove succession as leading to a terrestrial climax. That mangroves can play the role of land-formers under some environmental regimes is well known and not disputed, neither is it denied that terrestrial, mangrove and oceanic systems are coupled by both physical and biological transport processes. However, depending on sealevel fluctuations, sedimentation rates (which may be positive or negative) and decomposition rates of the mangrove detritus, the replacement over time of a mangrove community by a terrestrial community is by no means preordained (Thom 1967). In addition, it does not necessarily follow that there must be a shoreward sequence of succession within the mangal *per se*, even though such a species replacement shoreward has been the normal manner in which the succession of mangroves has been visualized.

The basic characteristic of an obligatory successional sequence (irrespective of the mechanism) is that species A (the colonizer) is replaced by species B, which is replaced by species C, etc., until the climax species is reached, which is the only species capable of self replacement. In a broader context

Tasks for vegetation science, Vol. 8 ed. by H.J. Teas
 ISBN 978-90-481-8526-9

one can think of community A being replaced by community B and so on. Whether Clemensian, Hornian or otherwise, given this sequence of events, the basic questions I propose are: do the mangrove species nearest the shore represent the final stage in a sequence of mangrove species and do terrestrial communities represent the climax for mangrove systems?

On the low rainfall areas of the Papuan coast, the sharp demarcation between terrestrial communities and the mangal both in space (a plantless zone frequently separates the two communities) and structure (terrestrial plants invading the shoreward fringe of the mangal are extremely rare) suggested that the mangal is a community in its own right (rather than just an assemblage of colonizing plants), and that replacement of the mangal by a terrestrial community is not obligatory. (In high rainfall areas freshwater runoff, by reducing the salinity in the back mangrove zone, can create a situation in which terrestrial species can survive in the affected part of the mangal as long as the depressed salinity conditions are maintained.)

If the mangal is a community in its own right, a climax mangrove assemblage might exist, which need not necessarily be at the shoreward fringe of the mangal. If this hypothesis is recognized, then, four patterns of succession within the mangal become possible (Table 1). It is also implied that if a climax assemblage can be identified, the existence of a developmental pattern would have to be ultimately reconciled with Thom's (1967) steady-state model of mangal development which is based on the assumption that biotic factors only play a minor role in the mangal and that every zone is fundamentally an environmentally cued climax.

The present paper is a preliminary attempt to identify the location of the climax community in zoned mangrove systems. Data are presented that suggest that the classical successional sequence (sea to land) may be an inappropriate model because parameters that are believed to correlate with ecosystem development fail to show that the most climax-like zone of the mangal is always at the landward fringe.

Materials and methods

The following environmental and ecological parameters were measured in the zoned mangal on the western side of Hood Lagoon (Central Province, Papua New Guinea) (Fig. 1):

(1) Zonation was determined by counting and identifying the plants in 100 m^2 contiguous quadrates along a transect (100° E of N) from the seaward to the landward fringe of the mangal. 0 to 600 m (landward) was completed in one day and 1000 to 600 m (seaward) the next day. The 600 m marks were not identical because the precise point marking the end of the first day's work could not be located on the second day, a problem which will be appreciated by those who have worked in such an extensive mangal.

(2) Soil samples were collected every 20 m along the transect.

(3) Soil nitrate, ammonia and phosphorus were determined using a Hach Soil Test Kit.

Table 1. Possible patterns of succession in a mangrove system.

Climax location	Zonation	Suggested successional mechanism
Seaward fringe	Present	Either: classical, obligatory sequential species replacement (Clemensian);
Landward fringe	Present	or: competition (the outcome being modulated by environmental factors);
At some point between the seaward and landward fringes	Present	or: a combination of Clemensian and competition mechanisms
No distinct community differences that permit the identification of a climax assemblage	Absent	Species replacement on a first arrival basis at disturbance created gaps

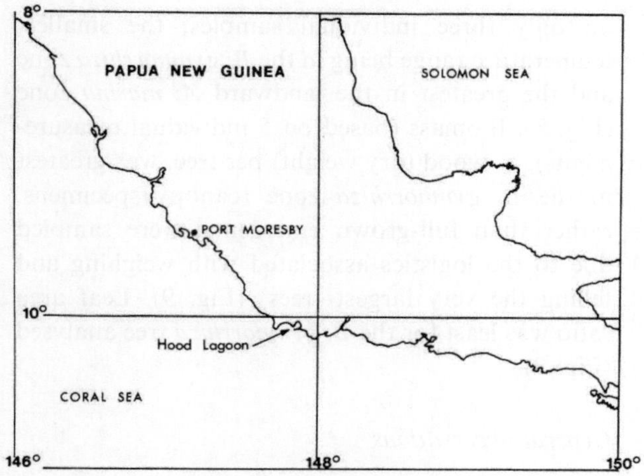

Fig. 1. Location of Hood Lagoon.

(4) Soil organic content was determined by combusting dried weighed samples at 550° for 12 hours.

(5) For the leaf area ratio, the leaves from a tree were weighed and the area of subsamples determined with an electronic area meter. The fresh-weight of the wood of the whole tree was determined (to the nearest kg) and the dry weight was determined on subsamples. Sampling sites are shown in Fig. 2.

(6) Air temperature (2 m above the ground) was measured with a max/min mercury thermometer over a 24 h period. Sampling sites are shown in Fig. 2.

In addition, general examples of species distributions and colonization patterns were drawn from observations made at numerous sites on the Papuan Coast.

Results

Hood Lagoon data

The zonation of mangrove species on the western shore of Hood Lagoon is well developed (Figs 2, 3 and 4). Species diversity (as species number) is least in the *Avicennia marina** (landward) zone and greatest in the shoreward fringe (Fig. 5). Plant density (excluding seedlings) is least in the *Bru-*

* That is, *A. marina sensu lato* as it exists from Arabia to New Zealand. The plants in New Guinea are possibly better considered as the separate species *A. eucalyptifolia.*

Fig. 2. Interface of *Ceriops tagal v. tagal* zone and *Avicennia marina* (rear) zone. Oblique view from 900′ altitude on IR film.

guiera gymnorrhiza zone and greatest in the *Ceriops tagal v. tagal* zone (Fig. 5). Soil nitrate concentrations were below detectable levels in all samples; ammonia and phosphate reach their highest levels in the *B. gymnorrhiza* and *C. tagal v. tagal* zones, decreasing both seaward and shoreward of these zones (Fig. 6). The content of organic matter in the soil is greatest in the *Bruguiera cylindrica*, *B. gymnorrhiza* and *C. tagal v. tagal* zones (Fig. 7). The diurnal air temperature variation results are based

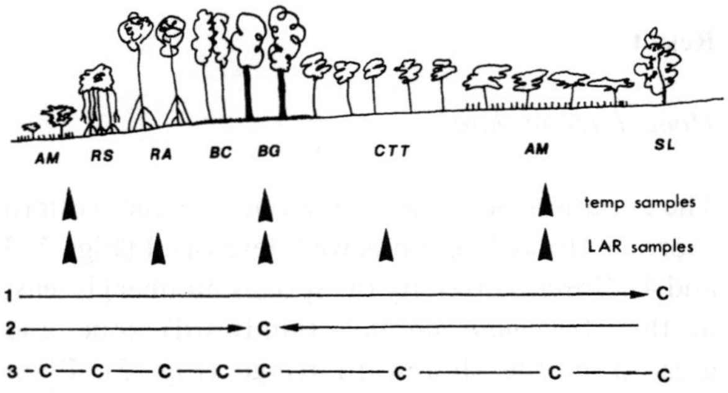

Fig. 3. Sketch of the main zonation pattern at Hood Lagoon. Zones: AM = *A. marina*, RS = *R. stylosa*, RA = *R. apiculata*, BC = *B. cylindrica*, BG = *B. gymnorrhiza*, CTT = *C. tagal v. tagal*, SL = strandline; 1 = classical seral model, 2 = proposed model, 3 = steady state model, C = climax.

on only three individual samples, the smallest temperature range being in the *B. gymnorrhiza* zone and the greatest in the landward *A. marina* zone (Fig. 8). Biomass (based on 5 individual measurements), as wood (dry weight) per tree, was greatest in the *B. gymnorrhiza* zone (canopy specimens, rather than full-grown emergents, were sampled due to the logistics associated with weighing and felling the very largest trees) (Fig. 9). Leaf area ratio was least for the *B. gymnorrhiza* tree analysed (Fig. 9).

General observations

(a) 'Double distributions' and distributions throughout zoned mangrove systems: many Papuan mangrove species have double distributions, reaching their greatest abundance both at the seaward and landward edges of the mangal (Johnstone and Frodin, in press). Most mangrove species are not restricted to a single zone, for isolated individuals can usually be found throughout the mangal, for example note the distribution of *Rhizophora stylosa* in Fig. 4.

(b) At Hood Lagoon, *C. tagal v. tagal* seedlings are frequent in the shade of *A. marina* on the

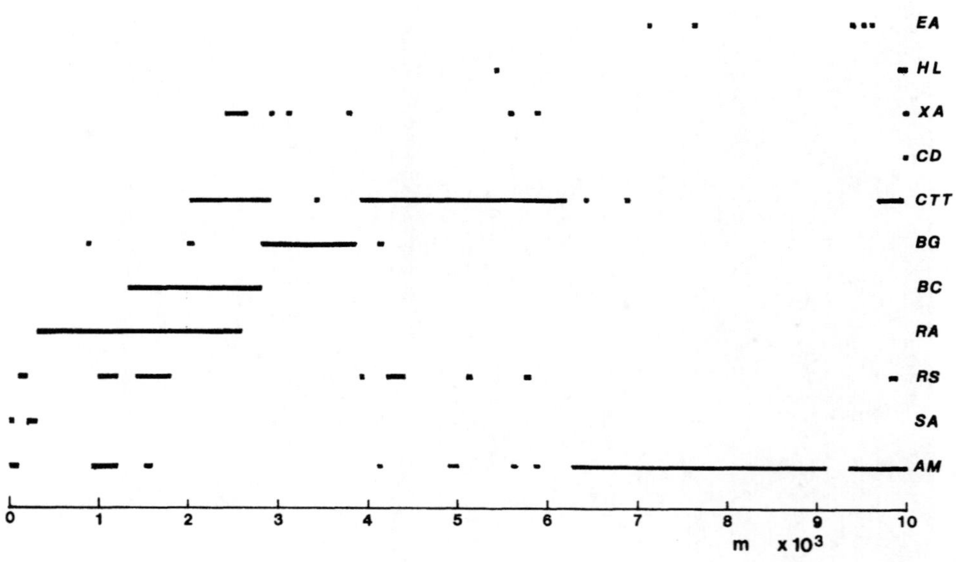

Fig. 4. Species distribution along transect from seaward fringe to shore.EA = *E. agallocha*, HL = *H. littoralis*, XA = *X. australasicus*, CD = C. decandra, CTT = *C. tagal v. tagal*, BG = *B. gymnorrhiza*, BC = *B. cylindrica*, RA = *R. apiculata*, RS = *R. stylosa*, SA = *S. alba*, AM = *A. marina*.

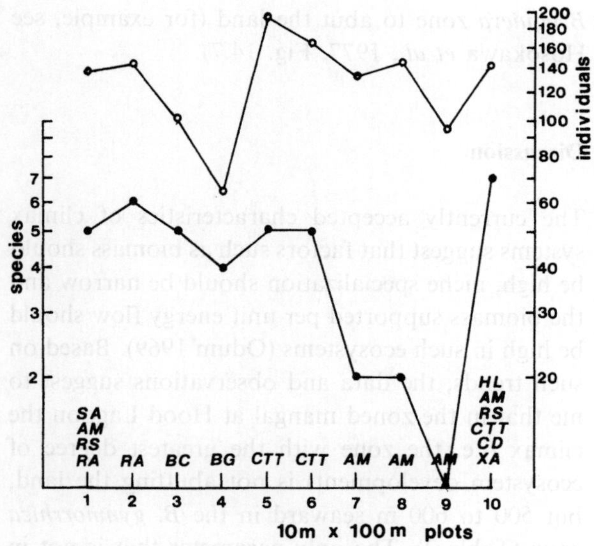

Fig. 5. Species number (●) and plant density (○) along transect. The species number of 5 plotted above the 1 indicates that 5 species were recorded in a 100 × 10 m band starting at the seaward fringe of the mangal; the subsequent plots are contiguous with the first. The 'dominant' species in each plot are noted (see legend Fig. 4 for abbreviations).

seaward side of the landward *A. marina* zone; *B. gymnorrhiza* seedlings are frequent in the *B. gymnorrhiza*, *B. cylindrica* and *Rhizophora apiculata* zones and, to a lesser extent, in the *C. tagal v. tagal zone*. Similar observations of the ability of *Bruguiera* and *Ceriops* seedlings to establish in the shade of a closed canopy have been made by Macnae (1968), but the seedlings of these species are by no means shade demanding, since it is not uncommon to find *B. gymnorrhiza* and *C. tagal v. tagal* seedlings growing in full sunlight.

(c) At Hood Lagoon overmature (dead and dying) *Rhizophora* are present in the *Bruguiera* zones, but very few seedlings and essentially no saplings of *Rhizophora* are present in the shade of the *Bruguiera* canopy. (Shade may not be the only factor excluding other mangroves from *Bruguiera* forest; the pavement of knee roots would seem to make the establishment of any seedlings relatively unlikely.)

(d) An abandoned village site in the *Bruguiera* zone at Hood Lagoon is overgrown with *A. marina*, *C. tagal v. tagal*, *Exocoecaria agallocha* and a few

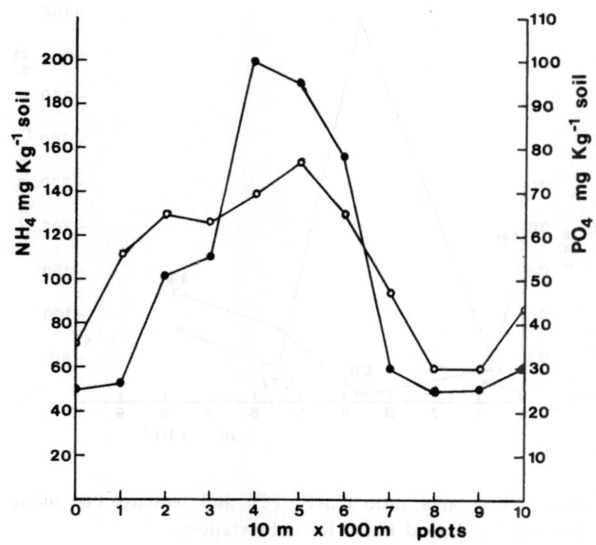

Fig. 6. Phosphate (●) and ammonia (○) concentrations along transect. The figures for plot 'O' represent samples from the bare sand seaward of the outer mangrove fringe. Each data point is the mean of replicated samples taken at 20 m intervals along each plot. See legend Fig. 5.

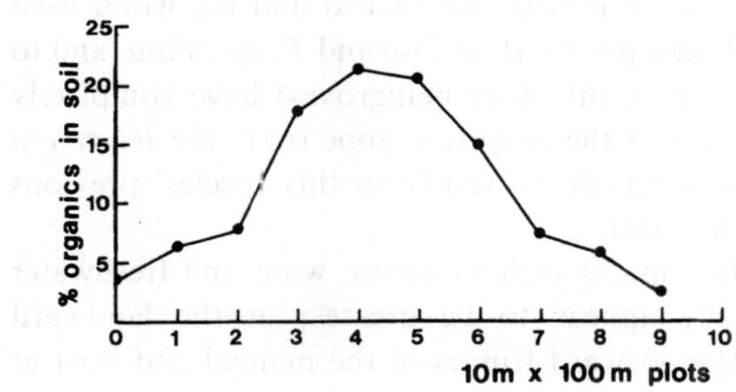

Fig. 7. Soil organic content. See legends Figs. 5 and 6.

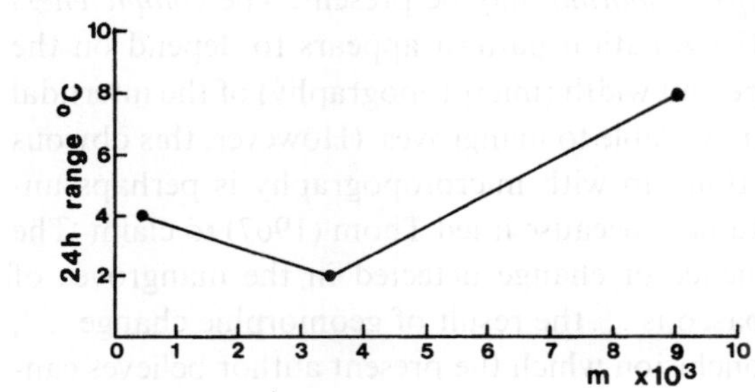

Fig. 8. 24 h temperature range (10/11.VII.79) along transect.

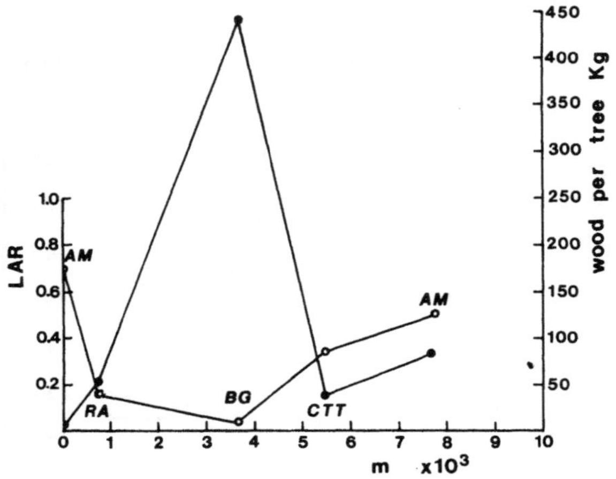

Fig. 9. Leaf area ratio (LAR) (○) and biomass (●) along transect. See legend Fig. 4 for abbreviations.

Heritiera littoralis. In undisturbed *Bruguiera* forests these species are very rare (suggesting competitive exclusion).

(e) In the Port Moresby area where *B. gymnorrhiza* has been eliminated in recent years from some swamps (it is selectively felled and the wood used for house posts), *R. stylosa* and *R. apiculata* (and to a lesser extent other mangroves) have completely taken over the *Bruguiera* zone (only the *Bruguiera* stumps remain to testify to this species' previous dominance).

(f) Impacts such as waves, wind and freshwater usually appear to be greatest at the landward and/or seaward fringes of the mangal and least at that point furthest from the open sea and the land.

(g) Sites exist on the Papuan coast where one or more of the back zones are absent – in the extreme case only a fringe of *Avicennia* and/or *Aegialites* and/or *Osbornia* may be present. The *completeness* of the zonation pattern appears to depend on the slope and width (microtopography) of the intertidal area available to mangroves. (However, this obvious relationship with microtopography is perhaps unfortunate, because it led Thom (1967) to claim 'The sequence of change detected in the mangroves of Tabasco is ... the result of geomorphic change ...', a conclusion which the present author believes cannot be extrapolated *in toto* to the Papuan mangal.) When back zones are absent, it is possible for the *Bruguiera* zone to abut the land (for example, see Hosokawa *et al.*, 1977, Fig. 14.7).

Discussion

The currently accepted characteristics of climax systems suggest that factors such as biomass should be high, niche specialization should be narrow and the biomass supported per unit energy flow should be high in such ecosystems (Odum 1969). Based on such trends, the data and observations suggest to me that in the zoned mangal at Hood Lagoon the climax (i.e. the zone with the greatest degree of ecosystem development) is not abutting the land, but 500 to 600 m seaward in the *B. gymnorrhiza* zone (Table 2). The only parameter that is not in agreement with this conclusion is 'species diversity'. However, this lack of agreement is probably an artifact resulting from the computation of species diversity as species number. Elsol and Saenger (1980) have computed the relative importance of each species, in a system similar to that at Hood Lagoon, from transect density, quadrat density and biomass data. Based on this measure no single species is dominant in the *Bruguiera* zone – as one might predict from theory if the *Bruguiera* zone is the mangal climax.

The preliminary observations reported in this paper are not new, only their interpretation would appear to be novel. Saenger and Robson (1977) report 'double distributions' (*Rhizophora*, *Aegiceras*, *Lumnitzera* and *Ceriops*), distributions throughout the mangal (*Avicennia*, *Rhizophora*), and clearly show that canopy height (and, by implication, biomass) is greatest towards the center of the zonation pattern for a mangrove system on the Queensland coast. The possibility of competition between mangrove species has long been recognized (Macnae 1968). The failure of abiotic environmental factors to completely account for zonation has been experimentally documented for new world mangroves (Rabinowitz 1978). That the *Bruguiera* zone is sometimes at the rear of the mangal is interpreted as being simply a result of local physical features of the shore and not that *Pandanus*, for example, is about to invade the *Bruguiera* zone (see

Table 2. Interpretation of results in the context of successional trends.

Parameter	Direction or intensity of parameter in a climax system	Location in Hood Lagoon mangal at which parameter shows greatest tendency towards a climax-like state
Total biomass[a]	Maximum	
• kg wood		*Bruguiera* zone
Energy flow/biomass[b]	Minimum	
• LAR		*Bruguiera* zone
Environmental control[c]	Well developed	
• Temperature range (min)		*Bruguiera* zone
Nutrient pools	Large	
• Soil ammonia		*Bruguiera-Ceriops* zone
• Soil phosphate		*Bruguiera-Ceriops* zone
Role of detritus	Important	
• Soil organic content (max)		*Bruguiera-Ceriops* zone
Niche specialization	Narrow	
• Restricted distribution in Hood Lagoon mangal[d]		*S. alba* *B. cylindrica* *B. gymnorrhiza* *H. littoralis* *E. agallocha*
Size of organisms	Large	*B. gymnorrhiza*
Loss of energy/nutrients	Low	
• Least wave action and freshwater erosion (removing detritus and nutrients from system)		*Bruguiera-Ceriops* zone
Symbiotic relationships	Developed	
• Ant plants[e]		*Ceriops* zone
Species diversity	High	
• Species number[f]		Strandline zone
Self replacement	Frequent	*Bruguiera-Ceriops* zone

[a] This parameter should include all life-forms and not just trees, but it seems reasonable, as a first approximation, to assume tree biomass represents the general biomass ratios between zones. Although data were not obtained on an area basis, this can be roughly calculated by multiplying density (Fig. 5) by biomass (Fig. 9). Such calculations support the conclusion that the *Bruguiera* zone supports the highest biomass.

[b] This is based on the assumption that leaf area is an indirect measure of energy input (LAR, like LAI, is related to a plant's growth rate), and, although it is known that different mangrove species photosynthesize at different rates (Lugo *et al.*, 1975), these differences are small when compared with the differences in the LAR. Although the data are expressed as LAR rather than LAI they can be extrapolated to a per area basis in a similar way to that for biomass as explained in note (a).

[c] Regulation of chemical cycles is generally used as the criterion when considering environmental regulation, but a system which modifies physical environmental factors is also usually considered to be more developed than one that does not.

[d] Any or all of these restricted distributions could be due to competition rather than niche specialization.

[e] The presence of ant plants (as highly developed mutualistic systems) has been used as a measure of the potential for symbiotic relationships to develop in the different zones. Ant plants are totally absent from the *Rhizophora* and *Avicennia* (both front and back) zones at Hood Lagoon. They are most abundant in the *Ceriops* zone with lesser numbers elsewhere. Their distribution was not quantified because of the difficulty in accurately counting them without climbing every tree.

[f] See text.

Hosokawa *et al.* 1977, Fig. 14.7). Similarly, with respect to succession in high rainfall areas, the presence of freshwater or terrestrial species in the back zones of the mangal is considered to be more indicative of the salinity regime than representing an active process of colonization from the land as a part of an integrated successional system.

The ideas expressed in the present paper suggest that succession in the mangal is neither entirely shoreward nor necessarily linked to terrestrial patterns of succession. Indeed, it might be quite valid to define a mangrove climax, rather than treating the mangal as a pioneer assemblage with no true climax. Such an approach would avoid situations such as encountered by Macnae (1968) where he describes *Bruguiera* as 'dominant' on one hand and simply as a 'stage in the development from mangrove to freshwater-swamp or terrestrial forest' on the other.

The analyses in this paper are based on the assumption that the presence of zonation implies succession (at least in terms of plant–plant competition). This is denied within the context of Thom's (1967) steady state concept of mangal development, an idea that is clearly expressed by Lugo and Snedaker (1974) who suggest that zonation and succession are not necessarily functionally related. This suggestion appears untenable, at least in the New Guinea context. That environmental factors play a role in the development of zonation patterns is not denied (at least insofar as they modify competitive interactions); but the virtual rejection, by these authors, of a possible complementary role by the biotic factors appears unreasonable in the mixed-species mangal of the New Guinea region. The rejection in this paper of the relative importance of the steady state model is based on the observation that many mangrove species can and do grow anywhere in the mangal from the seaward to the landward fringe and that many species are most abundant on the landward and seaward fringes – apparently being excluded from the mid-zone areas of the mangal. Because of this ability of many mangrove species to grow anywhere within the normal mangrove tidal range, zonation can only be due to competition (modified or not by the environment). Competition is the basis of succession in at least some contemporary forest development models (Horn 1976).

It is interesting that the same problem exists in studies on salt marsh zonation where it is usually claimed that the zonation pattern is correlated with environmental gradients such as salinity, and yet it is well known that most marsh plants have broadly overlapping salt tolerances (Niering and Warren 1980).

The foregoing discussion would appear to create a paradox in that the mangal has been thought of by many workers, in recent years, as a steady state system rather than as a seral state system. If one accepts the steady state concept, then one cannot delineate a climax within the mangal because this theory claims that every zone is an environmentally cued climax. That the zones in a mangrove swamp differ in their degree of development is, however, strongly suggested by the present paper and, therein is the paradox. The resolution of this problem would appear, however, not to just depend on the importance of plant–plant competition versus environmental sorting, but rather the paradox is apparently tripartite because time dependent processes, in the Clemensian sense, probably also play a role in the development of the mangal.

Literature cited

Chapman, V.J. 1976. Mangrove Vegetation. J. Crammer, Vaduz, 447 pp.

Elsol, J.A. and P. Saenger. 1980. A general account of the mangroves of Princess Charlotte Bay with particular reference to zonation of the open shoreline. 2nd Int. Symp. Biol. and Mgmt Mangroves and Trop. Shallow Water Comm. Port Moresby. Mimeo, 23 pp.

Horn, H.S. 1976. Succession. In: Theoretical Ecology Principles and Applications. Saunders, Philadelphia, pp. 187–204.

Hosokawa, T., H. Tagawa and V.J. Chapman. 1977. Mangals of Micronesia, Taiwan, Japan, The Philippines and Oceania. In: Ecosystems of the World. I. Wet Coastal Ecosystems. Elsevier, Amsterdam, pp. 271–291.

Johnstone, I.M. and D.G. Frodin. (In press). Mangroves of the Papuan subregion. In: Biogeography and Ecology of New Guinea. Junk, The Hague.

Lugo, A.E. and S.C. Snedaker. 1974. The ecology of mangroves. Ann. Rev. Ecol. Syst. 5:39–64.

Lugo, A.E., G. Evink, M.M. Brinson, A. Broce and S.C. Snedaker. 1975. Diurnal rates of photosynthesis, respiration, and transpiration in mangrove forests of South Florida. In: Tropical Ecological Systems. Springer-Verlag, Berlin. pp. 335–350.

Odum, E.P. 1969. The strategy of ecosystem development. Science, 164:262–270.

Macnae, W. 1968. A general account of the fauna and flora of mangrove swamps and forests in the Indo-West Pacific region. Adv. Marine Biol. 6:73–270.

Niering, W.A. and R.S. Warren. 1980. Vegetation patterns and processes in New England salt marshes. Bioscience 30:301–307.

Rabinowitz, D. 1978. Early growth of mangrove seedlings in

Panama, and an hypothesis concerning the relationship of dispersal and zonation. Jour. Biog. 5:113–133.
Saenger, P. and J. Robson. 1977. Structural analysis of mangrove communities on the central Queensland coastline. Mar. Res. in Indonesia 18:101–118.
Thom, B.G. 1967. Mangrove ecology and deltaic geomorphology: Tobasco, Mexico. Jour. Ecol. 55:301–343.

CHAPTER 16

Degradation of mangrove leaf and stem tissues *in situ* in Mgeni Estuary, South Africa

T.D. STEINKE, G. NAIDOO and L.M. CHARLES

Department of Botany, University of Durban-Westville, Durban, South Africa

Abstract. *In situ* rates of degradation of mangrove litter were obtained from regular harvests of submerged material contained in nylon-mesh bags. Litter components comprised leaves and stems of *Avicennia marina* (Forsk.) Vierh. and *Bruguiera gymnorrhiza* (L.) Lam. and stems of *Rhizophora mucronata* Lam. Leaves degraded more rapidly than stems and after six months *Avicennia* and *Bruguiera* leaf debris comprised only 9.5% and 15.3% respectively of their initial mass. At the end of this period approximately 80% of the initial mass of the stems of each species remained. The material from each harvest was analysed for N, P, K and total C. In the bags containing leaves of both species nitrogen percentages showed an initial sharp decrease before levelling off. Little difference was revealed in all analyses of stems and in total carbon values of leaves. The significance of leaf litter in the contribution of nutrients to this estuary is discussed.

Introduction

Mangroves have been shown to be important for the contribution which they make to the detrital food chain (Odum and Heald 1972; Teas 1979). Degradation studies of mangrove litter, which appear to support this concept, have followed chiefly two lines of investigation: chemical analysis of litter components and living plant parts in an attempt to quantify the contribution of breakdown products to an estuarine system (Clough and Attiwill 1974); and the identify and role of the micro-organisms responsible for the degradation of mangrove material (Kohlmeyer 1969; Lee and Baker 1972; Fell and Master 1973; Newell 1976; and others). However, relatively little is known of the rates of breakdown of mangrove litter components and the accompanying release of nutrients. Albright (1976) studied *in situ* rates of leaf, root and pneumatophore degradation of *Avicennia marina* var. *resinifera* in New Zealand, although no data were available on release of nutrients during decomposition. Fell *et al.* (1972) studied changes in the C:N ratio of decaying leaves of *Rhizophora mangle*. Most of the research of this nature has been carried out in terrestrial forests.

Research to assess the productivity of mangroves by recording litter fall was carried out at Beachwood, Mgeni Estuary, and it was felt that as much information as possible on the decomposition of this litter should be obtained. Consequently in an effort to start building up a more complete picture of the role of the mangrove litter in a food chain, a project on litter degradation was initiated at Beachwood during the spring of 1978 and continued for approximately six months. Although this work was intended to be only of a preliminary nature, useful information has been obtained. However, more research in this field is necessary before a fairly reliable estimate of the contribution of mangroves towards nutrient recycling can be formulated for this estuary.

Tasks for vegetation science, Vol. 8 ed. by H.J. Teas

 ISBN 978-90-481-8526-9

Methods

The study was conducted in the Beachwood mangrove swamp at the mouth of the Mgeni River. The mangrove and other plant communities in this area have recently been described by Padia (1980).

Yellow, senescent leaves, whose abscission was judged to be imminent, and dead twigs were picked from trees and taken to the laboratory within an hour. Leaves of *Avicennia marina* (Forsk.) Vierh. and *Bruguiera gymnorrhiza* (L.) Lam. were sorted into groups of approximately equal numbers for each species, weighed to equal weight, and placed in nylon bags with a very fine, but irregular, mesh. The diameter of the twigs of *B. gymnorrhiza* and *Rhizophora mucronata* Lam. was 3–5 mm and that of *A. marina* twigs was 2–4 mm. They were cut into Lengths of 50–100 mm before being weighed and placed in litter bags as above.

The litter bags were transferred to the Beachwood stream and tied to marker posts just below the level of low water ordinary spring tides (Site C, Fig. 1). The bags were almost continually submerged: they were exposed only on very low spring tides. Litter bags were collected at three-weekly intervals over a period of 27 weeks and at each of these harvests five bags were removed. There were two exceptions: the eighth harvest was delayed by one week and took place only on April 21, 1979; and the final harvest of *Avicennia* leaves was not recovered. The bags were removed to the laboratory where washing of the contents was carried out on a 30-mesh sieve using only a gentle stream of water to prevent damage to the material as far as possible. The material was then oven-dried for a week at 70° C and weighed. Milling of the dried material was carried out on a Cassella mill to pass a 1 mm mesh sieve.

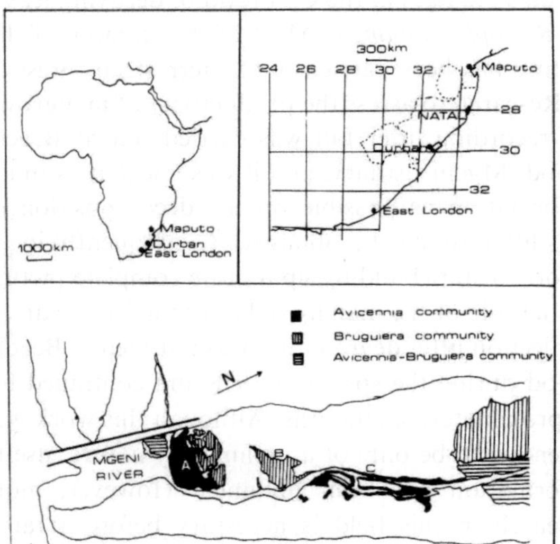

Fig. 1. Mgeni River, showing the southern section of the Beachwood mangrove swamp (after Padia, 1980).

Leaf and stem samples from each harvest were analysed for nitrogen, phosphorus, potassium and total carbon. Nitrogen was determined by the micro-Kjeldahl method (Horwitz 1965). Extracts for the determination of phosphorus and potassium were prepared by ashing according to the method described by Ewel (1976). Phosphorus was determined by the molybdophosphoric blue colour method (Jackson 1962), and potassium was determined by atomic absorption spectrophotometry using a Varian Techtron Model 1200. Carbon analyses were conducted with the use of a Coleman carbon-hydrogen analyser.

Temperature and salinity measurements in the estuary (site C) were taken at the time of harvest which was usually before 0900 h. Additional readings were made whenever possible at other times of the day. Measurements of water temperature at 100 mm below the surface varied from 28.5° C in the summer to 18° C in the cooler months (towards the end of the experiment), although more extreme temperatures would probably have been experienced. In salinity there were large differences in the values recorded, particularly during the rainy season. Values as low as 0‰ and as high as 34‰ were obtained, although generally the salinity was approximately 12‰.

Results and discussion

Dry matter

The decrease in mass of leaves and stems with decomposition is shown in Fig. 2. Leaf material

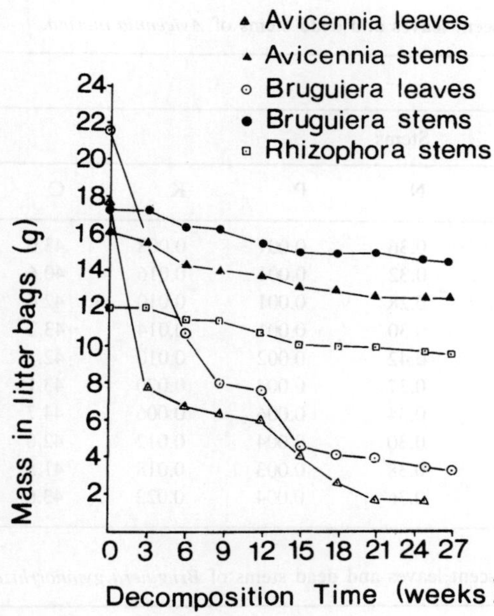

Fig. 2. Dry matter losses of mangrove leaves and stems during decomposition.

showed a more rapid decrease than stems, and leaves of *Avicennia* in turn degraded faster than those of *Bruguiera*. With the first harvest (i.e. after three weeks) *Avicennia* leaves had shown a 57% decrease in mass with the loss by the eighth harvest at 90.5%. At the first harvest the mass of *Bruguiera* leaves had abated by only 27.4% which had increased to 50.6% by the second harvest. The loss by the final harvest (i.e. after 27 weeks) was 84.7%. In both species the early losses were greater, followed by a slower, steady decrease for the remaining period of the experiment. From the fourth harvest (i.e. towards the end of January) the leaf material was difficult to identify to species, although it was still recognizable as leaf material. By the end of the experiment (i.e. beginning of May), with the exception of a few individual petioles, it was difficult to recognize the residue as leaf material. At this stage the residue was intermingled with sediment inside the bags. It would appear that under these conditions most litter was incorporated into the soil organic matter profile within six months.

Among the contents of the bags containing leaves were numerous small (±5 mm) crustaceans (Amphipoda, Isopoda, Tanaidacea) and polychaetes. Because mesh of the bags was too fine to permit entry of these animals in the sizes in which they were found, they must have entered either as eggs, which hatched inside the bags, or as very small juveniles. The polychaetes probably contributed significantly to the breakdown of the leaf material, although the importance of the crustaceans in this respect is not known. Whether they were active in actually degrading the leaf material or were present as filter feeders was not assessed.

The stems exhibited comparatively little loss of mass over the period of the experiment, with *Avicennia*, *Bruguiera* and *Rhizophora* showing decreases of 21.4, 15.9 and 19.6% respectively. By the third harvest (i.e. after nine weeks) the decreases were 13.1, 6.2 and 5.3% respectively. The stems were clearly identifiable until the end. The more rapid early decrease in mass in *Avicennia* was almost entirely the result of the loss of bark which flaked off the wood. These flakes of bark were apparently degraded only partially, because much of this material persisted until the end of the experiment, while the woody stem tissue remained hard and showed no obvious signs of breakdown. The stems of *Bruguiera* were most resistant to breakdown, as borne out by the fact that this material, after more than six months of submersion, remained generally firm and hard. In contrast, many of the stems of *Rhizophora* had a soft, pulpy feel by the end of the experiment.

Chemical analyses

The levels of nutrients were generally very low in this dead stem material (Tables 1, 2 and 3). The slow degradation of the stems is supported by the results of these analyses which show no consistent downward trends, suggesting minimal breakdown and consequent release of nutrients. As woody material forms a relatively small percentage of the litter at Beachwood (Steinke and Charles 1980), this material obviously contributes relatively little in the way of nutrients to this system.

While no differences in total carbon of leaves were observed, there were significant changes in potassium, phosphorus and nitrogen percentages of this material (Tables 1 and 2). The amount of

Table 1. Concentration of nutrients (percentage oven-dry weight) in senescent leaves and dead stems of *Avicennia marina.*

Decomposition time (weeks)	*Avicennia marina*							
	Leaves				Stems			
	N	P	K	C	N	P	K	C
0	0.66	0.026	0.432	48.7	0.36	0.007	0.008	43.0
3	0.91	0.015	0.104	49.1	0.32	0.001	0.016	40.6
6	1.10	0.012	0.028	48.4	0.28	0.001	0.010	42.4
9	1.14	0.011	0.028	52.4	0.30	0.001	0.014	43.2
12	1.18	0.013	0.030	44.4	0.42	0.002	0.010	42.1
15	1.22	0.014	0.038	48.8	0.37	0.004	0.020	43.1
18	1.45	0.013	0.036	47.6	0.34	0.006	0.006	44.7
21	1.45	0.014	0.016	45.7	0.30	0.004	0.012	42.6
25	1.26	0.012	0.038	43.9	0.38	0.003	0.018	41.5
27	—	—	—	—	0.36	0.004	0.022	43.6

Table 2. Concentration of nutrients (percentage oven-dry weight) in senescent leaves and dead stems of *Bruguiera gymnorrhiza.*

Decomposition time (weeks)	*Bruguiera gymnorrhiza*							
	Leaves				Stems			
	N	P	K	C	N	P	K	C
0	0.28	0.022	0.196	43.3	0.50	0.012	0.016	42.8
3	0.38	0.011	0.130	43.0	0.34	0.004	0.032	41.9
6	0.42	0.010	0.028	50.2	0.41	0.002	0.028	43.0
9	0.56	0.010	0.046	44.8	0.30	0.003	0.014	40.9
12	0.59	0.012	0.030	45.6	0.38	0.004	0.018	41.4
15	0.61	0.013	0.038	46.0	0.37	0.002	0.020	43.0
18	0.72	0.009	0.028	49.7	0.51	0.002	0.030	41.9
21	0.67	0.011	0.040	49.9	0.38	0.007	0.030	41.7
25	0.62	0.012	0.040	50.2	0.42	0.001	0.032	41.5
27	0.74	0.010	0.050	48.1	0.36	trace	0.028	40.1

Table 3. Concentration of nutrients (percentage oven-dry weight) in dead stems of *Rhizophora mucronata.*

Decomposition time (weeks)	*Rhizophora mucronata* (stems)			
	N	P	K	C
0	0.39	0.016	0.018	42.1
3	0.53	0.013	0.028	41.9
6	0.51	0.014	0.032	41.8
9	0.48	0.008	0.012	43.2
12	0.68	0.013	0.016	43.7
15	0.47	0.014	0.016	42.4
18	0.48	0.013	0.016	41.9
21	0.56	0.016	0.024	41.7
25	0.53	0.015	0.034	43.0
27	0.41	0.011	0.014	43.2

potassium in the leaves of both species showed a sharp initial decrease, although the loss was more rapid in *Avicennia* (Figs. 3 and 4). As this element appears not to have a structural role in plants, this rapid loss is probably due to leaching. The very low amounts that were recovered from the litter bags subsequently, appeared to confirm this. Similar results were obtained with leaf litter decomposition in a tropical forest in Guatemala (Ewel 1976) and hardwood leaves in Canada (MacLean and Wein 1978).

The loss of phosphorus from leaves also occurred at a more rapid rate initially with little difference between species (Fig. 5). However, after the initial

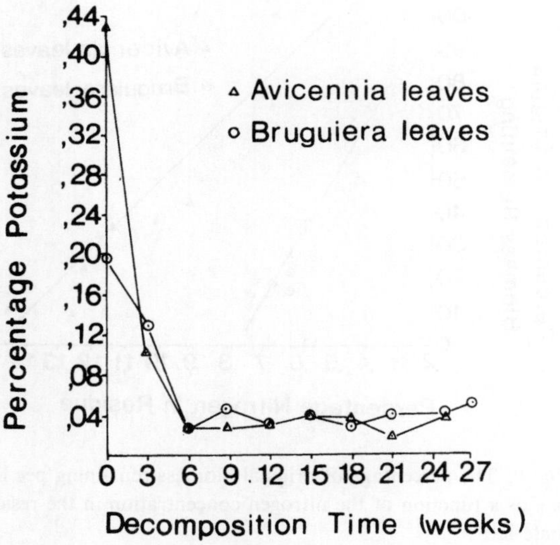

Fig. 3. Decrease in potassium percentages of *Avicennia* and *Bruguiera* leaves during decomposition.

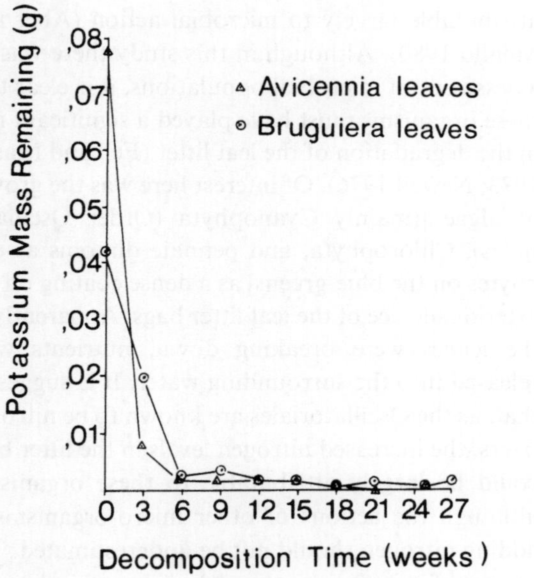

Fig. 4. Decrease in mass of potassium per litter bag during decomposition.

decrease, which was not as great as in the case of potassium, the phosphorus percentages appeared to stabilise at a relatively high level. A similar trend was shown by MacLean and Wein (1978). The phosphorus content of the bags followed a pattern similar to that of the dry matter loss (Fig. 6), suggesting rapid recycling under these conditions.

There was a steady increase in nitrogen percentage in the litter bags over the period of the experiment (Fig. 7), although the total amount of nitrogen decreased as a result of the decrease in dry matter (Fig. 8). Fell *et al.* (1972) similarly reported that the percentage of nitrogen in bags of *Rhizophora mangle* leaves doubled in 50 days. There was an inverse linear relationship between the percentage of original biomass remaining and nitrogen concentration in the residual material (Fig. 9) with a high correlation coefficient of $r = 0.92$ in *Avi-*

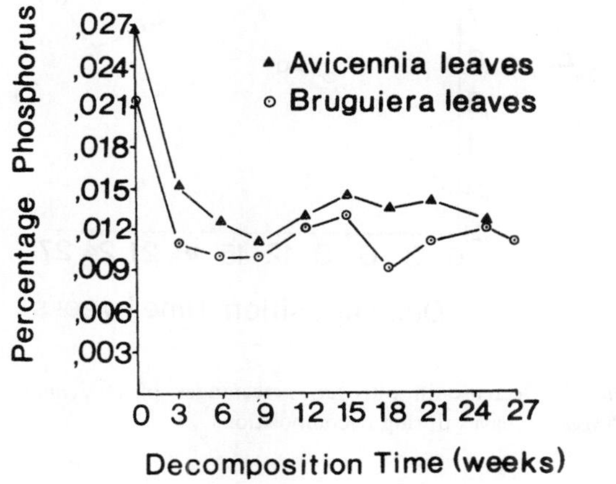

Fig. 5. Decrease in phosphorus percentages of *Avicennia* and *Bruguiera* leaves during decomposition.

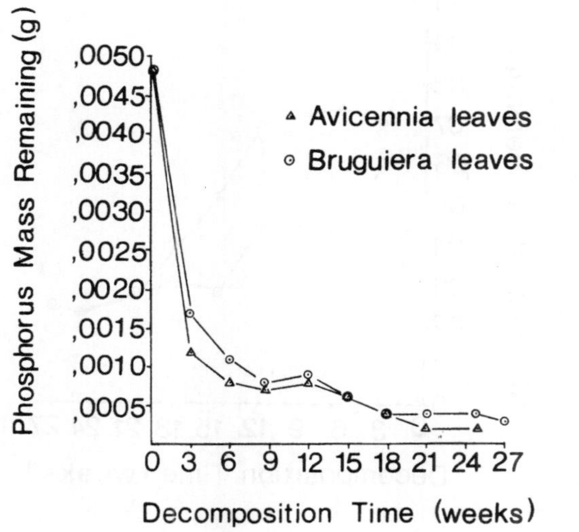

Fig. 6. Decrease in mass of phosphorus per litter bag during decomposition.

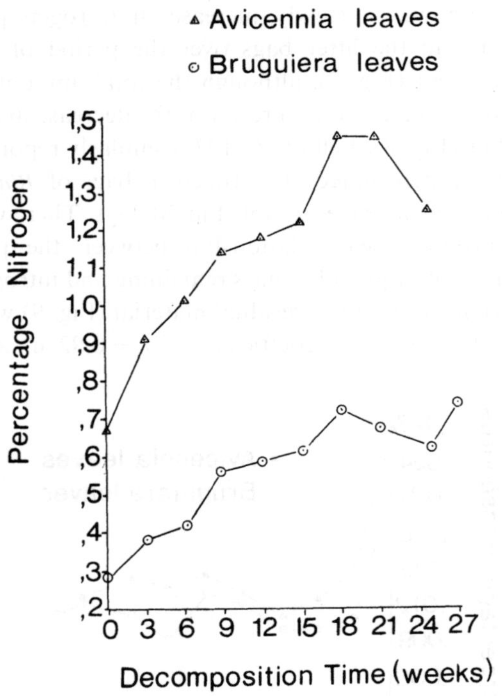

Fig. 7. Decrease in nitrogen percentages of *Avicennia* and *Bruguiera* leaves during decomposition.

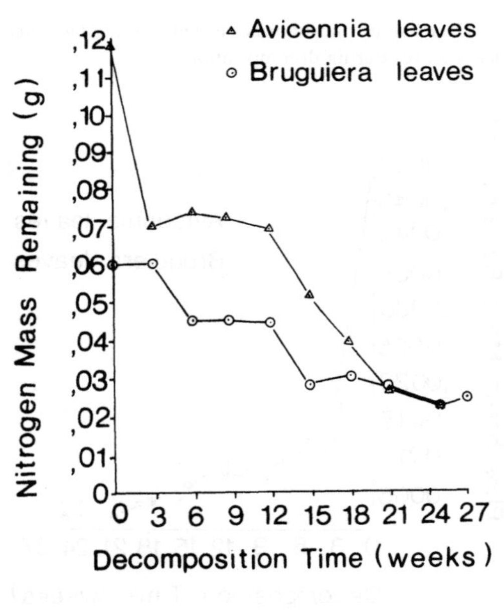

Fig. 8. Decrease in mass of nitrogen per litter bag during decomposition.

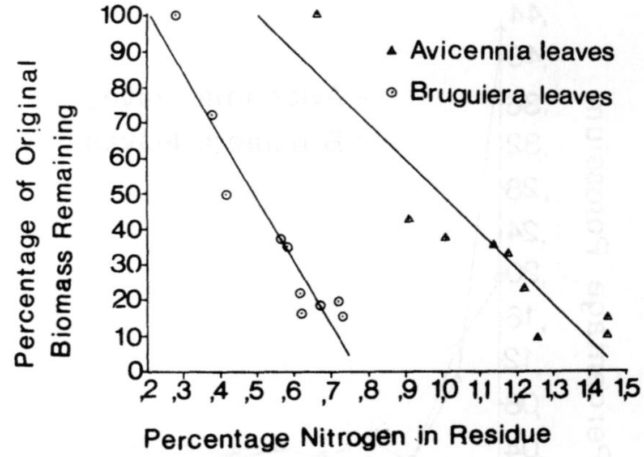

Fig. 9. The percentage of original biomass remaining per litter bag as a function of the nitrogen concentration in the residual material.

cennia and $r = 0.95$ in *Bruguiera* leaf litter. This inverse linear relationship is well established and is attributable largely to microbial action (Aber and Melillo 1980). Although in this study there was no assessment of microbial populations, it is clear that these organisms must have played a significant role in the degradation of the leaf litter (Fell and Master 1973; Newell 1976). Of interest here was the growth of algae [(mainly Cyanophyta (Order Oscillatoriales), Chlorophyta, and pennate diatoms as epiphytes on the blue-greens] as a dense coating on the exterior surface of the leaf litter bags. Apparently as the leaves were breaking down, nutrients were released into the surrounding water. It is suggested that, as the Oscillatoriales are known to be nitrogen fixers, the increased nitrogen levels in the litter bags could be largely attributable to these organisms, although the actions of other micro-organisms in adding nitrogen should not be underestimated. The uptake of nutrients, such as phosphorus and potassium, from decomposing litter by these organisms and their reciprocal contribution of nitrogen provides further proof, if this is necessary at all, of the importance of litter inputs in an estuarine ecosystem such as this. Steinke and Charles (1980) have estimated that the mangroves at Beachwood contribute approximately 274 tonnes of litter per year, of which leaves form almost 60%. Clearly this contribution is of tremendous value not only as a source of nutrients but also as a substrate for a variety of organisms.

General discussion

The differences in behaviour of the leaves of *Avicennia* and *Bruguiera* are interesting. Generally, the leaves of *Avicennia* decomposed faster with a more rapid release of nutrients. Furthermore, the leaves of *Bruguiera* are more buoyant than *Avicennia* leaves. Table 4 shows that at three salinities all *Bruguiera* leaves floated, while within one day almost all the *Avicennia* leaves had sunk. Approximately 50% of *Bruguiera* leaves remained floating for up to 8–9 days, while some remained buoyant for more than 15 days (Fig. 10). This study was conducted in large tanks where the build-up of decomposers and consequent breakdown and loss of buoyancy could be expected to be more rapid than in the field. Obviously there is a greater probability that leaves which float will be carried out to sea on an outgoing tide and therefore lost to that estuary. Conversely, leaves which sink are more likely to remain in that estuary, although nutrients in solution, derived from the degradation of these leaves, must also be lost at least in part.

Anatomical sections (Figs. 11 and 12) possibly indicate explanations for the differences between leaves of these two species. *Bruguiera* leaves have both surfaces covered with a thick cuticle which would impede the entry of water and degradative organisms. The presence of large intercellular air spaces, not readily waterlogged because of the relatively impervious cuticle, would also contribute significantly to the buoyancy of these leaves. Only the adaxial surface of *Avicennia* leaves has a thick cuticle, while the lower (abaxial) surface is covered

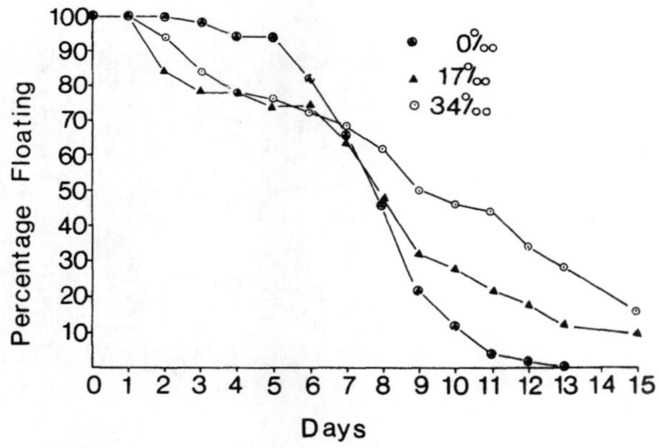

Fig. 10. Buoyancy of ***Bruguiera*** leaves at three salinities.

with numerous fine, non-glandular hairs. Fahn and Shimony (1977) have shown that most of the non-glandular hair is covered by a very thin cuticle which may even be absent in parts. It is suggested that through this vulnerable lower surface water may enter the leaves more readily, causing water-logging of intercellular air spaces, and carrying with it micro-organisms, or their propagative materials, to the more sensitive inner structures. In this way more rapid sinking, degradation, and loss of nutrients could be accounted for in *Avicennia* leaves.

This pilot study has provided some information on litter breakdown under Beachwood conditions, but has possibly posed more questions than it has been possible to answer. While stem material was not shown to form a significant component of the litter, reproductive material (i.e. buds, flowers, propagules) may contribute approximately 25% of the total litter (Steinke and Charles 1980), although

Table 4. Percentage buoyancy of senescent leaves of two mangrove species immediately after abscission and 24 hours later.

	Salinity					
	34‰		17‰		0‰	
	Immediate	24 h	Immediate	24 h	Immediate	24 h
Bruguiera gymnorrhiza	100	100	100	100	100	100
Avicennia marina	38	4	16	0	2	1

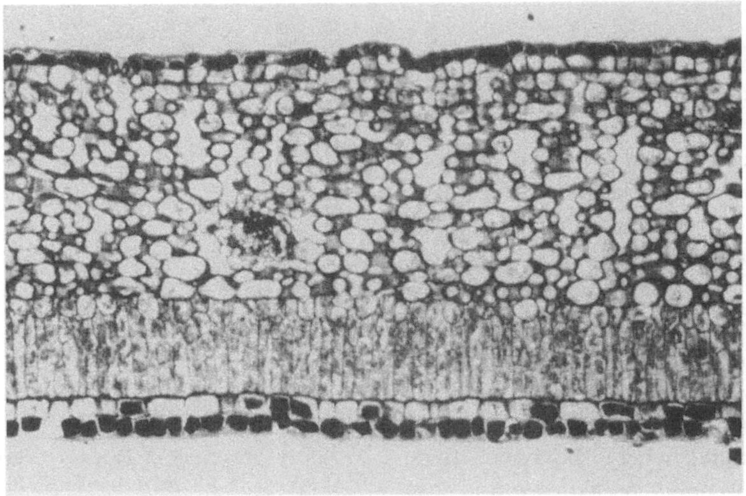

Fig. 11. Cross section of leaf of *Bruguiera gymnorrhiza* showing thick cuticle on upper and lower surfaces.

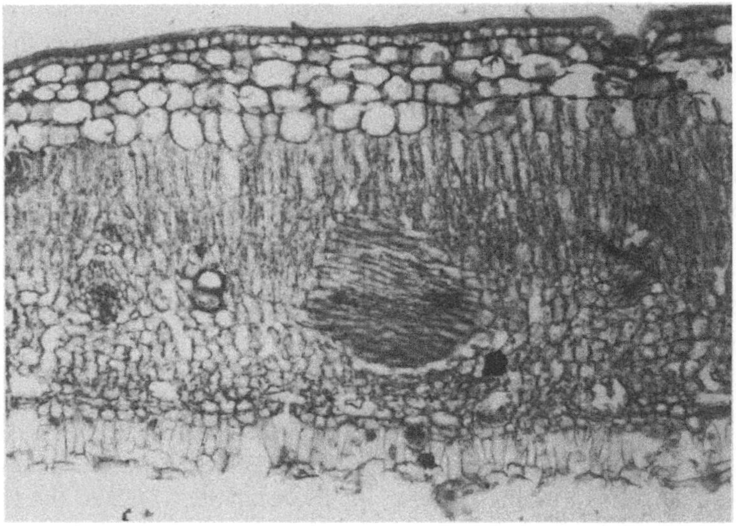

Fig. 12. Cross section of portion of leaf of *Avicennia marina* showing thick cuticle on upper surface and non-glandular hairs on lower surface.

much of this may be lost from an estuary too. Therefore, before meaningful estimates of nutrient cycling can be made, research in the future must include an assessment not only of the contribution of all components of the litter to the estuary (including debris which is carried in with an incoming tide), but also of the extent of loss of litter and/or soluble nutrients from the estuary. This research is in progress now.

Acknowledgments

The writers wish to thank Dr. H. Baijnath for his helpful comments and Mrs. E.L. Lawes for her assistance with the diagrams and photography.

Literature cited

Aber, J.D. and J.M. Melillo. 1980. Litter decomposition: measuring relative contributions of organic matter and nitrogen to forest soils. Can. Jour. Bot. 58:416–421.

Albright, L.J. 1976. *In situ* degradation of mangrove tissue. N.Z.Jour. Mar. Freshw. Res. 10:385–389.

Clough, B.F. and P.M. Attiwill. 1974. Nutrient cycling in a community of *Avicennia marina* in a temperate region of Australia. Proc. Internat. Symp. Biol. Mgmt Mangroves, pp. 137–146.

Ewel, J.J. 1976. Litter fall and leaf decomposition in a tropical forest succession in eastern Guatemala. Jour. Ecol. 64:293–308.

Fahn, A. and C. Shimony. 1977. Development of the glandular and nonglandular leaf hairs of *Avicennia marina* (Forsskal) Vierh. Bot. Jour. Linn. Soc. 74:37–46.

Fell, J.W., B. Hopper, R. Cefalu and I.M. Master. 1972. Mangrove detrital system. In: An ecological study of South Biscayne Bay and Card Sound, Florida. Report of Rosenstiel School of Marine and Atmospheric Science, Univ. of Miami, pp. VII-1 to VII-27.

Fell, J.W. and I.M. Master. 1973. Fungi associated with the degradation of mangrove (*Rhizophora mangle* L.) leaves in South Florida. In: Estuarine Microbial Ecology. Univ. S. Carolina Press, pp. 455–466.

Horwitz, W. 1965. Official methods of analysis of the Association of Official Agricultural Chemists (10th edition), published by A.O.A.C., Washington, pp. 744–745.

Jackson, M.L. 1962. Soil chemical analysis. Constable, London, pp. 144–146.

Kohlmeyer, J. 1969. Ecological notes on fungi in mangrove forests. Trans. Br. Mycol. Soc. 53:237–250.

Lee, B.K.H. and G.E. Baker. 1972. Environment and the distribution of microfungi in a Hawaiian mangrove swamp. Pacific Sc. 26:11–19.

MacLean, D.A. and R.W. Wein. 1978. Weight loss and nutrient changes in decomposing litter and forest floor material in New Brunswick forest stands. Can. Jour. Bot. 56:2730–2749.

Newell, S.Y. 1976. Mangrove fungi: the succession in the mycoflora of red mangrove (*Rhizophora mangle* L.) seedlings. In: Recent Advances in Aquatic Mycology. Elek Science, pp. 51–91.

Odum, W.E. and E.J. Heald. 1972. Trophic analyses of an estuarine mangrove community. Bull. Mar. Sc. 22:671–738.

Padia, R. 1980. A vegetation map of the Mgeni/Beachwood Nature Reserve, Durban, from aerial photographs. Unpublished B.Sc. Hons. project, Univ. of Durban-Westville.

Steinke, T.D. and L.M. Charles. 1980. Productivity and phenology of *Avicennia marina* (Forsk.) Vierh. and *Bruguiera gymnorrhiza* (L.) Lam. in Mgeni estuary, South Africa (in press).

Teas, H.J. 1979. Silviculture with saline water. In: The biosaline concept. Plenum Publishing Corporation, pp. 117–161.

CHAPTER 17

Mangroves and sewage: a re-evaluation

B.F. CLOUGH*, K.G. BOTO* and P.M. ATTIWILL**

* *Australian Institute of Marine Sciences, PMB NO. 3, Townsville, Q. 4810, Australia*
** *School of Botany, University of Melbourne, Parkville, Vic. 3052, Australia*

Abstract. The capacity of mangrove systems to remove and immobilise nutrients, heavy metals and other constituents of sewage effluent is strongly influenced by tidal and drainage characteristics, and by the redox state of the soil. Some mangrove systems may not remove sufficient phosphorus and nitrogen to reduce the quantity of these nutrients in sewage effluent to a level below that required for significant biological activity in offshore waters. In particular, the common practice of discharging treated sewage effluent directly into estuaries or tidal channels fringed by mangroves would appear to be a relatively ineffective scheme for improving the quality of sewage effluent.

Nutrient enrichment of mangrove systems in general does not appear to be harmful, and in some cases may have a beneficial effect on growth and productivity. Similarly, anaerobic mangrove soils seem likely to have the capacity to trap heavy metals and pesticides without harm to mangroves themselves, although these materials could have an adverse effect on the sediment fauna. The high organic carbon content of sewage effluent may lead to further reduction of soil redox potential. This may place additional stress on mangroves and their fauna which are already hard pressed to cope with anaerobiosis. In the long term this could be a major drawback to the use of mangrove systems as sites for the discharge of sewage effluent.

Introduction

Mangrove-dominated swamps and estuaries have long been regarded as convenient sites for the disposal of sewage and wastewater. The historical reasons for this seem obvious. Firstly, many urban and industrial centers are adjacent to rivers and estuaries fringed by mangroves and, secondly, mangroves traditionally have been viewed as an ecological and economic wasteland.

More recently, however, it was suggested by Nedwell (1974) that mangroves might be utilized and managed in a positive sense as the final stage in an overall sewage treatment facility. This suggestion was based on the view that mangrove ecosystems may be more resistant to nutrient enrichment and eutrophication than neighbouring lagoon and reef ecosystems. Furthermore, Nedwell (1974) also suggested that the discharge of treated sewage into mangrove ecosystems could provide a means of reducing eutrophication in coastal waters which otherwise would receive treated sewage effluent directly.

As a consequence of mounting pressures from population growth and from urban and industrial development it is likely that mangroves will be utilized increasingly throughout the world for a wide range of purposes, including sites for the discharge of sewage and wastewater. It therefore seems appropriate at this time to re examine the two basic questions raised by Nedwell's (1974) paper. These are, firstly, what is the effect of sewage and

Tasks for vegetation science, Vol. 8 ed. by H.J. Teas
 ISBN 978-90-481-8526-9

wastewater on mangroves and, secondly, how effectively can mangrove systems trap the various components of these effluents which would otherwise enter offshore systems?

Since very little work addressing these questions directly has been published, our discussion relies heavily on information derived from saltmarsh ecosystems and from more basic studies of mangrove ecology. This involves some degree of extrapolation and speculation. Nevertheless, we believe that many useful and valid predictions can be derived from this approach.

The components of sewage and wastewater which are most likely to have an impact on mangrove or nearby offshore systems are (i) dissolved and particulate-bound nutrients; (ii) heavy metals; (iii) pesticides – mainly from agricultural runoff; and (iv) dissolved and particulate organic matter.

The fate of these components in a mangrove system is determined chiefly by the chemical, biological and hydrodynamic characteristics of the system.

Nutrients

Relatively little information is available for the nutrient status of mangrove sediments. The few studies which have been done suggest that their fertility lies somewhere towards the middle to higher end of the range between native terrestrial ecosystems and fertilized agricultural systems. Nevertheless, there is evidence from work in southern Florida (Hicks and Burns 1975; Lugo *et al.* 1976; Onuf *et al.* 1977) and from our own studies (discussed below) that some mangrove ecosystems are nutrient limited, particularly with respect to nitrogen and phosphorus. For this reason, and because sewage effluent is generally rich in nitrogen and phosphorus, the present discussion will be restricted to these two nutrients.

Mangroves grow chiefly in saline, waterlogged, anaerobic soils. Anaerobiosis exerts such a profound influence on the chemistry of phosphorus and nitrogen that it is worth examining briefly here the sequence of oxidation-reduction reactions which occur in waterlogged soils. These are shown diagrammatically in Fig. 1. Initially, aerobic microorganisms deplete the soil of oxygen and the redox potential falls to around +350 mV. After the disappearance of oxygen, nitrate is reduced to gaseous N_2; this reaction is followed, in turn, by a sequence of reduction reactions involving manganese, iron, sulphur and, finally, CO_2. Sediment organic matter is the source of energy for these microbiologically mediated reduction reactions.

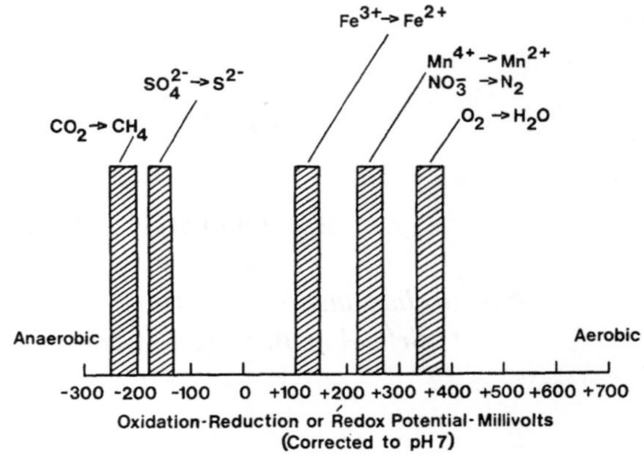

Fig. 1. Oxidation-reduction potential at which various inorganic substances in flooded soils are reduced. (After DeLaune *et al.*, 1976.)

The redox potentials of mangrove sediments vary from one area to the next, depending on the frequency and duration of tidal inundation, drainage, sediment organic matter content, and the availability of electron acceptors such as nitrate, Fe^{3+}, Mn^{4+} and sulphate. In general, redox potential would be expected to increase with decreasing frequency and duration of tidal inundation, with higher redox potentials at higher elevations where tidal inundation is infrequent. However, this trend may be reversed by the presence of mangroves or other vegetation. Boto and Wellington (in press) found that mangrove soils at more elevated sites, where tidal inundation was less frequent and where mangrove biomass was lower, were consistently more anaerobic than at lower sites, which were more frequently inundated and where the biomass of mangroves was much higher. They attributed this positive correlation between redox potential and mangrove biomass to enhanced oxygen transport to the sediment via mangrove roots. Similar findings in a salt marsh system have been reported by Howes *et al.* (1981). Notwithstanding these effects, the redox potential of mangrove sediments seldom exceeds +100 mV, and more commonly is much lower (Boto and Wellington, in press). Under these conditions the sediment is mostly devoid of nitrate and iron exists chiefly as Fe^{2+}. These characteristics have an important bearing on the chemistry of mangrove sediments.

Phosphorus

Total phosphorus concentrations in mangrove sediments appear to fall within the range 200–1500 μg g^{-1} oven dry weight (Hesse 1961, 1962, 1963; Attiwill and Clough 1978; Clough and Attiwill, unpublished data; Boto and Wellington, in press). Because of their high organic content mangrove sediments may contain relatively large amounts of organic phosphorus. Hesse (1962, 1963), for example, found that organic phosphorus accounted for 75–85% of the total phosphorus in mangrove sediments from Sierra Leone and Nigeria. Most of the organic phosphorus in mangrove sediments probably occurs as complexes with humic and fulvic acids, as has been found in lake sediments (Sommers *et al.* 1972).

Most of the inorganic phosphorus in mangrove sediments is adsorbed or otherwise incorporated within hydrated iron and aluminium colloidal sesquioxides (Table 1). Calcium-bound phosphorus may also be an important component of the inorganic phosphorus pool in some sediments, particularly where the sediment contains appreciable $CaCO_3$ and the pH is greater than 7.5. However, alkaline mangrove sediments are relatively uncommon.

In the general case, low redox potentials in mangrove and other waterlogged soils lead to the release of phosphate, particularly at low pH (<7). The mechanisms involved in this release have been summarised by Patrick and Mahapatra (1968) as: (i) Reduction of insoluble ferric phosphate to more soluble ferrous phosphate; (ii) the reduction of the hydrated ferric oxide coating on clay and silt particles resulting in the release of occluded phosphate; (iii) phosphate displacement from ferric and aluminium phosphates by organic anions; (iv) hydrolysis of ferric and aluminium phosphates; (v) anion (phosphate) exchange between clay and organic anions.

In view of the wide range of mechanisms involved in phosphorus transformations in reduced sediments it is probable that a state of thermodynamic disequilibrium exists between soluble interstitial phosphate, readily exchangeable phosphate and the much less soluble, organic and occluded or precipitated mineral forms. This state of disequilibrium should be more marked when phosphate is being removed by plants, and where tidal exchange and drainage lead to periodic gains and losses of interstitial and exchangeable phosphates. Boto and Wellington (in press) found that exchangeable phosphorus and total sediment phosphorus both decreased with distance away from the edge of a tidal creek in North Queensland. Weak acid-exchange-

Table 1. Forms of inorganic phosphorus in mangrove sediments as measured by the method of Chang and Jackson (1957).

Fraction	μg P g^{-1} oven dry weight	
	Westernport Bay, Victoria[a]	Sierra Leone[b]
Water-soluble phosphorus	0	2
Aluminium phosphate	5–22	6
Iron phosphate	40–100	29
Calcium phosphate	20–32	25
Reductant soluble iron phosphate	400–1100	0
Occluded iron and aluminium phosphates	2–15	6

[a] Attiwill and Clough (1978).
[b] Hesse (1962).

able phosphate fell from about 30 μg P g^{-1} oven dry weight at the streamside site to about 5 μg P g^{-1} oven dry weight 200 m further inland. This appeared to be related to increasing elevation with distance from the streamside site and the decreased frequency and duration of tidal influences.

When soluble phosphate is added to soils it is rapidly immobilised by adsorption reactions (Hsu 1964; Holford and Mattingly 1976; Holford and Patrick 1979), giving rise to an increase in exchangeable or labile phosphate. In a fertilization experiment in which 400 kg P ha^{-1} was added, as superphosphate, to mangrove sediment (Boto and Wellington, 1983), weak acid-extractable phosphate was found to increase from 5 to 65 μg Pg^{-1} oven dry weight.

However, the capacity of mangrove sediments to immobilise phosphorus is probably limited by the availability of exchange sites, particularly in sediments already rich in phosphorus. Adsorption isotherms (Olsen and Watanabe 1957) for several mangrove sediments from northern and southern Australia indicate adsorption maxima in the range 250–700 μg P g^{-1} over dry weight (Attiwill and Clough 1978, and unpublished data). A lower adsorption maximum (ca. 180 μg P g^{-1} oven dry weight) was found in a sediment which had received treated sewage effluent for about 20 years; in this sediment total phosphorus was 1720 μg g^{-1} oven dry weight and weak acid-extractable phosphate was 16.5 μg g^{-1} oven dry weight.

In a study of the effects of pH and redox potential on phosphate removal by soil, Holford and Patrick (1979) found that reduction led to a decrease in removal of phosphate from simulated wastewater, except at pH 6.5 where phosphate removal was enhanced, apparently due to increased adsorption on newly precipitated $Fe(OH)_2$. They noted that removal of phosphate in the short term was related to the phosphate buffer capacity of the soils, whereas in the longer term their capacity for continuing phosphate removal appeared to be dependent on the precipitation of $Fe(OH)_2$. These results led Holford and Patrick (1979) to the conclusion that soils have had a relatively short-lived capacity to reduce the phosphorus concentration of sewage effluent to a level below that required for significant biological activity. On the basis of the limited data now available we believe that this conclusion may be equally applicable to mangrove sediments.

Nitrogen

Total nitrogen levels in native mangrove sediments generally range from 600 to 2000 μg N g^{-1} oven dry weight (Boto and Wellington, in press; Clough and Attiwill, unpublished data). In mangrove sediments, as in other waterlogged soils (c.f. Keeney 1973 and references therein; DeLaune *et al.* 1976), much of this nitrogen probably occurs as organic nitrogen. Ammonium is the main form of inorganic nitrogen in anoxic mangrove sediments because nitrification of organic nitrogen stops at NH_4^+ due to the lack of oxygen to oxidise it further to nitrate (Ponnamperuma 1972). Moreover, under reducing conditions nitrate is rapidly denitrified by anaerobic bacteria to form gaseous nitrogen or nitrous oxide.

Like other cations, ammonium is capable of occupying cation exchange sites in soils. However, the high level of sodium in most mangrove sediments tends to swamp the cation exchange sites, thereby displacing ammonium. In consequence, ammonium in mangrove sediments occurs mainly in the interstitial phase, where our own measurements indicate that ammonium concentrations of up to 1000 μg l^{-1} may be found. Interstitial ammonium is highly mobile and is thus susceptible to being leached by heavy rain, or lost through drainage following tidal inundation. The permeability of mangrove sediments is generally quite low, particularly in sediments with a high clay and silt content, so that the loss of ammonium and other soluble nutrients through leaching and drainage may not necessarily be severe. To some extent, however, the low permeability of the sediment may be offset by drainage via channels which form as the result of the burrowing activity of animals and the decomposition of roots.

Denitrification can lead to significant losses of nitrogen from flooded, anaerobic soils and sediments (Tusneem and Patrick 1971; Keeney 1973; Bouldin *et al.* 1974; Van Cleemput *et al.* 1976).

Furthermore, the loss of nitrogen through denitrification appears to be higher in sediments which have been amended with large amounts of exogenous nitrogen (e.g. fertilizer or sewage effluent) than in unenriched, natural sediments (Buresh and Patrick, personal communication, 1978). These observations suggest that denitrification could significantly reduce the nitrogen levels in treated sewage effluent applied to mangrove sediments, particularly when the effluent contains appreciable quantities of nitrate.

Plant-nutrient interactions

Nutrient uptake by mangroves leads to the immobilization of significant amounts of nitrogen and phosphorus. The requirement of mangroves for nitrogen and phosphorus is not known exactly, but a reasonable estimate can be obtained from analyses of the nutrient status of foliage and wood. The data presently available indicate that leaves contain on average about 1% nitrogen and 0.1% phosphorus, and woody tissues about 0.5% nitrogen and 0.05% phosphorus (Clough and Attiwill 1975; Spain and Holt 1980). Well-developed *Rhizophora* forests in northern Australia produce about 5 tonnes of new leaves per hectare annually (Duke *et al.* 1981). No data are available for wood production by *Rhizophora* in northern Australia, but overseas estimates (c.f. Teas 1979) indicate a likely production of 20–40 tonnes ha^{-1} y^{-1}. Based on these figures, a well-developed *Rhizophora* forest would have an annual requirement for 150–250 kg N ha^{-1} and 15–20 kg P ha^{-1}, of which about 50 kg N and 5 kg P would be returned in the form of leaf litter. Most of this litter is probably lost from the mangrove systems which are regularly flushed by the tide (Boto 1982). Therefore, uptake of nitrogen and phosphorus by mangroves in northern Australia could lead to the immobilisation of 100–200 kg N ha^{-1} and 10–20 kg P ha^{-1} annually.

In a study of the effects of enrichment by sewage effluent on mangroves near Darwin in northern Australia it was found that enrichment over a period of about 20 years led to an almost two-fold increase in total nitrogen and phosphorus status of mangrove sediments, with corresponding increases in foliar N and P (Table 2). Similarly, Boto and Wellington (1983) observed that fertilization with up to 400 kg ha^{-1} of nitrogen or phosphorus resulted in significant increases in growth rate and foliar N and P levels in mangroves at some sites in North Queensland.

A significant enhancement of growth, and earlier onset of flowering for *Rhizophora mangle* in response to nutrient enrichment was reported by Onuf *et al.* (1977). These authors compared the nutrient status and growth of mangroves on two spoil islands near Fort Pierce in Florida – a control island and one which received about 3 kg guano m^{-2} y^{-1} from a colony of birds. They found higher concentrations of ammonium in the interstitial water from the site enriched by guano and also higher concentrations of nitrogen in plants at the enriched site than at the unenriched site.

These observations indicate that nutrient enrichment from fertilization or from sewage effluent is not likely to be deleterious to mangroves, and indeed, may well be beneficial, particularly in area where the nutrient status of mangrove sediments is

Table 2. Mean values for nitrogen and phosphorus in mangrove sediments and foliage from enriched and unenriched sites near Darwin, northern Australia.

	Site conditions	No. of observations	Total N ($\mu g\ g^{-1}$ ODW)[a]	Total P ($\mu g\ g^{-1}$ ODW)[a]
Sediment	Enriched	6	2839	1438
	Unenriched	12	1586	787
Foliage	Enriched	6	2.04	0.167
	Unenriched	18	1.15	0.097

[a] ODW, oven-dry weight

low. An enhancement of growth in concert with higher levels of nitrogen and phosphorus in both the foliage and the woody parts of mangroves would lead to much higher rates of nutrient immobilisation in vegetative biomass than our earlier calculations suggested.

Heavy metals

Due chiefly to stricter environmental legislation, heavy metal contamination from industrial effluent is becoming less of a problem. Other sources include leaching of mine tailings, and agricultural chemicals such as fungicides. Domestic sewage normally contains very little heavy metal contamination. It is worthwhile, however, to examine briefly the possible fates of heavy metals in mangrove systems and the factors governing their residence time. Much of the information discussed here is derived from salt marsh studies but is equally applicable to mangrove systems given that the mobility of heavy metals is governed mainly by the characteristics of anaerobic sediments.

Heavy metals are usually strongly bound in anaerobic sediments, either by adsorption on ion exchange sites, incorporation into lattice structures or by precipitation as insoluble sulphides. Metal ions may also be adsorbed within iron and manganese colloidal sesquioxides or complexed by large molecular weight organic compounds such as humic acid. Gambrell *et al.* (1977), and Khalid *et al.* (1978) studied the bioavailability of heavy metals in anaerobic marsh sediments as a function of Eh and pH. They found that alterations in the redox status and pH by activities such as dredging, or by natural alterations in the water table, can give rise to mobilisation of adsorbed heavy metals normally trapped in sulphide-rich sediments. In the absence of such alterations, however, mangrove soils could act as high capacity sinks for many metals.

Various workers have studied uptake of heavy-metals by salt marsh species and rice. Banus *et al.* (1975), for example, examined lead, zinc and cadmium in various components of a marsh which had been enriched experimentally with these metals. For lead, generally only 6–8% was taken up by the grass. The degree of cadmium uptake, on the other hand, was dependent on the dosage rate. For example, for an input rate of $\leqslant 15$ mg Cd m^{-2} yr^{-1}, $\geqslant 80\%$ of the Cd was retained in the sediments and $\leqslant 20\%$ taken up by the grasses. At dosage rates of >40 mg Cd m^{-2} yr^{-1}, however, up to 65% of the Cd was exported from the marsh system. Large zinc losses by undetermined mechanisms occurred. In a similar study with rice plants in controlled redox and pH environments, Reddy and Patrick (1977) found that a large proportion of cadium and lead was subject to plant uptake, the degree of uptake increasing with increasing Eh and decreasing pH. These results are not necessarily applicable to mangroves, but serve as a warning that heavy metal trapping by anaerobic sediments may not be permanent at high dosage rates or where severe disturbances take place.

Pesticides

The main source of pesticide contamination is from agricultural use. Organochlorine insecticides (DDT, lindane, heptachlor) and herbicides (2,4–D,2,4,5–T) give rise to most concern because of their persistence in most natural environments and their toxicity towards nontarget organisms. Studies by Pionke and Chesters (1973) and Parr *et al.* (1974) show that these compounds are strongly adsorbed onto some sediments, notably those with a high clay and humic acid content, and that they are readily degraded in anaerobic soils.

It is therefore reasonable to assume that mangroves could act as an effective buffer system, protecting nearby coastal waters from excessive pollution by these types of compounds. It should be kept in mind, however, that during riverine flood conditions, the pesticides adsorbed onto sediment particles can be transported well away from the coast and remain in suspension for long periods under aerobic conditions. This view is supported by findings of low but detectable levels of lindane in various organisms along the Great Barrier Reef, up to 100 km from the coastal sugar growing areas of north-eastern Australia (Olafson 1978). Under normal river flow conditions, however, it is more probable that a good deal of the sediment-borne pesticides are deposited in estuarine mangrove systems, where efficient trapping and rapid degradation will occur.

Organic matter

Sewage effluent usually contains moderate to high levels of dissolved and particulate organic carbon. Effluent which has received some form of initial treatment in settling ponds typically contains 50–70 mg organic carbon per litre (McPherson 1979) and has a BOD of 10–70 mg l^{-1} (McPherson 1979; Quin 1979; Reynolds *et al.* 1979). The BOD is closely correlated with the level of labile organic carbon. These components of sewage effluent represent a potential threat to mangrove systems.

Undisturbed mangrove waterways may contain up to 40 mg l^{-1} dissolved organic carbon and 3 mg l^{-1} particulate organic carbon (Boto and Bunt 1981a, b). Dissolved organic carbon, in particular, has been shown to result in consistently lower dissolved oxygen and pH in many mangrove waterways. Increased dissolved organic carbon input, acting as an energy source for bacteria, could give rise to further reduction in dissolved oxygen, possibly with harmful effects on aquatic fauna.

Similarly, inputs of dissolved and particulate organic carbon from sewage will enhance bacterial activity in the already anaerobic sediments. Although mangroves and other marsh plants are adapted to anaerobic soils, recent studies in marsh and mangrove systems (Morris, 1980; Howes *et al.* 1981; Mendelssohn *et al.* 1981; Boto and Wellington, in press) indicate that anaerobiosis is a major factor controlling plant growth and that extremely anaerobic conditions lead to a severe restriction in growth. Additional inputs of organic carbon could alter the precarious redox balance in these systems which are already hard pressed to cope with anaerobic stress.

On the positive side, additional inputs of nutrients from sewage effluent may help to ameliorate the problem of anaerobic stress. Because of the lack of quantitative data it is difficult to make firm predictions on the overall effect of increased organic matter loadings on mangrove systems; however, there is a strong possibility that these effects will be detrimental to both the mangroves and their fauna. This aspect deserves much more attention than hitherto it has received, preferably by the use of case studies.

Hydrodynamic considerations

The fate of nutrients and other constituents of sewage effluent discharged into mangrove systems is also influenced by hydrodynamic factors such as the frequency and duration of tidal inundation, and patterns of water movement and drainage. These factors determine effluent dispersal patterns and also influence exchange processes between effluent and sediment. In many cases these factors may have a greater influence on nutrient and heavy metal removal and immobilisation by mangrove systems than the chemical and biological characteristics of the sediment.

We will consider these hydrodynamic effects here in the context of three likely discharge situations. These are: (i) discharge directly into an estuary or creek bordered by mangroves; (ii) discharge into mangrove areas which are inundated regularly by tides; and (iii) discharge into 'dryland' mangrove communities near the upper tidal limit where tidal inundation is infrequent.

From an examination of the literature and from our own experience it appears that the most common practice to date has been to discharge sewage effluent directly into an estuary or creek bordered by mangroves. The work of Nedwell (1974, 1975) provides an example of this case. In the latter paper, Nedwell reported that 56% of the total soluble inorganic nitrogen, 30% of the nitrate, and 63% of the soluble ammonium discharge in sewage effluent was removed within the estuary. The greater loss of ammonium than of nitrate was attributed to its preferential assimilation by primary producers in the euphotic zone. Nedwell (1975) also found that the bottom sediments of the estuary had a considerable potential for denitrification. However, the loss of nitrate via this process appeared to be limited by exchange of nitrate across the land–water interface due to inadequate mixing of the effluent with estuarine water. This is unlikely to be an isolated case, as well-defined haloclines are common in many estuaries.

Estuarine waters often carry high levels of suspended materials derived from erosion in upland catchments. The extent of sediment suspension and deposition is governed by tidal velocities, turbulence, particle size distribution, and the nature of the clay minerals (Boto and Patrick 1978). Significant amounts of dissolved nutrients (particularly ammonium and phosphate), heavy metals and other potential toxins may be removed by ion exchange or by adsorption on suspended colloidal materials (Boto and Patrick 1978). Depending on hydrodynamic conditions this material may be carried out to offshore systems, or deposited within the estuary or adjacent mangrove wetlands. This 'scavenging' effect may lead to appreciable inputs of particulate-bound nutrients to adjacent mangrove wetlands where the dense assemblage of prop roots and pneumatophores can reduce tidal velocities and enhance sedimentation (Bird 1972; Wolanski *et al.* 1980).

Removal of dissolved and particulate materials from estuarine water by adjacent mangrove communities can only occur during the period that the mangroves are covered by tidal water, which, even with the most favourable tidal characteristics, would be restricted to perhaps no more than twelve hours daily. In effect this will reduce the impact of the effluent on adjacent mangrove communities; it also limits the capacity of the mangrove community to reduce nutrient loadings from sewage to a level below that required for significant biological activity.

Tidal influences also play a major role in determining the fate of sewage effluent discharged directly into mangroves. In systems subject to regular tidal action – the second discharge situation described earlier – periodic inundation gives rise to pronounced tidal asymmetry effects. At high tide, effective mixing of effluent with tidal water and relatively low tidal velocities (due to surface roughness and the effect of pneumatophores and prop roots) should facilitate exchange across the mud-water interface and the 'scavenging' effect of sedimentation. At low tide, by contrast, effluent is discharged directly onto the sediment surface, where the removal of nutrients, heavy metals and other soluble constituents of the effluent will be governed by exchange and adsorption processes, and by drainage characteristics.

The influence of tidal inundation on the fate of sewage effluent becomes less significant at higher topographical levels, where inundation is less frequent and of shorter duration. The most extreme example of this case is where effluent is discharged into 'dryland' mangrove systems which are rarely exposed to tidal influence. Such systems are often characterized by high soil salinity and low fertility, particularly with respect to phosphorus. Other chemical characteristics of the soils, such as pH and redox, may also differ from those in mangrove soils subject to more regular tidal action. In the absence of regular tidal influences, the removal of nutrients, heavy metals and other constituents from the effluent will be governed largely by exchange and adsorption processes, and by drainage characteristics.

The effect of sewage effluent on mangrove systems near the upper tidal limit is difficult to predict with certainty. From what little we know in general of these systems, it appears likely that application of sewage effluent would lead to a reduction in soil salinity and higher phosphorus status of the system. This would probably be beneficial and could lead to a significant increase in productivity. However, it could also give rise to a marked change in the species composition of the system.

Based on general considerations, these three situations represent a probable gradient from the least (discharge directly into an estuary) to most effective (discharge in 'dry land' mangroves) schemes for reducing the impact of sewage on offshore systems through the use of mangrove systems. However, it must also be emphasised that detailed quantitative information on the effect of sewage on any mangrove system is entirely lacking.

Conclusions

What little information we have suggests that inputs of nutrients, heavy metals and pesticides to mangrove systems from sewage effluent will not be harmful to mangroves themselves. Indeed, nutrient enrichment may well have a beneficial effect on the productivity of many mangrove systems. Furthermore, the productivity of mangroves growing on more saline substrate could be enhanced as a result of irrigation with treated effluent. On the other hand, significant inputs of dissolved and particulate organic materials may further compound the problem of severe anaerobiosis for both plants and animals alike, which are already hard pressed to cope with the problem of highly reduced substrates.

In the short term, the rate at which nutrients and other components are removed from sewage effluent discharged into mangrove systems is largely determined by hydrodynamic factors, such as tidal inundation and drainage, which affect effluent dispersal and mixing, and exchange rates across the mud–water interface. The common practice of discharging treated effluent directly into an estuary or tidal channel fringed by mangroves appears to be a relatively inefficient scheme for reducing the potential pollutant loading to offshore systems. By contrast, discharging sewage effluent directly into mangrove systems near the upper tidal limits appears to offer considerably greater potential for trapping nutrients and other components in sewage effluent and wastewater.

In the longer term, the capacity of mangrove systems to immobilise nutrients and other constituents in sewage effluent depends on sediment properties and on the biological characteristics of the system. The highly anaerobic nature of most mangrove sediments suggests that they have a considerable long-term capacity to immobilize heavy metals and detoxify pesticides with moderate loadings of these materials. Furthermore, under hydrodynamic conditions which favour effective mixing and rapid exchange across the mud-water interface the reduced nature of mangrove sediments can lead to appreciable losses of nitrate via denitrification processes. On the other hand, the capacity of sediments to immobilize both ammonium and phosphate chemically is impaired by low redox potential. Under these conditions, the long-term capacity of mangrove systems to immobilize ammonium and phosphate may depend on their incorporation into plant biomass.

The activity of crabs, polychaetes and other burrowing animals could have an important bearing on the fate of sewage discharged into mangrove systems. In addition to creating subsurface drainage channels, these activities may lead to significant sediment reworking. These effects are likely to be relatively more important in areas with large populations of burrowing animals, notably those areas subject to regular tidal inundation.

Mangrove systems vary widely in their physical, chemical and biological properties, often over quite a small local scale. We have attempted here to identify those characteristics which are important in determining the fate of sewage effluent discharged into mangrove systems and its effect on the productivity and stability of the system. Future proposals to discharge treated sewage effluent into mangroves require detailed studies of the relevant site characteristics in order to evaluate their potential as a suitable tertiary treatment facility. In the meantime, our understanding of the complex interactions between sewage effluent and mangrove systems could be improved measurably by detailed investigations of mangrove systems already exposed to significant sewage loads.

Literature cited

Attiwill, P.M. and B.F. Clough. 1978. Productivity and nutrient cycling in the mangrove and seagrass communities of Westernport Bay. Ministry for Conservation, Victoria, 86 pp.

Banus, M.D., I. Valiela and J.M. Teal. 1975. Lead, zinc and cadmium budgets in experimentally enriched salt marsh ecosystems. Est. Coast. Mar. Sci. 3:421–30.

Bird, E.C.F. 1972. Mangroves and coastal morphology in Cairns Bay, North Queensland. J. Trop. Geogr. 35:11–16.

Boto, K.G. 1982. Nutrient and organic fluxes in mangroves. In: Mangrove Ecosystems of Australia: Structure, Function and Management (B.F. Clough, ed.). Australian National University Press, Canberra, pp. 239–257.

Boto, K.G. and J.S. Bunt. 1981a. Dissolved oxygen and pH relationships in northern Australian mangrove waterways. Limnol. Oceanogr. 26:1176–1178.

Boto, K.G. and J.S. Bunt. 1981b. Tidal export of particulate organic matter from a northern Australian mangrove system. Estuarine Coastal and Shelf Science. 13:247–255.

Boto, K.G. and W.H. Patrick, Jr. 1978. Role of wetlands in the removal of suspended sediments. In Wetland Functions and Values: The State of our Understanding (P.E. Greeson, J.R.

Clark and J.E. Clark, eds.). American Water Resources Association, Minneapolis, Minnesota, pp. 479–489.

Boto, K.G. and J.T. Wellington. (In press). Soil characteristics and nutrient status in a northern Australian mangrove forest. Estuaries.

Boto, K.G. and J.T. Wellington. 1983. Nitrogen and phosphorus nutritional status of a northern Australian mangrove forest. Marine Ecology Progress Series 11:63–69.

Bouldin, D.R., R.L. Johnson, C. Burda and C.W. Kao. 1974. Losses of inorganic nitrogen from aquatic systems. J. Environ. Quality 3:107–114.

Chang, S.C. and M.L. Jackson. 1957. Fractionation of soil phosphorus. Soil Sci. 84:133–144.

Clough, B.F. and P.M. Attiwill, 1975. Nutrient cycling in a community of *Avicennia marina* in a temperate region of Australia. In: Proceedings of the International Symposium on Biology and Management of Mangroves (G.E. Walsh, S.C. Snedaker and H.J. Teas, eds.). Institute of Food and Agricultural Sciences, University of Florida, Gainesville. Vol. I, pp. 137–146.

DeLaune, R.D., W.H. Patrick, Jr. and J.M. Brannon. 1976. Nutrient transformations in Louisiana salt marsh soils. Sea Grant Publication No. LSU-T-76-009, Center for Wetland Resources, Louisiana State University, Baton Rouge, 38 pp.

Duke, N.C., J.S. Bunt and W.T. Williams. 1981. Mangrove litter fall in North-east Australia. I. Annual totals by component in selected species. Aust. J. Bot. 29:547–553.

Gambrell, R.P., R.A. Khalid, M.G. Verloo and W.H. Patrick, Jr. 1977. Transformations of heavy metals and plant nutrients in dredged sediments as affected by oxidation-reduction potential and pH – Volume II: Materials and Methods/ Results and discussion. Dredged Materials Research Program, Waterways Experiment Station, U.S. Army Engineers, Vicksburg, Mississippi, Contract Report No. D-77-4.

Hesse, P.R. 1961. Some differences between the soils of *Rhizophora* and *Avicennia* mangrove swamps in Sierra Leone. Plant and Soil 14:335–346.

Hesse, P.R. 1962. Phosphorus fixation in mangrove swamp soils. Nature 193:295–296.

Hesse, P.R. 1963. Phosphorus relationships in a mangrove-swamp mud with particular reference to aluminium toxicity. Plant and Soil 19:205–218.

Hicks, D.B. and L.A. Burns. 1975. Mangrove metabolic response to alterations of natural freshwater drainage to southwestern Florida estuaries. In: Proceedings of the International Symposium on Biology and Management of Mangroves (G.E. Walsh, S.C. Snedaker and H.J. Teas, eds.). Institute of Food and Agricultural Sciences, University of Florida, Gainesville. Vol. I, pp. 238–255.

Holford, I.C.R. and G.E.G. Mattingly. 1976. A model for the behaviour of labile phosphate in soil. Plant and Soil 44: 219–229.

Holford, I.C.R. and W.H. Patrick Jr. 1979. Effects of redox potential and pH on phosphate removal from wastewater during land application. Prog. Water Techn. 11:215–225.

Howes, B.L., R.W. Howarth, J.M. Teal and I. Valiela. 1981. Oxidation-reduction potentials in a salt marsh: Spatial patterns and interactions with primary production. Limnol. Oceanogr. 26:350–360.

Hsu, P.H. 1964. Adsorption of phosphate by aluminium and iron in soils. Soil Sci. Soc. Amer. Proc. 28:474–478.

Keeney, D.R. 1973. The nitrogen cycle in sediment-water systems. J. Environ. Quality 2:15–29.

Khalid, R.A., W.H. Patrick, Jr. and R.P. Gambrell. 1978. Effect of dissolved oxygen on chemical transformations of heavy metals, phosphorus and nitrogen in an estuarine sediment. Est. Coast. Mar. Sci. 6:21–36.

Lugo, A.E., M. Sell and S.C. Snedaker. 1976. Mangrove ecosystem analysis. In: Systems Analysis and Simulation in Ecology (B.C. Patten, ed.). Academic Press, New York. Vol. IV, pp. 113–145.

McPherson, J.B. 1979. Land treatment of wastewater at Werribee past, present and future. Prog. Water Techn. 11:15–31.

Mendelssohn, I.A., K.L. McKee and W.H. Patrick, Jr. 1981. Oxygen deficiency in *Spartina alterniflora* roots: Metabolic adaptation to anoxia. Science 214:439–441.

Morris, J.T. 1980. The nitrogen uptake kinetics of *Spartina alterniflora* in culture. Ecology 61:1114–1121.

Nedwell, D.B. 1974. Sewage treatment and discharge into tropical coastal waters. Search 5:187–190.

Nedwell, D.B. 1975. Inorganic nitrogen metabolism in a eutrophicated tropical mangrove estuary. Water Research 9:221–231.

Olafson, R.W. 1978. Effect of agricultural activity on levels of organochlorine pesticides in hard corals, fish and molluscs from the Great Barrier Reef. Mar. Environ. Res. 1:87–107.

Olsen, S.R. and F.S. Watanabe. 1957. A method to determine a phosphorus adsorption maximum of soils as measured by the Langmuir isotherm. Soil. Sci. Soc. Amer. Proc. 21:144–149.

Onuf, C.P., J.M. Teal and I. Valiela. 1977. Interactions of nutrients, plant growth and herbivory in a mangrove ecosystem. Ecology 58:514–526.

Parr, J.F., G.H. Willis, L.L. McDowell, C.E. Murphree and S. Smith. 1974. An automatic pumping sampler for evaluating the transport of pesticides in suspended sediments. J. Environ. Quality 3:292–294.

Patrick, W.H. Jr. and I.C. Mahapatra. 1968. Transformation and availability to rice of nitrogen and phosphorus in waterlogged soils. Adv. Agromony 20:323–359.

Pionke, H.B. and G. Chesters. 1973. Pesticide-sediment-water interactions. J. Environ. Quality 2:29–45.

Ponnamperuma, F.N. 1972. The chemistry of submerged soils. Adv. Agromony 24:29–96.

Quin, B.F. 1979. Surface irrigation with sewage effluent in New Zealand – a case study. Prog. Water Techn. 11:103–126.

Reddy, C.N. and W.H. Patrick, Jr. 1977. Effect of redox potential and pH on the uptake of cadmium and lead by rice plants. J. Environ. Quality 6:259–262.

Reynolds, J.H., M.O. Braun, W.F. Campbell, R.W. Miller and L.R. Anderson. 1979. The long term effects of land application of wastewater. Prog. Water Techn. 11:283–300.

Sommers, L.E., R.F. Harris, J.D.H. Williams, D.E. Armstrong and J.K. Syers. 1972. Fractionation of organic phosphorus in

lake sediments. Soil Sci. Soc. Amer. Proc. 36:51–54.
Spain, A.V. and J.A. Holt. 1980. The elemental status of the foliage and branchwood of seven mangrove species from northern Queensland. Division of Soils Divisional Report NO. 49. CSIRO, Australia, 8 pp.
Teas, J.H. 1979. Silviculture with saline water. In: The Biosaline Concept (A. Hollaender, J.C. Aller, E. Epstein, A. San Pietro and O.R. Zaborsky, eds.). Plenum Press, New York, pp. 117–161.
Tusneem, M.E. and W.H. Patrick, Jr. 1971. Nitrogen transformations in waterlogged soil. Bulletin No. 657. Department of Agronomy, Louisiana State University, Baton Rouge, 75 pp.
Van Cleemput, O., W.H. Patrick, Jr. and R.C. McIlhenny. 1976. Nitrite decomposition in flooded soil under different pH and redox potential conditions. Soil Sci. Soc. Amer. J. 40:55–60.
Wolanski, E., M. Jones and J.S. Bunt. 1980. Hydrodynamics of a tidal creek – mangrove swamp system. Aust. J. Mar. Freshwater Res. 31:431–450.

CHAPTER 18

Ecology of a mangrove swamp near Juhu Beach, Bombay with reference to sewage pollution

S.N. DWIVEDI and K.G. PADMAKUMAR*

Central Institute of Fisheries Education, Bombay-400 058, India

Abstract. The mangrove areas in the vicinity of Bombay are the waste dumping grounds for the city. Large scale destruction of the *Rhizophora* spp. has resulted in the domination of the tidal swamps by dwarf shrubs of *Avicennia alba.*

Three stations in a three km stretch between Seven Bungalows and Juhu, Bombay were studied from January to September 1978, for physico-chemical and biological factors.

The first station, at Juhu Beach, has sandy substratum, high energy wave action and is devoid of mangroves. The range of dissolved oxygen was 0.00–6.1 ml/l. The BOD varied between 1.2 and 77.0 mg/l and the phytoplankton counts varied between 0.48×10^4 and 37.0×10^4 cells/l.

The second station, at Sluice Gate, the confluence of open drains which discharges domestic effluents and sewage, had dissolved oxygen values from 0.00 to 4.7 ml/l. The BOD ranged from 29.6 to 121.0 mg/l. The Phytoplankton counts varied from 0.78 to 43.9×10^4 cells/l.

The third station, at Seven Bungalows, which is about two km from the point of sewage discharge, was least polluted. Dissolved oxygen ranged from 1.78 to 7.4 ml/l and BOD from 3.3 to 38.0 mg/l. The Phytoplankton counts fluctuated from 0.09 to 101.6×10^4 cells/l. Fishes (*Boleophthalmus sp.*) were abundant in this region but not at the other two stations.

The study indicates that mangrove swamps have a high capacity for degrading organic wastes if sufficient dilution is available.

Introduction

The city of Bombay situated between lat. 18° 55′ and 19° 19′ N and long. 72° 47′ and 72° 59′ E has grown from a group of islands, with extensive low lying areas, creeks and tidal swamps which are rich in mangrove vegetation. Bombay and adjacent Colaba, West Thana, Janjira and part of Ratnagiri districts have been estimated to have a total of 24,870 ha of mangroves (Qureshi 1959). The mangroves, the so-called builders and guardians of coastal land (Sidhu 1963), are an open system continually adjusting to changes in the environment. To understand the ways in which human activities affect mangrove swamps, we must understand the ways by which man's activities change that environment, which in turn may bring about changes in the biological structure of this ecosystem (Burns 1977). The mangroves of Bombay have been affected by human activities. Due to large scale reclamation of mangrove areas and cutting of the larger trees, the mangrove vegetation is fast dis-

*Present address: Kerala Agricultural University, Faculty of Fisheries, Coconut Research Station, Kumarakam, Kottayam, Kerala, India.

Tasks for vegetation science, Vol. 8 ed. by H.J. Teas
 ISBN 978-90-481-8526-9

appearing from the shores of these islands. The remaining mangrove swamps near Bombay have become waste dumping ground. *Avicennia alba*, which is the dominant mangrove species of this area today, occurs as dwarf shrubs that grow only to about 1 to 1.5 m high. Navalkar (1951) reported that sewage is an important biotic factor which contributes to the gradual disappearance of this vegetation.

The food web relationships in the mangrove ecosystems have been worked out by Heald (1971), Odum (1972) and Odum and Heald (1975) in Florida, U.S.A. Recent contributions to the study of mangroves on the West Coast of India have been published by Navalkar and Bharucha (1949), Navalkar (1953), Untawale *et al.* (1973) and Dwivedi *et al.* (1975a, b). However, very little information is available on the impact of the discharge of sewage and domestic wastes on the environment of the mangroves in the Bombay region.

Materials and methods

The general object of this study was to investigate the effects of sewage discharges into the area of mangrove swamps. Along the western part of the city of Bombay, facing the Arabian sea, a 3 km coastal stretch between Versova and Juhu was

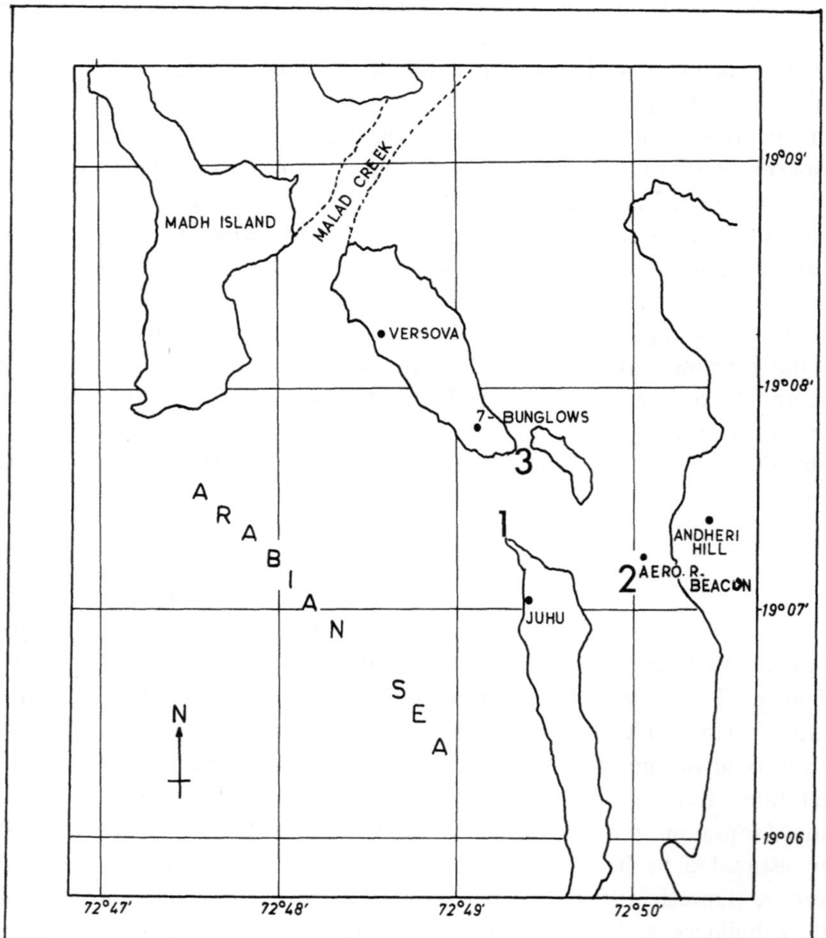

Fig. 1. Map of Juhu mangrove swamp showing the stations.

studied from January 1978 to September 1978. Fortnightly observations were made for the physico-chemical factors: temperature, salinity, dissolved oxygen and BOD of the channel waters of the mangrove swamp. Quantitative and qualitative composition of phytoplankton, benthic macrofauna and mangrove flora were also studied. Mohr's method was used for salinity and Winkler's method was used for dissolved oxygen (DO). The direct method was used for biochemical oxygen demand (BOD) on samples having a BOD of less than 3 mg/l; the unseeded dilution method was used for samples with higher BOD (Martin 1970). The BOD was measured after 5 days incubation. Benthic fauna were collected by the procedures of McClusky (1971). Station No. 1 at Juhu is an open sand beach. Station No. 2 at sluice gate is situated near of confluence of an open drains that discharges domestic wastes and raw sewage. Station 3 at Seven Bungalows is situated at the northern edge of the same mangrove swamp as station 2 but is nearer to the sea; it is separated from the sea by a sand bar. The locations of the three stations are shown in Fig. 1.

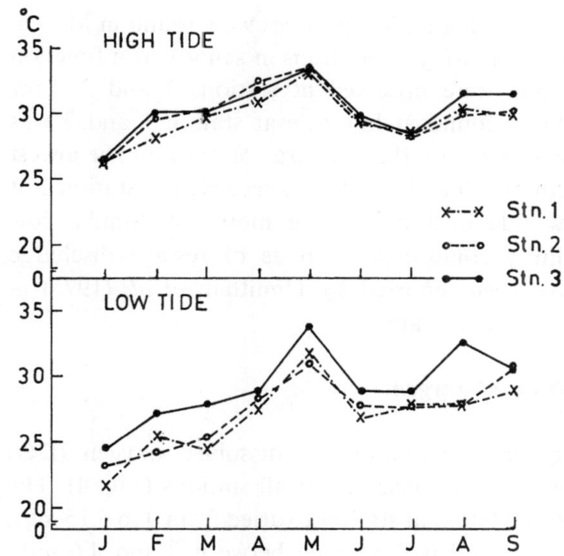

Fig. 2. Changes in temperature of the mangrove swamp waters.

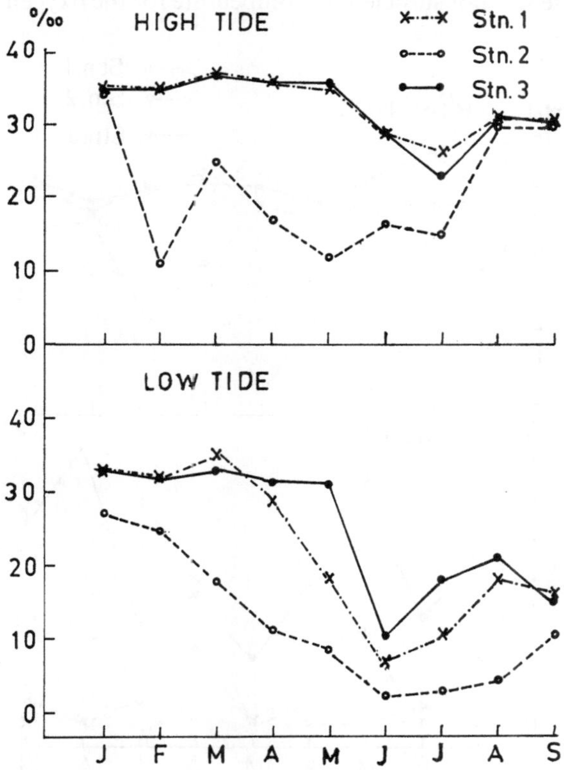

Fig. 3. Seasonal changes in salinity of the swamp waters.

Results and discussion

Temperature

Temperature of the channel waters of the mangrove swamp showed marked variation with station and season (Fig. 2). Temperatures varied from 21.8° C to 33° C as at station 1, 23.2° C to 33.6° C at station 2 and 24.6° C to 33.4° C at station 3. At all the three stations the temperature minima were observed in January and maxima in May. The thermal conditions of the swamp water appear to be influenced principally by the atmospheric temperature.

Salinity

In the mangrove swamp under investigation, the tides were predominantly of the semidiurnal type. Seasonal and tidal variations in salinity were pronounced (Fig. 3). Low salinity values were observed at all stations during the monsoon months of June and July; high salinity values were found in March, April and May. Variations in salinity as a function of tide were observed at stations 1 and 2. Low salinity found at low tide at station 1 and 2 was correlated with the discharge of sewage. The lowest salinity value of 2.43% was recorded at station 2 at low tide in June, during monsoon. Similar low salinity conditions at areas of sewage discharge have been reported by Unnithan *et al.* (1975) in Cochin backwaters.

Dissolved oxygen

Significant variations in dissolved oxygen (DO) content were observed at all stations (Fig. 4). The DO at station at station 1 varied from 0 to 6.15 ml/l; at station 2 it fluctuated between 0 and 4.6 ml/l; and at station 3, the DO varied from 1.17 ml/l to 7.4 ml/l. At station 1 and 2, DO was 0 at low tide during most of the months; even the dilution provided by the oxygen-rich flooding water at high tide was not sufficient to compensate for the oxygen

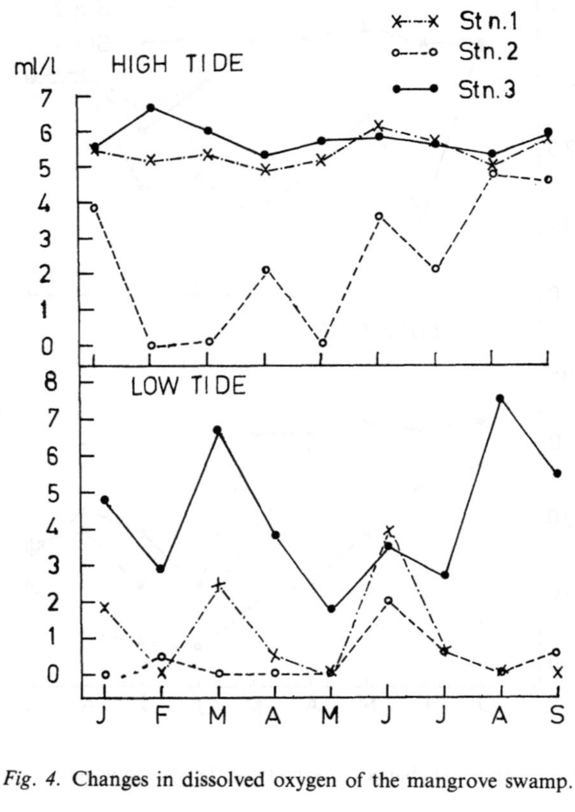

Fig. 4. Changes in dissolved oxygen of the mangrove swamp.

depletion at station 2.

Apparently decomposition of the heavy load of organic matter brought in by the sewage depleted the available oxygen at this station; the higher temperature during the summer months enhanced the decomposition process. Ganapati and Raman (1973) at Visakhapatnam Harbour and Vijayan *et al.* (1976) who worked in the Cochin backwaters have also reported similar oxygen depletion in sewage discharge areas. The DO values were always high at station 3 which indicates the relatively healthy nature of this area. The low values for DO observed at low tide at station 1, even though it is an open beach, are because the sewage water from station 2 flows into the sea through this area. However, during high tide the DO levels at this station were high.

Biochemical oxygen demand

The importance of BOD in pollution monitoring has been well established (McClusky 1971; Reish 1972). BOD vaues of more than 8 mg/l is indicative of moderate pollution (Martin 1970). The high BOD values obtained, especially at station 2, indicate severe pollution. BOD values were higher at low tide than at high tide (Fig. 5), probably because of the dilution provided by the flooding waters at high tide. The highest BOD value recorded in the present study was 120 mg/l at station 2; however, Zingde and Desai (1980) have recorded a BOD of 190 mg/l in the sewage-polluted areas of Mahim creek, Bombay, and Unnithan *et al.* (1975) have reported BOD values as high as 280 ppm in the organically polluted areas of Cochin backwaters. Low values of DO and high values of BOD observed at station 1 and 2 are indicative of organic pollution in these areas (Campbell 1978).

Benthic fauna

The use of benthic animals to assess the degree of pollution is well established. Benthic organisms reflect not only conditions at the time of sampling but also conditions that have been operative for some time previous to sampling (Reish 1972). The benthic animal biomass showed wide variations

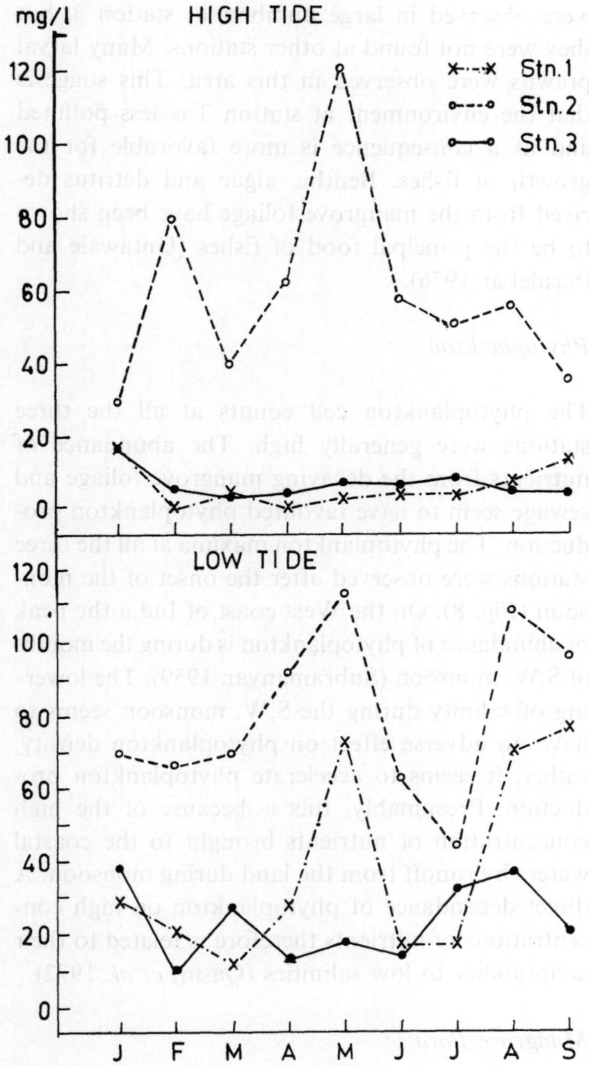

Fig. 5. Seasonal variation in BOD_5 in the swamp waters.

between stations and seasons (Fig. 6). The average biomass values were 0.15 gm/m^2 at station 1, 34.8 gm/m^2 at station 2 and 20.4 gm/m^2 at station 3. The benthic biomass values were low at station 1. Intertidal samples collected at a beach near station 1 showed a higher benthic biomass (11.2 g/m^2) (Dwivedi and Sreeramamurty 1979). The biomass values in their study were largely contributed by a filter feeding trochid, *Umbonium vestiarium*. However, in the present study *Umbonium vestiarium* occurred much less frequently, which may be due to the impact

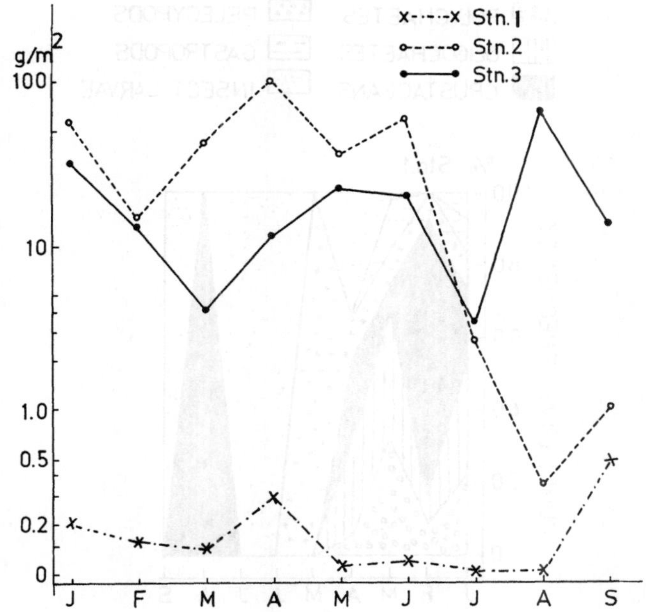

Fig. 6. Changes in the benthic biomass at the three stations of the mangrove swamp during January–September, 1978.

of sewage on these animals. The sewage influx coupled with high energy waves and current velocity at this open beach station appear to responsible for the low benthic biomass at station 1. The benthos at station 1 is dominated by crustaceans whereas at station 2 gastropods and oligochaetes were the dominant fauna. At station 3 gastropods were the dominant groups in the benthic fauna (Fig. 7). The oligochaete *Tubifex tubifex* is considered to be an indicator of organic pollution (Welch and Lindell 1980); large numbers were found at station 2 but were absent from station 3. The increase in number of tubificids is partly due to their capacity to tolerate pollution. Although tubificids are also affected adversely by low DO, they are able to maintain large biomass in the presence of a large food supply because of their reproductive capability (Brinkhurst 1965). The significant factors which may influence the distribution of bottom fauna are salinity and the nature of the bottom deposit (Kurian 1974). The reduced faunal biomass observed during the southwest monsoon may be due to the wide fluctuations in salinity.

Fishes such as mudskippers (*Boleophthalmus* sp.)

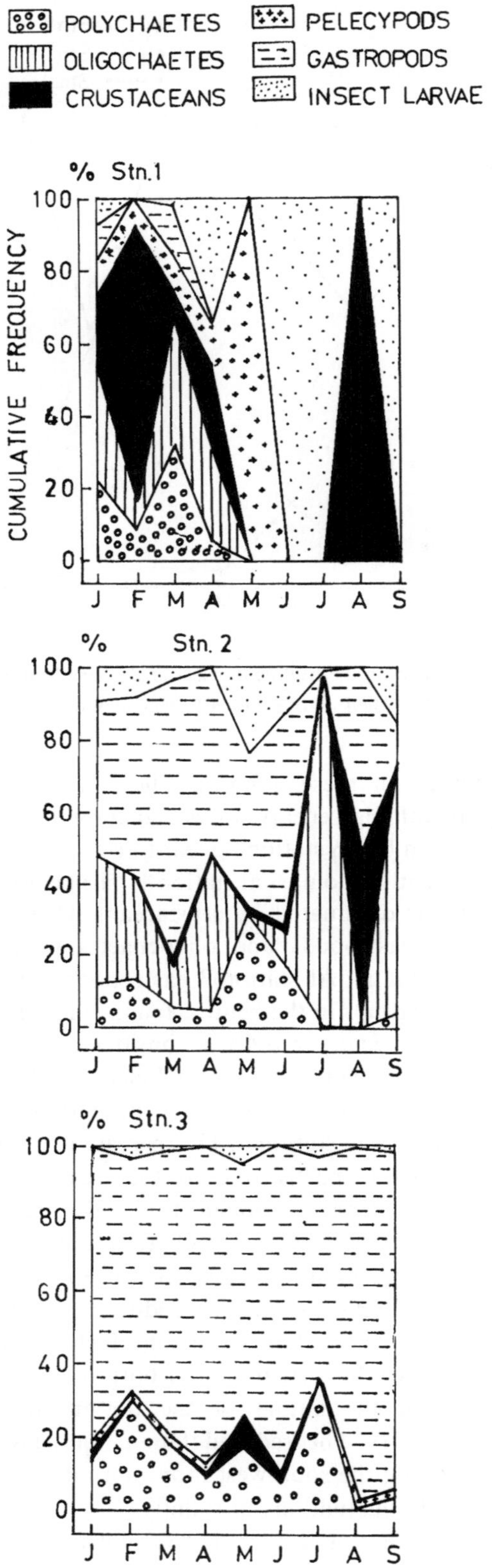

were observed in large numbers at station 3, but they were not found at other stations. Many larval prawns were observed in this area. This suggests that the environment at station 3 is less polluted and as a consequence is more favorable for the growth of fishes. Benthic algae and detritus derived from the mangrove foliage have been shown to be the principal food of fishes (Untawale and Parulekar 1976).

Phytoplankton

The phytoplankton cell counts at all the three stations were generally high. The abundance of nutrients from the decaying mangrove foliage and sewage seem to have favoured phytoplankton production. The phytoplankton maxima at all the three stations were observed after the onset of the monsoon (Fig. 8). On the West coast of India the peak of abundance of phytoplankton is during the months of S.W. monsoon (Subramanyan 1959). The lowering of salinity during the S.W. monsoon seems to have no adverse effect on phytoplankton density, rather, it seems to accelerate phytoplankton production. Presumably, this is because of the high concentration of nutrients brought to the coastal waters by runoff from the land during monsoon. A direct dependance of phytoplankton on high concentrations of nutrients therefore is related to their adaptability to low salinities (Qasim *et al.* 1972).

Mangrove flora

The mangrove flora of the study area is represented mainly by two species, *Avicennia alba* and *Acanthus ilicifolius*. These occur as dense formations with an average height of 1.25 m. This height is typical for *Acanthus* everywhere but here even the *Avicennia alba* is shrubby, whereas it normally grows as a tree. The indiscrimate cutting and grazing has obliterated any sign of the original mangrove zonation. Mangrove were totally absent at station 1, which is probably because of the sandy substratum and high

←

Fig. 7. Cumulative frequency distribution of major groups of benthic macrofauna.

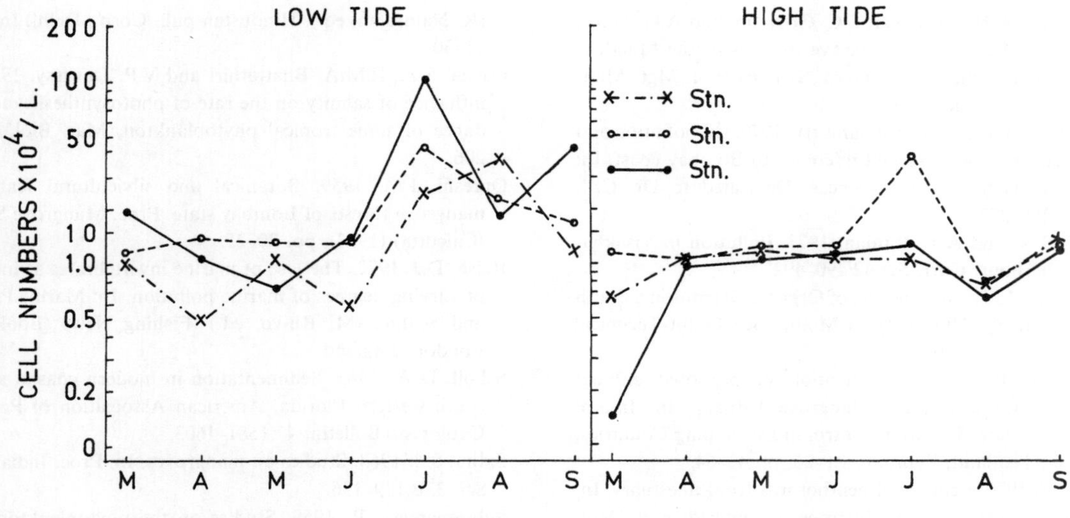

Fig. 8. Seasonal distribution of phytoplankton in the swamp waters.

wave energy in the area. A role for substrate is not unexpected since coastal sedimentation is known to be important in determining where mangroves can grow (Scholl 1963). At stations 2 and 3 the mangrove formations were dense and thick so that in certain places the average density of plants, including seedlings and saplings, was 23/m^2. The average height of the mangroves was 1.5 m at station 2 and 1 m at station 3. Discharge of sewage did not seem to have any serious effect on the mangrove flora. As compared to other estuarine organisms, mangrove flora and fauna can withstand large variations in environmental factors (Parulekar and Dwivedi 1975). Henley (1978) reported no marked adverse impace of sewage effluents on the mangrove communtly at Darwin, Australia. Mangrove swamps seem to have a high capacity for degrading organic wastes. However, if sewage nutrients are increased beyond a certain value and adequate dilution does not occur, these discharges may have an adverse effects on the mangroves.

The substratum and current apparently have a greater effect on the mangroves than does nutrient enrichment since the mangrove flora has not suffered greatly on account of sewage discharge. However, the benthic animal population is markedly affected by the influx of sewage. Tides play an important role in determining the environmental conditions which in turn influence the numbers and species of animals.

Acknowledgements

The authors are thankful to Dr. B.N. Desai and Dr. S.A.H. Abidi, National Institute of Oceanography, Regional Centre, Bombay for critically reviewing the manuscript. Thanks are also due to Shri V. Ravindranathan, Assistant Professor, Oceanography, Central Institute of Fisheries Education, for suggestions and Shri K.B. Sreeramamurthy for assistance.

Literature cited

Brinkhurst, R.O. 1965. Observations on the recovery of a British River from gross organic pollution. Hydrobiol. 25:9–51.

Burns, L.A. 1977. Mangrove swamps. In: Coastal Ecosystem Management – A technical manual for the conservation of Coastal Zone resources. (J.R. Clarke ed.), John Wiley and Sons, New York. 660–665.

Campbell, N.C. 1978. A biological investigation of an organically polluted urban stream in Victoria. Aust. J. Mar. Freshwater Res. 29:275–91.

Dwivedi, S.N., R.M.S. Bhargava and A.G. Untawale, 1975a. Marine living resources and ecosystem along the west coast of India. The 3rd Intern. Ocean Develop. Confern., Tokyo, Vol. III:31–41.

Dwivedi, S.N., A.H. Parulekar, S.C. Goswami and A.G. Untawale. 1975b. Ecology of mangrove swamps of the Mandovi estuary, Goa, India, Proc. Intern. Sympos Biol. Mgt. Mangroves, Univ. Florida Press.

Dwivedi, S.N. and K.B. Sreeramamurty. 1979. Oil pollution and its consequences with special reference to Bombay coast. In: Contributions to Marine Sciences. Dedicated to Dr. C.V. Kurian. 266–273.

Ganapati, S.N. and A.V. Raman. 1973. Pollution in Visakhapatnam harbour. Curr. Sci 42:490–492.

Heald, E.J. 1971. The production of Organic Detritus in a South Florida Estuary, University of Miami, Sea Grant Technical Bulletin No. 6, 162 pp.

Henley, D.A. 1978. An investigation of proposed effluent discharge into a tropical Mangrove Estuary. In: Intern. Confer. on Water Pollution control in Developing Countries, Bangkok, Thailand, February 21–25, pp 43–64.

Kurian, C.V. 1970. Ecology of benthos in a tropical estuary. In: INSA Bulletin No. 47. Proc. Sympos mar. intertdl. eco., Dept. Zool. Andhra Univ., Waltair, January 22–24, pp. 156–163.

Martin, D.F. 1970. Marine chemistry, vol. I. Marcel Dekker, Inc., New York, 281 pp.

McClusky, D.S. 1971. Ecology of estuaries, Heinemann Educational Books, London, 144 pp.

Navalkar, B.S. 1951. Succession of the mangrove vegetation of Bombay and Salsette Islands, J. Bombay Nat. Hist. Soc. 50:157–161.

Navalkar, B.S. and F.R. Bharucha. 1949. Studies in the ecology of mangroves, V. chemical factors of mangrove soils. J. Univ. Bombay 17:17–35.

Odum, W.E. 1972. Pathways of Energy Flow in a South Florida Estuary. University of Miami, Sea Grant Technical Bulletin No. 7, 162 pp.

Odum, W.E. and E.J. Heald. 1975. The detritus based foodweb of an estuarine mangrove community. In: Estuarine Research, Vol. I (Ed.) L.E. Cronin. Academic press, Inc., New York, pp. 265–286.

Parulekar, A.H. and S.N. Dwivedi. 1975. Ecology of benthic production during south west monsoon in an estuarine complex of Goa. In: Recent Researches in Estuarine Biology. (R. Natarajan, ed.). Hindustan pub. Corp., Delhi, India, pp. 21–30.

Qasim, S.Z., P.M.A. Bhattathiri and V.P. Devassy. 1972. The influence of salinity on the rate of photosynthesis and abundance of some tropical phytoplankton, Mar. Biol. 12:200–206.

Qureshi, I.M. 1959. Botanical and silvicultural features of mangrove forests of Bombay state. Proc. Mangrove Sympos. (Calcutta) (1957), pp. 20–25.

Reish, D.J. 1972. The Use of marine invertebrates as indicators of varying degrees of marine pollution. In: Marine Pollution and Sealife. (M. Ruivo, ed.). Fishing News (Books) Ltd, London, England.

Scholl, D.W. 1963. Sedimentation in modern coastal swamps, south western Florida. American Association of Petroleum Geologists Bulletin. 47:1581–1603.

Sidhu, S.S. 1963. Studies on mangroves. In: Proc. Indian Acad. Sci. 33B:129–136.

Subramanyan, R. 1959. Studies on the phytoplankton of the west coast of India, Parts I and II. Proc. Indian Acad. Sci. 59B:113–252.

Unnithan, R.V., M. Vijayan and K.N. Remani. 1975. Organic pollution in Cochin backwaters. Indian J. Mar. Sci. 4:39–42.

Untawale, A.G., S.N. Dwivedi and S.Y.S. Singbal. 1973. Ecology of mangroves in Mandovi and Zuari estuaries and the interconnecting Cumbarjua canal of Goa. Indian J. Mar. Sci. 2:47–53.

Untawale, A.G. and A.H. Parulekar. 1976. Some observations on the ecology of an estuarine mangrove of Goa. Mahasagar-Bull. Nat. Inst. Oceanogr. 9:57–62.

Vijayan, M., K.N. Remani and R.V. Unnithan. 1976. Effect of organic pollution on some hydrographic feature of Cochin backwaters. Indian J. Mar. Sci. 5:196–200.

Welch, E.B. and T. Lindell. 1980. Ecological effects of waste water. Cambridge University Press, Cambridge, London, 337 pp.

Zingde, M.D. and B.N. Desai. 1980. Waste water discharge and its effect on the quality of water of Mahim creek and Bay. Mahasagar Bull. Nat. Inst. Oceanogr. 13:205–213.

CHAPTER 19

Impact of oil spills on mangrove forests

ROY R. LEWIS, III

Mangrove Systems, Inc., Post Office Box 24748, Tampa, Florida 33623, U.S.A.

Introduction

Mangrove forests are the dominant intertidal plant community along most low wave energy shorelines in the tropics (Macnae 1968; Lugo and Snedaker 1974). Their value as habitat and detrital food sources for marine organisms as well as their direct commercial value as lumber, firewood and tanning agents are well documented (Odum and Heald 1972; Chapman 1976).

Increasingly large amounts of petroleum are entering the marine environment, primarily from three routes: natural seeps, 0.6 million metric tons per annum (mta); offshore drilling and production, 0.08 mta; and transportation losses, 0.77 mta (National Academy of Sciences 1975). These figures are presently being updated by the National Academy of Sciences and certain figures may increase substantially due to such events as the IXTOC I oil well blowout of 1979–80.

Due to the fact that floating oil is transported by waves and currents and strands on shorelines, low wave energy ecosystems such as mangrove forests are routinely sites where oil accumulates after a spill. In addition, the inaccessibility of most mangrove forests makes oil removal very difficult, if not impossible, and the fine grained anaerobic sediments characteristic of mangrove forests severely reduce the rate of microbial breakdown of oil (Lee 1980).

These factors plus the burrowing activities of crustaceans characteristic of mangrove forests can lead to persistent, high levels of oil contamination, not only on the soil's surface but also deep in the sediments in the mangrove root zone.

Table 1 lists thirteen oil spills, worldwide, where some documentation of impact to mangroves has been observed and reported. Most of the earlier reports are anecdotal with little or no quantification as to the levels of oil contamination, the actual type of oil stranded in the forest, the areal extent of mangroves oiled, or the actual numbers, sizes, or areal coverage of mangroves killed or stressed by the oil spill. Often, the species of mangroves involved was not mentioned. In addition to impacts to the trees themselves, the faunal component of the ecosystem is usually treated by noting dead or stressed animals at the time of the spill, or attempting to quantify losses or changes by one-shot, non-seasonal sampling. In particular, the sampling of infauna, in many of these cases, has not taken into account the normal seasonal variations in infaunal species and numbers. All these problems may lead to widely different observations of the 'impact' of a spill by different investigators, as will be discussed.

Materials and methods

Within 24 hours of the *Howard Star* (5 Oct 1978) spill, personnel were on site documenting the extent of the spill and any immediate mortalities of invertebrates, fish, or birds. Samples of dead inverte-

Tasks for vegetation science, Vol. 8 ed. by H.J. Teas

 ISBN 978-90-481-8526-9

Table 1. Comparison of oil spills impacting mangroves (modified from Baker *et al.*, 1981).

Source of spill and date	Type of oil	Amt. of oil	Mangrove species affected	Location	Impact on mangroves	Authors
Argea Prima, vessel, 7/16/62	Crude	10,000 tons	Unidentified	Guanica, Puerto Rico	. . . this habitat was virtually destroyed'	Diaz-Piferrer, 1962
Whitewater, vessel, 12/13/68	Diesel oil and bunker C	20,000 barrels	*Rhizophora mangle* *Avicennia* sp.	Galeta I., Panama	Death of young mangroves, loss of sessile animals and algae on prop roots (loss still visible 66 months after spill)	Rutzler and Sterrer, 1970; Birkeland *et al.*, 1976
Pipeline break, 1970	Light crude	100,000 barrels	Unidentified	Tarut Bay, Saudi Arabia	Defoliation, but many survived	Spooner, 1970
St. Peter, vessel, 2/76	*Crude*	*243,442 barrels* carried; quantity spilled unknown	*Rhizophora* sp. *Avicennia* sp.	Colombia, Ecuador	No 'noticeable long-term biological effects . . .'; temporary decline in fishery harvests and clam harvesting	Jernelöv and Linden, 1980; Hayes, 1977
Garbis, vessel 7/18/75	Crude oil and water emulsion	1500–3000 barrels	*Rhizophora mangle* *Avicennia germinans* (referred to as *A. nitida*)	Florida Keys, U.S.A.	Death of young red mangrove seedlings and some dwarf black mangroves	Chan, 1977
Zoe Colocotroni, vessel, 3/18/73	Venezuelan crude	37,000 barrels	*Rhizophora mangle* *Avicennia germinans* (referred to as *A nititda*)	Cabo Rojo, Puerto Rico	Death of adult trees (red and black) over an area of 1.0–2.7 ha within 3 years	VAST/TRC 1975; Nadeau and Berquist, 1977; Page *et al.*, 1979; Lewis, 1979a; Martinez *et al.*, 1979; Gilfillan *et al.*, 1981
Showa Maru, vessel, 1/6/75	Arabian light Berri and Murban crude	54,000 barrels	*Sonneratia* sp. *Rhizophora* sp.	Indonesia	Some dead trees (both species) unquantified; areas of greatest impact in sheltered bays; low numbers of crabs and snails associated with oil in sediments	Baker, 1981; Baker *et al.*, 1981
Pipeline rupture 10/13/76	Crude	377 barrels	*Avicennia germinans*	Corpus Christi, Texas	Mangroves burned to remove oil died; uncleaned mangroves recovered after minor defoliation	Holt *et al.*, 1978
Unidentified vessel, 3/77	Venezuelan Crude	1000 barrels	*Rhizophora mangle*	Guayanilla Bay, Puerto Rico	Damage to mangrove root community, trees survived	Lopez, 1978

Santa Augusta, vessel, 1971	Crude	12.5 million liters	*Rhizophora mangle*	St. Croix, U.S. Virgin Islands	5 ha completely destroyed, little or no recolonization after 7 years	Lewis, 1979b; Lewis and Haines, 1980
Funiwa 5, offshore oil well, 1/17/80 to 2/1/80	Crude	8.4 million gallons	Unidentified	Nigeria	Under study	Baker, 1981; OSIR 1980a; OSIR, 1980b
Peck Slip, barge, 12/19/78	Bunker C	440–460,000 gallons	*Rhizophora mangle*	Between Punta San Agustin and Yabucoa, Puerto Rico	Significantly affected mangrove, crab, snail, and epiphyte populations	Gundlach *et al.*, 1979; Getter *et al.*, 1981
Howard Star, vessel, 10/5/78	20% diesel 80% Bunker C	40,000 gallons	*Rhizophora mangle* *Avicennia germinans* *Laguncularia racemosa*	Tampa, Florida	Death of all three species of mangroves, death of molluscs and polychaetes, root abnormalities	Lewis, 1979b; Grundlach *et al*, 1979; Lewis, 1980a; Getter *et al.*, 1981; Snedaker *et al.*, 1981

brates were either identified to species on site or collected and preserved for later identification. Estimates of the number of dead invertebrates were made from random surface counts of dead animals in m^2 quadrats. Dead animals were particularly evident near the origin of the spill.

Twenty-five mangroves of all three species (black, *Avicennia germinans;* white, *Laguncularia racemosa;* and red, *Rhizophora mangle*) that had been oiled were tagged, and DBH (Diameter Breast High), height, and amount of oiling recorded (e.g. height and relative oiling on pneumatophores, prop roots, and stems). A group of twenty-five control trees (not oiled) were also tagged and measured.

All of the stations were visited at approximately monthly intervals between October 1978 and August 1980. Individual areas were regularly photographed and the health of individual experimental and control trees noted. Vertical, false-color, infrared aerial photographs of the stations were taken in November 1978, April 1979, and November 1979.

On 21 March 1980 ten three-inch PVC cores eight inches long were taken at each of the eight stations. These were sieved in the field through a 0.5 mm mesh and the residue and organisms retained on the sieve were relaxed in a propylene phenoxytol/seawater mixture, preserved with 10% formalin, and stained with rose bengal.

These cores were transported to the Mote Marine Laboratory in Sarasota, Florida. Due to the very high cost of sorting organisms from the plant material retained on the sieve, only five of the cores were analyzed from each station. All organisms were removed from the samples, identified to the lowest practical taxonomic level and enumerated.

On 2 April 1980 four two-inch sediment cores were taken at each of the eight stations by the Bowdoin College Hydrocarbon Research Center. The cores were taken to a depth of approximately 20 cm. Each core was taken with a stainless steel coring device that was rinsed with solvent between each station. After the core was taken, it was cut longitudinally into four quarters. Three of these were discarded and the last used as a subsample. Four subsamples from a given station were combined as the final sample for analysis of hydrocarbons. Samples were transported on ice from

Tampa to Bowdoin College, Brunswick, Maine for analysis.

In addition to the sediment samples taken in April 1980, samples of oil collected by the Coast Guard during the oil spill in October 1978 were transferred from the Coast Guard Central Oil Intelligence Lab (COIL) to the Bowdoin College Hydrocarbon Research Center for analysis and comparison with the sediment samples. These included samples of oil taken from fuel tanks on board the *Howard Star* as well as freshly spilled oil. Oil analysis was performed using the same techniques as described in Gilfillan *et al.* (1981).

The Bahia Sucia site was visited on 17–18 March 1979, six years to the day after the *Zoe Colocotroni* oil spill.

Previously established transects (Page 1979) were walked and standard measurements (DBH, height, surface salinity and interstitial salinity) taken at locations along the transects. These are the same transects that had been established previously for chemical analysis of the sediments for oil (Page 1979). Surface and interstitial salinities were taken with an A.O. temperature-corrected refractometer.

The intent of the survey was to examine the mangrove community for any evidence of (1) oil damage to mature trees; (2) stress or death in young mangroves, and (3) salinity stress indicated by high salinities in unoiled areas. Heights of 1–2 m indicate young trees if the diameter (DBH) of the stem is small (1–2 cm) and the canopy (leaf cover of stems) is also small. Scrubby mangroves also are 1–2 m tall, but the diameters are larger (greater than 2 cm) and the canopy more extensive. Due to recent rains, the soil was extremely soggy in some areas and could not be reached on foot.

In addition to oblique color aerial photographs of the site taken by Dr. David Page in November, 1978, five other vertical aerial photographs (1966–1977) of the area were photointerpreted.

The site is dominated by red mangroves (*R. mangle*), with fewer black mangroves (*A. germinans*) found either mixed with the red mangroves or as uniform stands in the more saline areas behind the red mangrove fringe. White mangroves (*L. racemosa*) were very uncommon.

Case histories

Three oil spills in particular will be discussed in order to amplify the problems of describing oil spill impacts to mangrove forests and the attendant legal and scientific arguments that result.

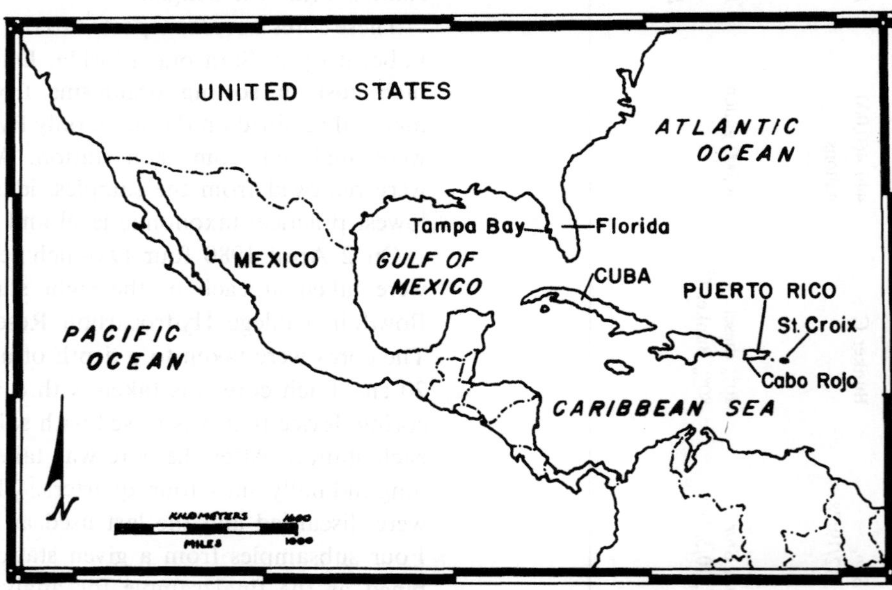

Fig. 1. Location map for oil spill sites at St. Croix (*Santa Augusta*), Cabo Rojo (*Zoe Colocotroni*) and Tampa Bay (*Howard Star*).

The first spill to be discussed is the *Santa Augusta* spill of 12 June 1971. This spill occurred when a tanker ruptured several of its tanks after striking an unidentified submerged object at the Hess Oil Refinery on the island of St. Croix, United States Virgin Islands (Fig. 1). An estimated 12.5 million liters (3.3 million gallons) of crude oil was spilled and much of it stranded on the south shore of the island. No reports of biological damage were made at the time of the spill. Lewis (1979b) and Lewis and Haines (1980) concluded after examination of historical aerial photography and recovery of weathered oil in concentrations of 50,000 ppm that an area of approximately 5 ha of mostly red mangroves, *Rhizophora mangle*, had been killed by the spill (Fig. 2). Natural regeneration seven years after the spill had been minimal and successful restoration efforts were conducted during 1978–1979 (Lewis and Haines 1980).

Fig. 2. Mangrove forest dominated by red mangroves, *Rhizophora mangle*, seven years after being impacted by spilled oil from the vessel *Santa Augusta*, St. Croix, U.S. Virgin Islands.

The second spill is that of the vessel *Zoe Colocotroni* on 18 March 1973 offshore of Cabo Rojo, Puerto Rico (Fig. 1). This spill was caused when the ship grounded on a reef and the captain decided to lighten the vessel by pumping 5.7 million liters (1.5 million gallons) of crude oil overboard. The oil was carried by currents and winds into Bahia Sucia, a semi-enclosed bay where much of it stranded along sandy shorelines and mangrove forests. The primary area of concern was located on the western side of the bay where a 4.9 ha mangrove forest was heavily oiled (Fig. 3), although oil also stranded in mangroves just south of this site and on the eastern sided of Bahia Sucia (Nadeau and Berquist 1977; Page *et al.* 1979; Gilfillan *et al.* 1981), most of the

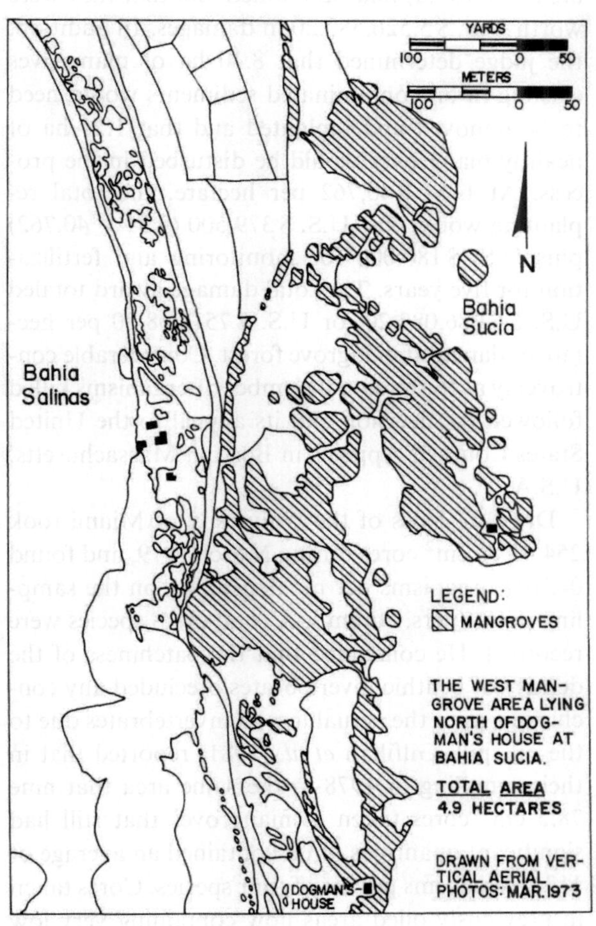

Fig. 3.

observed damage was in this area.

The effects of the spill have been the subject of litigation since the time of the spill, and which is still ongoing (1982). A number of precedent-setting legal decisions have been rendered and numerous conflicting reports have been prepared (VAST/TRC 1975; Nadeau and Berquist 1977; Page *et al.* 1979) and many volumes of legal testimony generated.

The main subject of these disagreements revolves around two points. The first is how many benthic invertebrate organisms were killed by the oil and the second is how many hectares of mangroves died as a result of the oiling they received. In the first legal decision on the case (Commonwealth of Puerto Rico *vs.* the SS Zoe Colocotroni, 456 F. Supp. 1327, 1978), it was determined by the judge that 92,109,720 marine organisms (11,371,570 per hectare over an area of 8.10 ha) had been killed and that they were worth U.S. $ 5,526,583.20 in damages. In addition, the judge determined that 8.10 ha of mangroves existing in oil-contaminated sediments would need to be removed and replanted and that 1.21 ha of healthy mangroves would be disturbed in the process. At U.S. $ 40,762 per hectare, the total replanting would cost U.S. $ 379,500 (9.31 × 40,762) plus U.S. $ 180,000 for monitoring and fertilization for five years. The total damage award totaled U.S. $ 6,086,083.20, or U.S. $ 751,368.30 per hectare of damaged mangrove forest. Considerable controversy as to the actual number of organisms killed followed the decision and its appeal to the United States Court of Appeals in Boston, Massachusetts, U.S.A.

Dr. G.L. Voss of the University of Miami took 254 41.85 cm^2 cores during March 1979, and found 0–3107 organisms per m^2 depending on the sampling point (pers. comm.). A total of 18 species were recorded. He concluded that the patchiness of the density of benthic invertebrates precluded any conclusions as to the actual loss of invertebrates due to the oil spill. Gilfillan *et al.* (1981) reported that in their sampling in 1978 in the same area that nine 78.5 cm^2 cores taken in mangroves that still had significant quantities of oil contained an average of 113.1 organisms per m^2 of four species. Cores taken in previously oiled areas now containing very low oil residues ranged from 0–20,001 organisms per m^2, again depending on the station. Gilfillan *et al.* (1981) concluded that only 0.81 ha of previously oiled red mangroves still showed a reduced number of infaunal organisms due to remaining oil residues.

The original decision that 92,109,720 organisms were killed was based upon sampling at just four stations at Bahia Sucia where six 41.85 cm^2 cores were taken at each station and compared with four control (unoiled) stations. One of the pair of stations was not used in the final calculations for questionable reasons. In the three remaining paired stations, the results were: first station, 0 organisms/m^2 oiled site versus 955/m^2 at the control; second station, 0/m^2 at the oiled site versus 1846/m^2 at the control; third station, 21/m^2 at the oiled site versus 637/m^2 at the control site. This is an average difference of 1138 organisms/m^2 which was the basis for the 92,109,720 number. A number of questions have been raised (Gilfillan *et al.* 1981) about the blasing of these sample locations since they were chosen in areas where '...there was a large quantity of hydrocarbons present...' (testimony of Dr. David Ballantine) and may not have been representative of the entire 8.1 hectare under question.

The second point of disagreement involved the actual area of mangroves damaged by the oil spill. All parties seem to agree that approximately 8.1 ha of mangroves on both the *east* and *west* sides of Bahia Sucia were oiled, but no claim was made by the Commonwealth of Puerto Rico for damages to mangroves on the *east* side of the bay. Within the mangroves on the *west* side of the bay where oil residues remain (Fig. 3) a total of 1.0 ha of mixed red (*Rhizophora mangle*) and black (*Avicennia germinans*) mangroves were reported to have died by 1976 (Nadeau and Berquist 1977), three years after the spill. After examining both pre-spill and post-spill aerial photography and inspecting the site, it was determined that 1.4 ha of dead mangroves attributable to oil contamination were present in March of 1979, six years after the spill (Fig. 4). Martinez *et al.* (1979) reported that 2.7 ha of mangroves were killed by the oil. Support for the figure of 8.1 ha of damaged mangroves used in litigation is lacking.

The major problem with determining the area of

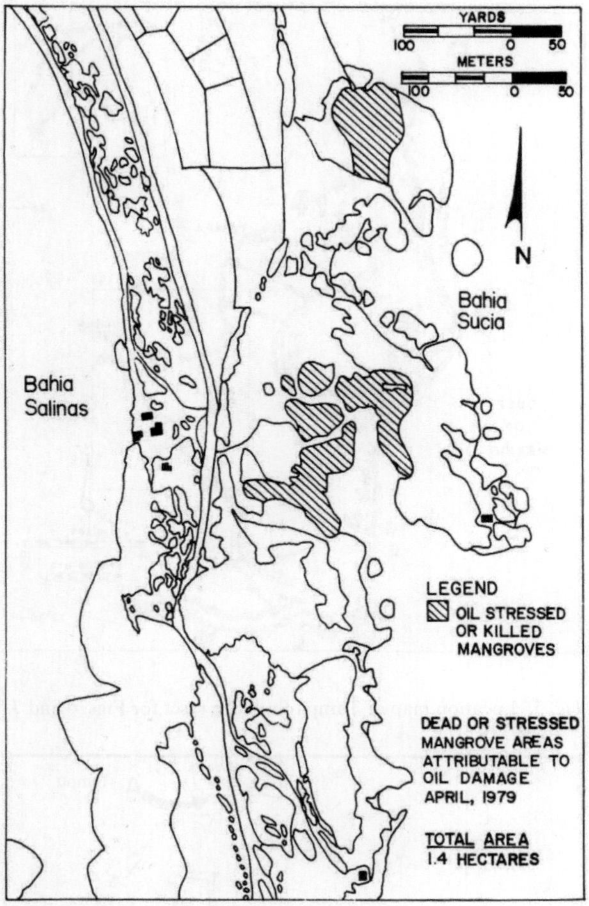

Fig. 4.

damage is that the mangroves at Bahia Sucia were not thriving *prior* to the spill and large areas of saline lagoons devoid of mangroves and stands of dead mangroves due to salinity stress were present at the time of the spill. Martinez *et al.* (1979) describe the Bahia Sucia mangroves as follows:

> In general, fringe mangroves are the dominant physiographic type . . . and basins are poorly developed. Poor development of the basins bay be due to at least two factors: (1) the lack of runoff to prevent salt accumulation; and (2) wave exposure . . . wave action in this particular coastal strip is high enough to contribute to the deposition (within the mangrove fringe) of berms of sand and *Thalassia* debris. These berms may be low, 10–15 cm, but effectively reduce the exchange of water in the inner fringe and basin. Salinity behind the berm (interstitial) is much higher and rapidly approaches more than 60‰ . . . In most sectors, salinity increases so rapidly that only a narrow fringe of red mangrove is found. Behind this fringe, the typical profile contains a *zone of dead mangroves* and a shallow hypersaline lagoon which end in a wide salitral or salt flats [emphasis added].

The arid nature of this portion of Puerto Rico creates natural stresses on the mangroves and combined with storms, is responsible for the natural cycle of death and regrowth of mangroves. As noted by Cintron *et al.* (1978) in describing mangroves of arid environments in Puerto Rico:

> The cycle mortality and expansion of mangrove forests in response to cyclic climatic events appears to be a common feature of arid coastlines. The climatic cycles may be associated with hurricanes (periods of 10–30 years) or with rainfall (periods of 5–10 years). This association has at least two implications for mangrove management. First, managers should consider open water areas and salt flats as part of the mangrove ecosystem and not as separate ecosystems. Failure to recognize this fact may result in incompatible land uses that may affect the normal expansion of the mangrove forest during periods of high rainfall and lower soil salinities. Secondly, periods of high mangrove mortality are normal occurrences in these environments, and care should be taken before attributing this mortality to other factors, including man and isopods.

Apparently, much of the confusion about the 8.1 ha of 'oil damaged' mangroves resulted from the inclusion of normal areas of mangrove mortality in the calculations.

In any case, U.S. Court of Appeals ruled on 12 August 1981 that (1) the Commonwealth of Puerto Rico could not collect damages for the loss of the 92 million organisms since there was no plan to actually purchase the organisms and use them to restore the area and (2) rejected the claim of U.S. $ 559,500 for replanting mangroves in the oiled areas due to previously mentioned conflicts as to exactly how large an area of mangroves was damaged and the fact that replanting mangroves in oiled sediments seemed 'pointless'.

The court remanded the case back to the District Court in Puerto Rico with the suggestion that an alternate site (unoiled) be considered for restoration. Such a suggestion had been made by Dr. Howard Teas during the court case and is presently being negotiated by the parties to the case based upon restoration of tidal flow and planting of mangroves in a large saline lagoon due south of the oil damaged site.

The final example is the *Howard Star* oil spill of October, 1978 (Fig. 1). The ship apparently accidentally pumped 151,000 liters (40,000 gallons) of a fuel oil mixture (80% Bunker C, 20% diesel) overboard in the upper harbor area of Tampa Bay. The resulting slick moved south stranding oil in at least four sites that contained mangroves (Lewis 1979a; Gundlach *et al.* 1979; Lewis 1980) (Fig. 5–7).

One of the common problem with observing the acute, short term impacts of oil in mangroves is that trained observers rarely are present due to the unplanned nature of an accidental spill. Lewis (1979a) noted that during the *Howard Star* spill, large numbers of molluscs (*Melongena corona*) and polychaetes (*Laeonereis culveri*) were killed during the first 24–48 hours of the spill, decayed rapidly due to the warm temperatures, and were not obvious as oil spill mortalities as soon as four days after the spill. Thus an observer arriving as soon as one week or even two months after a spill is handicapped by the rapid decay of soft-bodied invertebrates and general unfamiliarity with the area. Nadeau and Berquist (1977) arrived on the scene of the *Zoe Colocotroni* spill site 24 hours after the spill and reported 'large numbers of sea cucumbers, conchs, prawns, sea urchins, and polychaete annelids were being washed ashore' on the third day after the spill. Chan (1977) noted mortalities of crustaceans and molluscs soon after the *Garbis* spill.

In contrast to these, Jernelöv and Linden (Jernelöv *et al.* 1976) visited the *St. Peter* oil spill site three months after the spill, and Baker *et al.* examined the *Showa Maru* spill site two years after the spill (Baker *et al* 1981). The *Zoe Colocotroni* site was visited six years after the spill (Voss, pers. comm. 1979). These time frames limit the ability of the investigator to describe the impacts of a given oil spill.

The chronic impacts of the *Howard Star* oil spill have been reported by Getter *et al.* (1981) and Lewis (1979a; 1980). These reports document death of three species of mangroves (*R. mangle*, *A. germinans*, *L. racemosa*) due to oiling at a number of sites along the path of the oil slick. Estimates of the area of mortality are 0.5 ha with 1.0 ha of sublethal damage for a total of 1.5 ha requiring restoration.

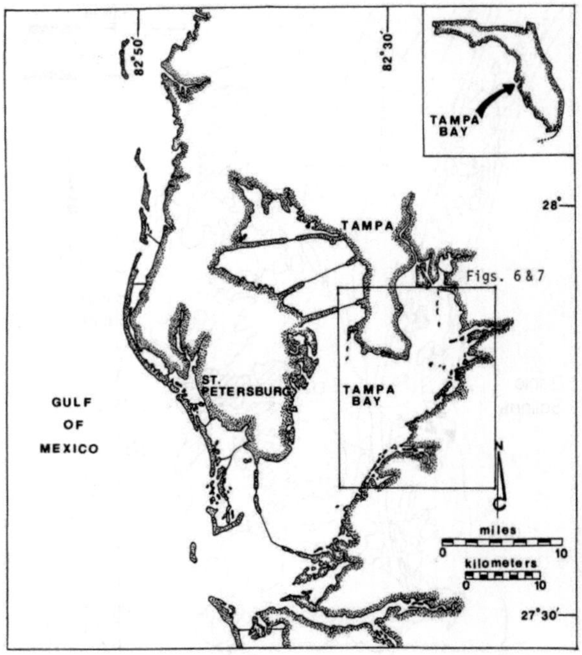

Fig. 5. Location map of Tampa Bay with inset for Figs. 6 and 7.

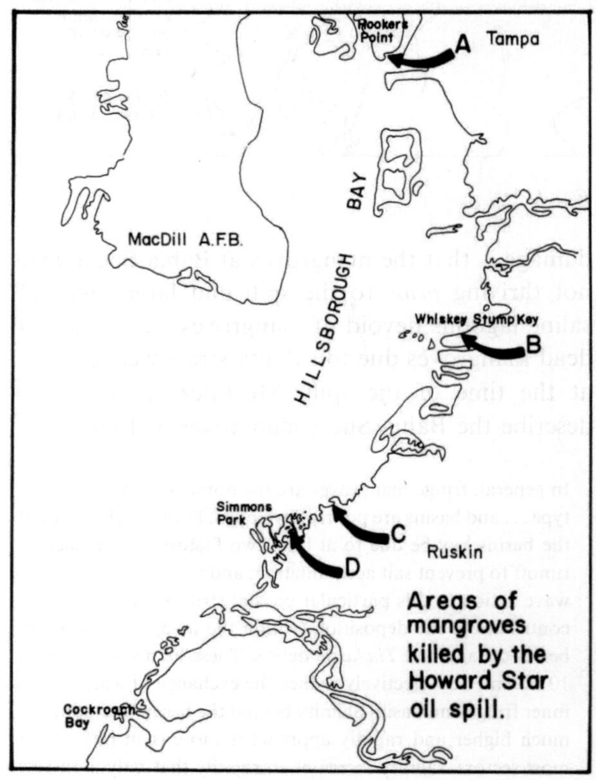

Fig. 6. Tampa Bay.

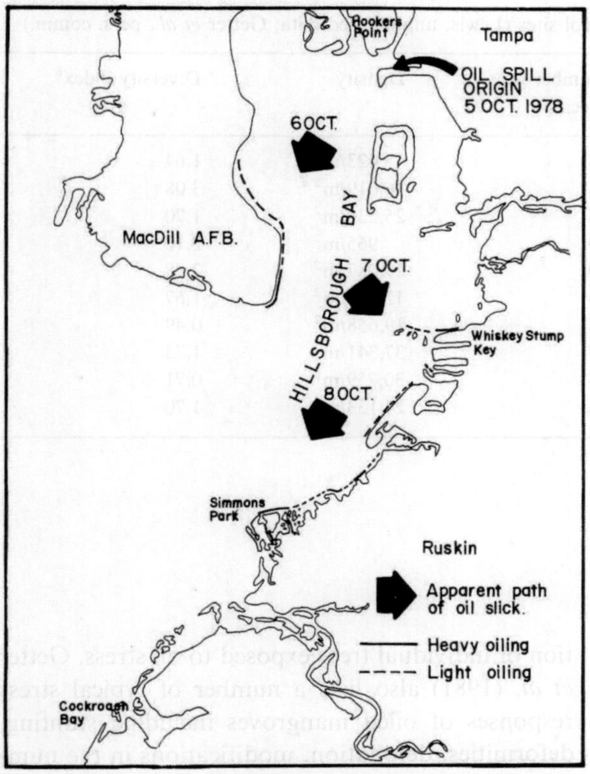

Fig. 7. Upper Tampa Bay and Hillsborough Bay showing the origin of the spill and apparent path of the slick with stranding sites.

Estimated costs for acquiring land, construction, and planting of an inland site to be converted to intertidal mangroves was U.S. $ 60,000 per hectare for a total restoration cost of U.S. $ 90,000. This damage estimate was based upon the concept of 'alternate site restoration' arising from the *Zoe Colocotroni* litigation. Any additional removal of remaining oil remaining oil residues from damaged mangrove areas was felt to be too costly and unlikely of success. A U.S. $ 750,000 clean up effort had been undertaken by the U.S. Coast Guard at the time of the spill and further expenditure of clean up money was not felt warranted.

Getter *et al.* (1981) did not estimate the areal extent of damage. Snedaker *et al.* (1981) report the death of pneumatophores of black mangroves (*A. germinans*) as a result of the *Howard Star* spill and the apparent resultant growth of anomalous aerial roots.

Getter (pers. comm. 1980) and Lewis (unpublished data) attempted a comparison of benthic infaunal density at oiled and unoiled sites. A comparison of the results of both of these analyses is presented in Table 2. It should be noted that Lewis (unpublished data) used five 7.65 cm diameter cores per station and sieved the sediments through a 0.5 mm mesh to recover infauna. Getter (pers. comm. 1980) used three 7.3 cm diameter cores per station and sieved the sediments through a 1.0 mm mesh.

Table 2 shows a comparison of the benthic infaunal data of Getter (pers. comm. 1980) and Lewis (unpublished data). The most obvious differences are in the numbers of species and densities, regardless of whether it was an oiled or a control site. The higher figures of Lewis are no doubt due to the smaller sieve size which allows fewer of the smaller macrobenthos to pass through. There is no consistent pattern in any of the data, all of which were collected at approximately the same time after the spill (17 months post spill, Lewis; 15 months post spill, Getter *et al.*)

Dr. David S. Page of the Bowdoin College Hydrocarbon Research Center, who performed the hydrocarbon analyses on samples collected at the *Howard Star* spill site, indicates that stations 1 and 5 show high levels of relatively unweathered oil that is closely related to oil collected by the Coast Guard from the *Howard Star* for comparative analysis. A sample of the same oil was used in Dr. Page's analysis. Stations 2 and 7 have very low levels of highly weathered oil and any adverse long term effects of the spilled oil would be expected to occur only at stations 1 and 5. Station 1 was the site of the observed deaths of polychaetes and molluscs, reported in Lewis (1979a). There were also large numbers of black and white mangroves and some cordgrass, *Spartina alterniflora*, that died at station 1. Additional oil analyses at this station showed values of 3,100–56,800 ppm total oil and grease ($\bar{x}$ = 27,900) in the area where mangroves died. Station 5 was the only other area where an oil sample was taken and mangroves were observed to have died as a result of the oiling.

As mentioned before, the benthic community data (Table 2) indicates no distinct pattern. Densities range from 965 to 37,341 organisms per meter

Table 2. Comparison of benthic infaunal densities at oiled and control sites (Lewis, unpublished data; Getter *et al.*, pers. comm.)

Station	Seive size	Oil residue present[a]	Number of species	Density	Diversity index[b]
RM-1 (oiled)[c]	1.0 mm	Not reported	4	5,423/m^2 [d]	1.64
CP-1 (control)[c]	1.0 mm	Not reported	6	6,619/m^2 [d]	3.08
1A (oiled)[e]	0.5 mm	3,640 ppm	14	25,230/m^2	1.70
2 (control)[e]	0.5 mm	177 ppm	9	965/m^2	2.10
3 (oiled)[e]	0.5 mm	1,120 ppm	29	19,087/m^2	2.28
4 (oiled)[e]	0.5 mm	3,790 ppm	15	12,681/m^2	1.67
5 (oiled)[e]	0.5 mm	36,400 ppm	8	19,658/m^2	0.49
6 (control)[e]	0.5 mm	3,920 ppm	17	37,341/m^2	1.73
7 (oiled)[e]	0.5 mm	2,930 ppm	14	30,759/m^2	0.71
8 (control)[e]	0.5 mm	686 ppm	18	24,133/m^2	1.70

[a] Total oil and grease.
[b] H′.
[c] Getter *et al.*
[d] Average of three cores.
[e] Lewis (1980).

square, with the lowest value at an unoiled station (no. 2) and the highest at another unoiled station (no. 6). Numbers of organisms at the two most heavily oiled stations (nos. 1 and 5) were midway to high. This appears to be due to very high numbers of pollution tolerant polychaete worms (e.g. *Capitella capitata*). Polychaetes of the family Capitellidae were also the most abundant benthic infauna reported by Gilfillan *et al.* (1981) for oiled sites at Bahia Sucia, Puerto Rico, and are common in areas of chronic oil input (Sanders *et al.* 1972). Although large numbers of dead polychaetes of the species *Laeonereis culveri* were observed soon after the spill at station 1, samples taken at this same station 17 months after the spill yielded 10,311 individuals of this species per m^2. It thus appears that at least some of the benthic invertebrate species have returned to normal densities at some oiled sites. Station 5 is the only oiled station where both species numbers and densities are reduced, probably due to the very high level of oil (36,400 ppm). *Capitella* is again the dominant polychaete, oligochaetes are numerous, and molluscs are completely absent.

Getter *et al.* (1981) also used a semi-quantitative 'compartmental method' of describing the presence of oiling at a site and applied a Mangrove Stress Index which uses an index number of 1–5 (1 = cryptic stress, 5 = mortality) to describe the condition of individual trees exposed to oil stress. Getter *et al.* (1981) also lists a number of typical stress responses of oiled mangroves including stunting, deformities, defoliation, modifications in the number and shape of lenticles, fissuring of root epidermis, and anomalous root formation, all of which were seen in trees impacted by the *Howard Star* spill.

Discussion

As noted by Getter *et al.* (1981), although there are many differences between oil spill sites involving mangroves, generalized responses and recovery processes have been observed to be common to many areas. This pattern of damage and recovery is dependent in time upon the type of oil spilled, the amount reaching the mangroves, the amount remaining after any clean up effort, and the various physiographic types of mangroves affected.

Getter *et al.* (1981) describe an 'inner fringe impact', an 'outer fringe impact', and an 'inner basin impact' depending on what portion of the forest shows stress symptoms and defoliation. Although mangrove overwash islands are described as being 'less sensitive to oil-induced damage', if they are isolated from normal waves, as happened in

Table 3. Generalized responses of mangrove forests to oil spills.

Stage	Observed impact	Reference
Acute		
0–15 days	Deaths of birds, turtles, fish, and invertebrates	Chan, 1977; Jernelöv *et al.*, 1976; Nadeau and Berquist, 1977; Lewis, 1979a
15–30 days	Defoliation and death of small (<1 m) mangroves – loss of aerial root community (bostrychietum)	Rutzler and Sterrer, 1970.
Chronic		
30 days – 1 year	Defoliation and death of medium (<3 m) mangroves – tissue damage to aerial roots	Chan, 1977; Lewis, 1980
1 year – 5 years	Death of larger (>3 m) mangroves, loss of oiled aerial roots and regrowth of new ones (sometimes deformed)	Lewis, 1979b; Snedaker *et al.*, 1981.
	Recolonization of oil damaged areas by new seedlings	Getter *et al.*, 1981.
1 year – 10 years?	Reduction in litter fall, reduced reproduction, and reduced survival of seedlings	Lugo, 1977.
	Death or reduced growth of young trees colonizing spill site?	
	Increased insect damage?	
10 – 50 years?	Complete recovery	

certain overwash islands at a public park in Tampa Bay, the overwash mangroves can then be more severely impacted due to oil being stranded in them. This occurred at station 5 after the *Howard Star* spill.

Tidal regime is also very critical in controlling the amount of damage to mangroves. Microtidal conditions, as seen in most of the Caribbean (tidal range less than 30 cm), would disperse the oil over a smaller lateral distance while macrotidal (3–5 m) conditions in parts of the Pacific could spread the oil farther inland.

Table 3 lists generalized responses of mangroves to oil spills in two general categories, acute and chronic exposure. Acute exposure would be short term (1–4 weeks) and may end with no further impact if the oil is removed either mechanically or by natural wave and current activity. An example of the former is the intense effort to spray seawater with portable pumps used in the *Howard Star* spill in portions of a public park accessible by road. This technique for cleaning oil from mangroves was suggested by Castle (1977). Monitoring of mangroves for two years after the *Howard Star* spill indicated no damage to the trees beyond loss of some oil soaked leaves. Uncleaned mangroves in the same park showed greater numbers of stress symptoms, and some died. Unfortunately, this technique is not applicable to most mangrove forests due to their inaccessibility.

Chronic exposure to oil and oil residues results in many of the stress symptoms previously noted. The lighter fraction of the oil, generally considered to be the most toxic, generally evaporates or degrades most rapidly (Lee 1977; Lee 1980; Page *et al.* 1979) leaving the heavier fraction as the cause of most of the chronic impacts. Exceptions to this can occur where the lighter fraction is incorporated into sediments and is thus degraded much more slowly.

Natural recovery of oil damaged mangroves can occur through recolonization of damaged areas by floating mangrove seedlings and planktonic larvae of benthic invertebrates. If the oil concentrations are still very high (30,000–80,000 ppm), however, the seeds may grow into deformed seedlings or die. Recolonization appears to be successful in many areas of Bahia Sucia after the *Zoe Colocotroni* spill, although both Gilberto Cintron (Puerto Rico Department of Natural Resources) and Ariel Lugo (Institute of Tropical Forestry, Puerto Rico) have testified that they believe these trees will either remain stunted or eventually die. Only continued monitoring will answer this question.

Active attempts to replant an oil damaged site

Fig. 8. A portion of the *Santa Augusta* oil spill site on St. Croix, 18 months after manual installation of 86,000 red mangrove (*Rhizophora mangle*) seedlings.

have been reported by Lewis (1979b) and Lewis and Haines (1980) for the *Santa Augusta* spill site. Natural recolonization had been prevented due to debris from the dead trees blocking movement of seeds into the damaged areas and mortality of seeds that were able to colonize more heavily contaminated areas. A total of 86,000 red mangrove seedlings were planted and 36,000 black mangrove seeds broadcast into the damaged area. Survival and growth of the planted seeds has been reported (Lewis and Haines 1980) (Fig. 8).

A key research project that would help clarify many of the questions raised here would be a controlled experimental oil spill into a mangrove forest, similar to that procedure that has been used in experimental oil spills in temperate tidal marshes (Bender *et al.* 1980; Lee *et al.* 1980). Hopefully, such an experiment will be undertaken in the near future.

Literature cited

Baker, J.B. 1981. The investigation of oil industry influences on tropical marine ecosystems. Mar. Poll. Bull. 12:6–10.

Baker, J.M., I.M. Suryowinoto, P. Brooks and S. Rowland. 1981. Tropical marine ecosystems and the oil industry; with a description of a post oil spill survey in Indonesian mangroves. In: Petroleum and the Marine Environment – Petromar 80, pp. 679–703 Graham and Trotman, Ltd., London, England.

Bender, M.E., E.A. Shearls, L. Murray and R.J. Huggett. 1980. Ecological effects of experimental oil spills in eastern coastal plain estuaries. Env. Int. 3:121–133.

Birkeland, C., A.A. Reimer and J.R. Younge. 1976. Survey of marine communities in Panama and experiments with oil. Ecological Research Series (NTIS PB-253–409), U.S. Environmental Protection Agency, Naragansett, Rhode Island, U.S.A., 176 pp.

Castle, R.W. 1977. Restoration of oil-contaminated shorelines. In: Proceedings of the 1977 Oil Spill Response Workshop. P.L. Fore, ed. Fish and Wildlife Service. FWS/OBS/77-24, 153 pp.

Chan, E.I. 1977. Oil pollution and tropical littoral communities: biological effects of the 1975 Florida Keys oil spill. In: Proceedings of the 1977 Oil Spill Conference. J.O. Ludwigson ed. pp 539–542. American Petroleum Institute, U.S. Environmental Protection Agency and U.S. Coast Guard. API Pub. No. 4284.

Chapman, V.J. 1976. Mangrove Vegetation. J. Cramer, Leutershausen, 499 pp.

Cintron, G., A.E. Lugo, D.J. Pool and G. Morris. 1978. Mangroves of arid environments in Puerto Rico and adjacent islands. Biotropica 10:110–121.

Diaz-Piferrer, M. 1962. The effects of an oil spill on the shore of Guanica, Puerto Rico. pp. 12–13. Assoc. Island Mar. Labs, 4th meeting, Curacao.

Getter, C.D., G.I. Scott and J. Michel. 1981. The effects of oil spills on mangrove forests: a comparison of five oil spill sites in the Gulf ot Mexico and the Caribbean Sea. pp. 535–540. In: Proceedings of the 1981 Oil Spill Conference, American

Petroleum Institute, Washington, DC.
Gilfillan, E.S., D.S. Page, R.P. Gerber, S. Hansen, J. Cooley and J. Hothan. 1981. Fate of the *Zoe Colocotroni* oil spill and its effects on infaunal communities associated with mangroves. In: Proceedings of the 1981 Oil Spill Conference, pp. 353–360. American Petroleum Institute, Washington, DC.
Gundlach, E.R., J. Michel, G.I. Scott, M.O. Hayes, C.D. Getter and W.P. Davis. 1979. Ecological assessment of the *Peck Slip* oil spill in eastern Puerto Rico. In: Proceedings, Ecological Damage Assessment Conference, pp. 303–317. Society of Petroleum Industry Biologists, 12–14 Nov 1979, Arlington, Virginia, U.S.A.
Hayes, T.M. 1977. Sinking of tanker *St. Peter* off Colombia. In: J.O. Ludwigson ed. Proceedings of the 1977 Oil Spill Conference, pp. 289–291. American Petroleum Institute, U.S. Environmental Protection Agency, and U.S. Coast Guard, API Pub. No. 4284. 640 pp.
Holt, S., S. Rabalais, N. Rabalais, S. Cornelius and J.S. Holland. 1978. Effects of an oil spill on salt marshes at Harbor Island, Texas. I. Biology. In: The Proceedings of the Conference on Assessment of Ecological Impacts of Oil Spills, pp. 344–352. American Institute of Biological Sciences, Washington, DC.
Jernelöv, A., O. Linden and I. Rosenblum. 1976. The *St. Peter* oil spill – an ecological and socio-economic study of effects. Columbia and Ecuador, May–June 1976. Swedish Water and Air Pollution Research Institute, Stockholm, Sweden, 34 pp. + Appendix.
Lee, R.F. 1977. Accumulation and turnover of petroleum hydrocarbons in marine organisms. In: Fate and effects of petroleum hydrocarbons in marine ecosystems and organisms. D.A. Wolfe, ed. pp. 60–70. Pergamon Press, New York.
Lee, R.F. 1980. Processes affecting the fate of oil in the sea. Chapter 12. In: Marine Environmental Pollution, 1. Hydrocarbons, R.A. Geyer, ed. pp. 337–351. Elsevier Scientific Publishing Co., Amsterdam, The Netherlands.
Lee, R.F., B. Dornseif, F. Gonsoulin, K. Tenore, R. Hanson and J. Blanton. 1980. Fate and effects of oil on Georgia coastal waters and marshes. Technical Report Series No. 80-7. Georgia Marine Science Center, Skidaway Island, GA.
Lewis, R.R. 1979a. Oil and mangrove forests: the aftermath of the *Howard Star* oil spill. Florida Scientist 42 (suppl.):26. (Abst.)
Lewis, R.R. 1979b. Large scale mangrove restoration on St. Croix, U.S. Virgin Islands. In: Proceedings of the Sixth Annual Conference on the Restoration and Creation of Wetlands. D.P. Cole, ed. pp. 231–242. Hillsborough Community College, Tampa, Florida.
Lewis, R.R. 1980. Oil and mangrove forests: observed impacts 12 months after the *Howard Star* oil spill. Florida Scientist 43 (suppl.):23. (Abst.)
Lewis, R.R. and K.C. Haines. 1980. Large scale mangrove restoration on St. Croix, U.S. Virgin Islands, II. Second Year. In: Proceedings of the Seventh Annual Conference on Restoration and Creation of Wetlands, pp. 137–148. Hillborough Community College, Tampa, Florida.
Linden, O. and A. Jernelöv. 1980. The mangrove swamp – an ecosystem in danger. Ambio 9:81–88.
López, J.M. 1978. Ecological consequences of petroleum spillage in Puerto Rico. In: The Proceedings of the Conference on Assessment of Ecological Impacts of Oil Spills, pp. 894–908. American Institute of Biological Sciences, Washington, DC.
Lugo, A.E. and S.C. Snedaker. 1974. The ecology of mangroves. Ann. Rev. Ecol. Syst. 5:39–64.
Macnae, W. 1968. A general account of the fauna and flora of mangrove swamps and forests in the Indo-West Pacific Region. In: Advances in Marine Biology, Volume 6. F.S. Russell and M. Younge, eds. pp. 74–270. Academic Press, New York.
Nadeau, R.J. and E.T. Berquist. 1977. Effects of the March 18, 1973 oil spill near Cabo Rojo, Puerto Rico and tropical marine communities. In: Proceedings of the 1977 Oil Spill Conference. J.O. Ludwigson, ed. pp. 535–538. American Petroleum Institute, U.S. Environmental Protection Agency, and U.S. Coast Guard. API Pub. No. 4284, 640 pp.
National Academy of Sciences. 1975. Petroleum and the Marine Environment, 107 pp.
Odum, W.E. and E.J. Heald. 1972. Trophic analyses of an estuarine mangrove community. Bull. Mar. Sci. 22:671–738.
Odum, W.E. and R.E. Johannes. 1975. The response of mangroves to man-induced environmental stress. In: Tropical Marine Pollution. E.J.F. Wood and R.E. Johannes, eds. pp. 52–62. Elsevier, New York, 192 pp.
Oil Spill Intelligence Report. 1980a. Nigeria plans clean up inspection following well blow-out. 7 March 1980. Cahners Publishing Company, Boston, Massachusetts.
Oil Spill Intelligence Report. 1980b. Texaco supplies food to Nigerian villages. 18 April 1980. Cahners Publishing Company, Boston, Massachusetts.
Page, D.S., D.W. Mayo, J.F. Cooley, E. Sorenson, E.S. Gilfillan and S.A. Hanson. 1979. Hydrocarbon distribution and weathering characteristics at a tropical oil spill site. In: The Proceedings of the 1979 Oil Spill Conference, pp. 709–712. American Petroleum Institute, Washington, DC.
Rutzler, K. and W. Sterrer. 1970. Oil pollution damage observed in tropical communities along the Atlantic seaboard of Panama. Bioscience 20:222–224.
Sanders, H.L., J.F. Grassel and G.R. Hampson. 1972. The West Falmouth oil spill. I. Biology. Woods Hole Oceanographic Institution, Woods Hole, Massachusetts. Technial Report No. 72-20.
Snedaker, S.C., J.A. Jimenez and M.S. Brown. 1981. Anomalous aerial roots in *Avicennia germinans* (L.) L. in Florida and Costa Rica. Bull. Mar. Sci. 31:467–470.
Spooner, M. 1970. Oil spill in Tarut Bay, Saudi Arabia. Mar. Poll. Bull. 1:166–167.
VAST/TRC. 1975. Oil spill – Bahia Sucia, Puerto Rico, 18 March 1973 – environmental effects. EPA contract 68-10-0542, 59 pp.
Westree, B. 1977. Biological criteria for the selection of clean up techniques in salt marshes. In: Proceedings of the 1977 Oil Spill Conference. J.O. Ludwigson, ed. pp. 231–235. American Petroleum Institute, U.S. Enviromental Protection Agency, and U.S. Coast Guard. API Pub. No. 4284, 640 pp.

CHAPTER 20

The effects of oil pollution on mangroves and fisheries in Ecuador and Colombia

ARNE JERNELÖV and OLOF LINDÉN

Swedish Water and Air Pollution Research Institute, Box 21060, S-100 31, Stockholm Sweden

Abstract. The damage to mangrove communities and fisheries were studied following extensive oil pollution of the coastal areas of northern Ecuador and southern Colombia in 1976. During the acute phase of the oil spill oil covered 2–3 m vertically on the mangrove trees along the sea front, and had penetrated 20–70 m horizontally. Acute effects on the mangrove communities included defoliation of trees, mortality of sessile organisms, and migration of semi-sessile and mobile crustaceans and molluscs. Dead fishes, sea snakes and sea birds were also found. The relatively large tidal range caused considerable washing off of the deposited oil from roots and trunks, and four months later the major part of the oil on the mangrove trees had disappeared. Previously defoliated mangroves had, with some exceptions, recovered, and mobile organisms had re-entered the affected area. In some areas where mangrove had died there was erosion of the substrate.

The oil spill affected the local fishery in a number of ways; for example, the absence of tuna in the region during that year suggested oil avoidance reactions in this group.

Introduction

In the beginning of February 1976, the 34,000-tonnes Liberian tanker Saint Peter with a cargo of 243 000 barrels of crude oil sunk at 01° 35′ N, 79° 13′ W in almost 400 fathoms of water. Oil was observed bubbling to the surface and spreading out into a large slick over the sunken ship.

In the middle of February oil slicks of medium to heavy thickness covering approximately 5,200 hectares were observed 5 km from the Colombian coast. The slicks which had spread southwards into Ecuadorian water drifted toward the coast. By 17 February, oil had reached beaches and mangrove swamps in Tumaco.

Description of the affected area

The coastal area in Colombia and Ecuador which was affected by the Saint Peter oil spill extended from Buenaventura, Colombia in the north to Punta Galera, Ecuador in the south. In this area the shoreline consists variously of mangrove swamps, sandy beaches and rocky shores, and is characterized by a relatively large tide water amplitude (2–4 m).

On the affected coastline, the heaviest contamination occured in the Tumaco-area (Colombia) and on the stretches of the shore southwards, towards the border of Ecuador. The shoreline in this area consists of two main biotopes: sandy beaches and mangrove swamps.

The main part of the shore-line in the Tumaco bay and the area southwards from Tumaco is occupied by mangrove swamps. In the north part of

Tasks for vegetation science, Vol. 8 ed. by H.J. Teas
ISBN 978-90-481-8526-9

the bay, the coast is very shallow and at low tide extensive areas become dry. The inner part of the Tumaco Bay and the coastal area southwards are not as shallow as the northern part of the bay.

Damage caused by the oil

Large parts of the mangrove swamps in the Tumaco area and southwards were severely oil-contaminated during the late part of February. A thick oil coat covered the roots and trunks of the mangrove trees. The covered part was often 2–3 m vertically and 20–70 m horizontally.

The acute effects of the oil in the worst affected parts of Colombia were investigated by a group from the Centro de Investigaciones Oceanogràficas de la Armada National. The results of their survey showed that the mangrove trees in large parts of the Tumaco area were partly defoliated. This applied also to epiphyte flora. The fauna of the mangrove-swamps was also disturbed in several ways. The sessile organisms like barnacles, mussels (*Anadara*) and oysters suffered massive mortality, and were absent or rare about 2 mounths after the spill. Semisessile organisms like gastropods migrated to a large extent to uncontaminated zones above the area coated by oil. A reduction of the number of fiddler crabs was observed and particularly the number of younger individuals decreased. Dead and decaying fishes was also observed at several places in the mangrove swamps during the survey at the beginning of April. Furthermore, dead or dying snakes and sea birds were observed on several places in the Tumaco area connected with the oil spill. From May 25 to June 10, 1976, Jernelöv *et al.* (1976) had the opportunity to study the more long term effects caused by the oil spill, at about 30 stations in Colombia and 10 in Ecuador. These May–June investigations revealed that at most stations which had been heavily oil-contaminated there remained only a relatively narrow belt of oil, close to the highest tidal level. Apparently the wave action had removed the main part of the oil coat from roots and trunks of the mangrove trees. The areas that had been less heavily oiled were largely undamaged and previously defoliated areas had recovered with new leaves, blooms and seedlings on the trees.

During the period of observation few clear residual effects on the mangrove fauna could be observed. The organisms (crustaceans, molluscs, fishes, insects and spiders) had returned to most stations to such an extent that these were indistinguishable from unoiled localities in the vicinity. This suggests that the mechanism for recovery is one of migration from unaffected parts of the mangroves rather than through new generations of organisms on site. When the acute toxicity of the oil had killed the organisms in an area, an empty space was created. Individuals from neighbouring areas moved into the vacant area as soon as the oil toxicity had disappeared.

At two stations (Comaco, San Jorge) which were sheltered from wave action, somewhat more oil was present on the mangrove. Here the abundance of gastropods (Piacuiles) was lower than in unaffected areas and barnacles were missing. At one station close to Resurrecciòn, a landslide was observed along approximately 100 m of shore. The mangrove trees at this locality appeared to be less viable, which could have been an effect of the oil cover. Deterioration and impairment of the trees could also have resulted from the mud and silt that collected around the dense mass of roots and trunks as a consequence of soil erosion from the landslide.

Studies of the impact of the oil spill on the fishery showed that the toxic and physical properties of the oil prevented people searching for clams and the oil had caused mortality among the clams that were collected in the mangroves. Otherwise it appeared that the oil did not physically hinder fishing or cause fish mortality on any large scale.

However, a change in fisheries which can be attributed to the oil spill concerns the tuna. Extraordinarily good catches of skipjack and yellowfin tuna were brought into Guayaquil, Ecuador, in December and January of 1976. The yearly migration route of those three-to-four-year-old tuna would normally have brought them to the Tumaco area in March or April, but this year they did not appear. Instead they apparently bypassed the northern Ecuador and Colombian coast and were found farther north in Costa Rica, Central America. Previous experience in other areas suggests that migratory tuna avoid oil contaminated areas (Jernelöv 1974).

Discussion

Mangroves occupy the coastal zone between high and low tide marks, a particularly exposed position for oil contamination. Oil pollution of mangrove swamps has been reported at several locations (Diaz-Piferrer 1962; Cerame-Vivas 1968; Rutzler and Sterrer 1970; Spooner 1970; Nadeau and Bergquist 1977; Chan 1977; and Lopez 1978). Chan (1977) reported that oil pollution in the mangroves of the Florida Keys resulted in the death of red mangrove seedlings that had greater than 50 percent oiling of their leaves. Black mangroves with greater than 50% oiling of pneumatophores also died, as did some trees where the substrate remained oiled one year later. A number of animals associated with the mangroves migrated away (*Uca* crabs, *Melampus* gastropods) and some mortality was observed among oysters and fish. The black coat of oil had a secondary effect of increasing absorption of solar energy and thus elevating the temperature above the lethal limits for many invertebrates and possibly mangrove plants.

Defoliation and mortality among mangroves were observed by Nadeau and Bergquist (1977) during three years following an oil spill that affected the mangrove swamps in southwestern Puerto Rico. The mangrove 'prop root community' showed a sparse and impoverished fauna, and mortality was observed among several major faunal groups. Observations of a very similar nature are reported by Rützler and Sterrer (1970) following an oil spill in the mangroves along the Atlantic coast of Panama where seedlings of red mangroves were killed together with the inter-tidal invertebrates, birds and turtles.

The biological investigation in Colombia following the Saint Peter oil spill demonstrated that at most stations which had earlier been heavily oil-contaminated traces of oil could only be detected close to the extreme high tide mark, where washing had least chance to occur. The mangroves themselves were largely undamaged and previously defoliated parts had recovered with new shoots blooms and seedlings on the trees. The organisms (crabs, molluscs, barnacles, fishes, shrimps and insects) had returned in abundance and sizes at most stations that made them indistinguishable from unaffected localities in this respect. However, at some stations the mangroves appeared less viable or were dying, and erosion effects were observed.

Partial defoliation but little mortality among the mangroves themselves have also been reported following an oil spill in Tartut Bay, Saudi Arabia (Spooner 1970).

There is no doubt that oil contamination can have several serious consequences in mangrove communities. The damage to the mangroves themselves is caused mainly by the direct smothering of lenticels in the pneumatophores or prop roots, resulting in oxygen starvation of the trees' roots. Harmful effects may also be produced by a number of fractions in oils, particularly the aromatics, which are very phytotoxic. Furthermore the dark oil film may dangerously elevate temperatures by absorbing solar radiation. However, some factors do tend to ameliorate the damage. Thus, high temperatures generally result in a rapid loss of the most toxic fractions through vaporization. It appears also that tidal action can rapidly wash away much of the oil deposited on roots and trunks of mangroves.

Literature cited

Cerame-Vivas, M.J. 1968. The wreck of the *Ocean Eagle*. Sea Frontiers 15:224.

Chan, E.I. 1977. Oil pollution and tropical littoral communities: Biological effects of the 1975 Florida Keys oil spill. 1977 Oil Spill Conference. American Petroleum Institute, Washington DC, USA 19–7, pp. 539–542.

Diaz-Piferrer, M. 1962. The effects of an oil spill on the shore of Guanica, Puerto Rico. Deep-Sea Research 11:855.

Jernelöv A., R. Rosenberg and Å. Hagström. 1974. Effects of Oil-Spill from the Tanker 'Trans Huron' on Kilton Islands, Laccadives, India. Swedish Water and Air Pollution Research Institute (IVL), Stockholm, Sweden.

Jernelöv, A., O. Linden and I. Rosenblum. 1976. The Saint Peter Oil Spill – An Ecological and Socioeconomic Study of Effects. Swedish Water and Air Pollution Research Institute (IVL), Stockholm, Sweden.

Lopez, J.M. 1978. Ecological consequences of petroleum spillage in the coastal waters of Puerto Rico. Conference on the Assessment of Ecological Impacts of Oil spills. American Institute for Biological Sciences, Washington DC, USA, pp. 895–908.

Nadeau R.J. and E.T. Bergquist. 1977. Effects of the 18 March 1973 oil spill near Cabo Rojo, Puerto Rico on tropical marine communities. 1977 Oil Spill Conference. American Petroleum Institute, Washington DC, USA 1977, pp. 535–538.

Rutzler, K. and W. Sterrer. 1970. Oil Pollution. Damage observed in tropical communities along the Atlantic seaboard of Panama. Bioscience 20:222.

Spooner, M. 1970. Oil spill in Tarut Bay, Saudi Arabia Marine Pollution Bulletin 1:166.

MIX
Papier aus verantwortungsvollen Quellen
Paper from responsible sources
FSC® C105338

If you have any concerns about our products,
you can contact us on
ProductSafety@springernature.com

In case Publisher is established outside the EU,
the EU authorized representative is:
Springer Nature Customer Service Center GmbH
Europaplatz 3, 69115 Heidelberg, Germany

Printed by Libri Plureos GmbH
in Hamburg, Germany